BRITISH MOSSES AND LIVERWORTS

BRITISH MOSSES AND LIVERWORTS

AN INTRODUCTORY WORK, WITH FULL DESCRIPTIONS
AND FIGURES OF OVER 200 SPECIES, AND KEYS
FOR THE IDENTIFICATION OF ALL EXCEPT
THE VERY RARE SPECIES

WRITTEN AND ILLUSTRATED BY

E. VERNON WATSON, B.Sc., Ph.D.
Lecturer in Botany, University of Reading

WITH A FOREWORD BY

PAUL RICHARDS, M.A., Sc.D.
*Professor of Botany, University College of
North Wales, Bangor*

CAMBRIDGE
AT THE UNIVERSITY PRESS
1955

PUBLISHED BY
THE SYNDICS OF THE CAMBRIDGE UNIVERSITY PRESS
London Office: Bentley House, N.W.1
American Branch: New York

Agents for Canada, India, and Pakistan: Macmillan

Printed in Great Britain at the University Press, Cambridge
(Brooke Crutchley, University Printer)

To my wife

CONTENTS

List of Plates . . . . . . *page* ix

Preface xi

Foreword, by Professor P. W. Richards . . xv

Introduction I

Glossary 25

Key to Mosses 36

Key to Liverworts 76

Synopsis of Classification 93

MUSCI

SPHAGNALES 100

ANDREAEALES 107

BRYALES 109

POLYTRICHALES 109

BUXBAUMIALES 121

FISSIDENTALES 122

DICRANALES 127

POTTIALES 154

GRIMMIALES 184

FUNARIALES 198

SCHISTOSTEGALES 203

TETRAPHIDALES 205

EUBRYALES 206

ISOBRYALES 234

HOOKERIALES 257

HYPNOBRYALES 258

HEPATICAE

JUNGERMANNIALES . . . . *page* 321

MARCHANTIALES 387

ANTHOCEROTALES 397

Habitat Lists 400

Index 413

Field Key . . . . . *in pocket at end*

LIST OF PLATES

I	*Polytrichum juniperinum*	*facing page*	116
II	*Polytrichum commune*		117
III	*Fissidens adianthoides*		152
IV	*Leucobryum glaucum*		153
V	*Grimmia pulvinata*		188
VI	*Rhacomitrium lanuginosum*		189
VII	*Mnium undulatum*		224
VIII	*Thamnium alopecurum*		225
IX	*Thuidium tamariscinum*		260
X	*Pseudoscleropodium purum*		261
XI	*Plagiothecium undulatum*		306
XII	*Hypnum cupressiforme*		307
XIII	*Ctenidium molluscum*		358
XIV	*Plagiochila asplenioides*		359
XV	*Marchantia polymorpha*		386
XVI	*Marchantia polymorpha*		387
XVII	*Conocephalum conicum*		390
XVIII	*Lunularia cruciata*		391

PREFACE

The object of this book is to provide an introduction to the study of British mosses and liverworts in a form which will be acceptable and helpful to those botanists, professional and amateur, who are concerned with the determination of the bryophytes commonly met with in the course of field work in this country. An effort has been made to draw up each description in a simple, readable form, which will avoid all unnecessary technicalities. Such technical terms as have been introduced are explained in the glossary.

After careful consideration it was decided to give full descriptions and figures of only the commonest or most notable species, amounting to a total of just over 150 mosses and 50 liverworts. As it was realized that this would make the work too limited in scope to be of much service as a bryophyte Flora, it was decided to include, in addition, a much less detailed treatment of a very considerable number of less common species. These additional species are included in the general Keys, but their treatment in the text is limited to brief notes designed to enable the student to separate them from commoner allied species.

The arrangement and nomenclature* for mosses are those of Richards and Wallace (Check-List, *Trans. Brit. Bryol. Soc.* I (1950), p. 426); for liverworts, the arrangement is that of Evans (*Bot. Rev.* V (1939), p. 90) and the nomenclature very largely that of Buch, Evans and Verdoorn (*Ann. Bryol.* X (1937), p. 3). Where, however, the name differs from that in the respective handbooks of Dixon and MacVicar the obsolete name has been appended in brackets. In this way it is hoped that students may be enabled gradually to acquire familiarity with the currently accepted name. That so many name changes should have become necessary is unfortunate, but it would seem to be inevitable.

It is with great diffidence that the original Keys in this book are offered. Although a serious attempt has been made to give them a thorough trial in advance of publication, I am well aware that more exhaustive use will bring many flaws to light, and any information as to such defects will be most gratefully received.

* At the suggestion of Professor Richards a few slight departures from the Check-List have been made, chiefly in the spelling of names.

At every stage in the planning and writing of this book the biggest problem has been how to provide an essentially simplified treatment of what is, after all, a technical subject, and yet at the same time to include sufficient precision of detail to promote more than a purely superficial acquaintance with bryophytes. Inevitably the treatment has become more detailed in some directions than in others. Thus, in the accounts of the species cell measurements have been freely used; yet, with one or two exceptions, no attempt has been made to treat of the delimitation and diagnostic characters of families and genera, as it was felt that this would unnecessarily lengthen and complicate the book without materially assisting in the task of determining species of moss and liverwort. How far the book has succeeded in its purpose only its users can decide. In the nature of things it cannot succeed completely, but in so far as use reveals error of fact or suggests improvement of presentation (as, indeed, it must), I would ask that my attention be duly drawn to these flaws.

All the drawings have been prepared specially for this work. My object has been to include, for each species, a series of studies (made from actual specimens) which will enable the beginner to form a useful overall picture of the plant. Particular care has been given to the details of cell structure, so often of critical importance in the determination of species. I am more than fortunate in having been able to include a series of photographs by the distinguished plant photographer Mr R. H. Hall.

Professor P. W. Richards has very kindly read through the entire manuscript and has been responsible for numerous improvements. He has also given unsparingly of his time in the difficult matter of making the initial selection of species for inclusion, and in assisting with the preparation of the ecological notes that accompany each description. I am most grateful to him for all his help and, not least, for the gracious Foreword he has written.

Dr Barbara Maxted was good enough to read through the descriptions of the mosses from the standpoint of the professional, but non-bryological, botanist, and I am much indebted to her for some useful constructive criticism. Mr A. D. Banwell's help with the liverworts has been unstinting in many directions, whilst Miss U. K. Duncan, Dr E. W. Jones and Dr P. Greig-Smith have very kindly assisted me respectively with *Sphagnum*, *Cephaloziella* and *Lejeunea*. Dr Jones has also put his unrivalled knowledge of *Barbula* freely at my disposal and has helped in yet another direction, by making numerous valuable suggestions for the improvement of the Field Key. To Mr R. E. Parker I am indebted for several excellent ideas for the improvement of the Habitat Lists. To all these, and to many other bryologist friends who have helped me in various ways and by their constant interest in the progress of the work, I am duly grateful.

Finally, I would thank Professor T. M. Harris, without whose initial suggestion this book might never have been written, both for that suggestion and for his unfailing encouragement during the years of its preparation. And to the staff of the Cambridge University Press a special word of thanks is due for all the skill and care they have bestowed upon the production.

E. V. WATSON

DEPARTMENT OF BOTANY
READING
November 1954

FOREWORD

In the British Isles there are over 600 mosses and nearly 300 liverworts, as compared with some 1500 flowering plants, but though the two groups of bryophytes together form a considerable part of our flora, they seldom receive the attention they deserve; even among professional botanists the names of any beyond the commonest species are known to comparatively few. Yet the study of these plants, small and inconspicuous, but wonderful in their beauty and variety, has a long history in this country. As long ago as 1741 the Oxford University Press published the first book about mosses ever written, the *Historia Muscorum* of Dr John Jacob Dillenius, a German from Giessen who was at that time the Sherardian Professor of Botany at Oxford. Since then the mosses and liverworts have been a favourite study of many British botanists of enduring fame, Robert Brown, William Wilson, the Hookers, father and son, Richard Spruce, William Mitten and, in our own century, H. N. Dixon and Symers MacVicar, to mention only a few.

Very recently there has been an outburst of new interest in our native bryophytes, which has been reflected in the founding in 1947 of a journal, the *Transactions of the British Bryological Society*, entirely devoted to them. The number of those studying the group would doubtless be even larger, but for the lack of a readily accessible introductory book and the belief that the study of such obscure plants demands exceptional skill or expensive equipment. Dixon's *Student's Handbook of British Mosses* and MacVicar's companion volume on the liverworts, admirable as they are, have long been difficult to obtain, and there is also a growing demand for a book which will smooth the path of the beginner by including only the commoner and more readily recognized species.

The book which Dr E. V. Watson has now written is thus assured of a wide and friendly welcome, and it should do much to dissipate the feeling that bryology is a science which can be learned only by a long and laborious apprenticeship. Since most bryophytes are very small plants, they cannot be studied by the naked eye alone, but with Dr Watson as a guide nobody who has the use of a reasonably good microscope need be deterred by the difficulties. Those who wish to know the bryophytes must begin their study out of doors, consequently this book rightly lays much emphasis on the appearance of the living plant. No aspect of the bryophytes is more fascinating than their ecology, and their habitat preferences are as varied, as individual and as un-predictable as a man's taste in food; it is therefore a welcome feature of the book that these preferences are treated more fully than in any other work of this kind.

Dr Watson, as well as being an expert bryologist, is also an artist, and his clear and accurate drawings add greatly to the value of his book. Not content with illustrating it himself, he has been fortunate in enlisting Mr R. H. Hall's co-operation; his delightful photographs are another attractive and unusual feature of the book.

In warmly commending to the inexpert and the not so inexpert alike, a work on which I know Dr Watson has spent much time and trouble, I should like to express the hope that it will make many new converts to the study of a neglected province of the plant kingdom.

PAUL RICHARDS

BANGOR
February 1954

INTRODUCTION

1. GENERAL CHARACTERS OF BRYOPHYTES

The mosses (Musci) and the liverworts (Hepaticae) together constitute the major group known as Bryophyta. All these bryophytes possess many features in common, and, in general, no moss or liverwort is likely to be confused with any member of any other major group of the plant kingdom. Filmy ferns (*Hymenophyllum* spp.) may, however, be mistaken for bryophytes at first glance, their fronds suggesting the shoots of some robust leafy liverwort; but their forked venation will separate them immediately. Occasionally the inexperienced may be for a moment in doubt whether to refer certain plants to liverworts or to the lichens. Lichens, of course, are fundamentally unlike bryophytes in their structure, and even those which most resemble liverworts in form have a tough consistency and opaque appearance very different from that of any liverwort. Moreover, even when green above, the lichen thallus is almost always quite different in colour on the underside; this is never so in a liverwort.

Bryophytes range in size from forms in which the whole plant is barely visible to the naked eye to those at the opposite extreme which grow to a height (or length) of 20 cm. or more. The plant body may consist of a thallus showing no differentiation into stem and leaves, as in some of the liverworts, or it may show a definite axis, or stem, on which delicate leaves are borne. The stem is always slender and, with a few exceptions, is much more weakly constructed than in higher groups—the so-called vascular plants. Even when tough and wiry, as in some mosses, the stem is always without the elaborate internal tissue differentiation found in vascular plants. The leaves, moreover, are always small (very rarely above about 12 mm. in length) and, except for the midrib, are normally only one cell thick. They are never stalked. In all liverworts and many mosses a thickened midrib (or nerve) is lacking, and the whole leaf thus commonly consists of a single fairly uniform expanse of cells. Whether 'thallose' or 'leafy', the plant is normally attached to the ground or other substrate by delicate, colourless or brown threads termed rhizoids. These are to be compared with the root-hairs, rather than with the roots of higher plants, and the details of their structure (different in the two groups—see below) can be seen only under the compound microscope.

Several mosses and a large number of liverworts regularly reproduce asexually, by means of small clusters of cells or plates of tissue which break away and readily germinate to give new plants. These minute agents of propagation are termed gemmae. The normal, sexual method of reproduction, however, involves special organs, the antheridia and archegonia. No close study of bryophytes can be made without familiarizing oneself with these two important organs. The antheridium varies considerably in detail, but is always a delicate sac whose contents form the spermatozoids (or male gametes). Antheridia will thus usually appear under the microscope as greyish or brown ovoid, ellipsoidal or globose objects, on stalks which vary much in length in different kinds of bryophytes. Among mosses they are normally accompanied by numerous club-shaped structures, or short filaments of cells, termed paraphyses. The archegonium varies comparatively little in shape among bryophytes, and is an easy structure to recognize, being shaped like a bottle or flask, with a wide, rounded body and a long, narrow neck. The cells in the interior of the neck break down, and it is down the canal so formed that a male gamete (from the antheridium) passes on its way to effect fertilization. In the centre of the body (or venter) of the archegonium lies the egg with which one male gamete unites, thus fertilizing it and making possible the development of a capsule and associated structures.

The fertilized egg cell develops into a rather elaborate structure which represents a distinct and important stage in the life cycle. In all but a few exceptional cases, this consists of a spore-containing capsule that is linked to an absorbing organ by a stalk termed the seta. In certain cases the seta is lacking, and—more rarely—the absorbing organ also. At maturity the capsule, which is popularly known as the fruit, sheds its spores as a fine dust, the mechanism of this process of spore dispersal being associated with different special structural features in the different groups of bryophytes. The spores are carried readily on currents of air (sometimes also by water or insects); and each, on germination, gives rise to a new moss or liverwort plant.

Mosses and liverworts thus possess two distinct stages in their life cycles: a green, or vegetative stage, characterized by bearing the antheridia and archegonia, ending at fertilization; and a dependent stage, not green at maturity, which is concerned with the elaboration and shedding of the spores. Important taxonomic characters, employed in classification and inevitably used in identification, are found to an almost equal extent in these two stages. It will be seen too that, as regards both the vegetative and the spore-producing phases of the life cycle, mosses and liverworts differ from one another in several important respects.

2. DIFFERENCES BETWEEN MOSSES AND LIVERWORTS

There are no transitional forms between mosses and liverworts, but the differences between them are not easy to state clearly, chiefly because so few characters are common to all mosses, or to all liverworts. One of the few such characters is that of the rhizoids, which are multicellular in mosses, whereas they are nearly always unicellular in liverworts. In deciding whether a particular plant is to be referred to one class or the other, however, one relies rather on a general impression that is gained by taking account of a number of characters collectively; and the sum of the evidence, viewed in this way, usually leaves little room for doubt. Even if a superficial inspection in the field may lead at times to a wrong conclusion (such as taking *Herberta* spp. for mosses or taking *Hookeria lucens* for a liverwort), a microscopic examination will almost always enable one to place the plant in the class to which it belongs.

In addition to the constant character of the rhizoids already mentioned, the following considerations may usefully be borne in mind:

(i) Any bryophyte that lacks differentiation into stem and leaves must be a liverwort, since all mosses show such differentiation.

(ii) Any bryophyte in which the leaves are deeply lobed or segmented must be a liverwort, since no moss shows this character.

(iii) All leafy liverworts have the leaves arranged in a characteristic way. Normally there are three ranks, two of them usually being laterally placed and consisting of leaves of normal size, the third being next to the substratum and consisting of the underleaves which are generally smaller than the lateral leaves. In many species the underleaves are lacking altogether so that the leaves appear in two ranks only; and in a few cases all three ranks of leaves are of nearly equal size. In mosses (except in some non-British species) it is in no case possible to distinguish a recognizable rank of underleaves, of reduced size, comparable with those of so many leafy liverworts. Furthermore, in very few genera of mosses do the leaves appear in regularly discernible ranks on the mature shoot. Most of the mosses with a regularly two-ranked leaf arrangement (e.g. *Fissidens*, *Distichium*) may be distinguished at once from liverworts by the presence of a nerve; for no liverwort has a true, thickened nerve or midrib in the leaf.

(iv) Mosses with nerveless leaves may invariably be distinguished from liverworts by their very different cell structure, or 'areolation'. Thus, in no leafy liverwort are the ordinary leaf cells markedly elongated—only very rarely are they as much as twice as long as broad. In many mosses with nerveless leaves they are much longer than this in relation to their breadth. In others, where the leaf cells are short and rounded, they are almost always very much smaller than the cells composing the leaf of a liverwort.

A few mosses, such as *Homalia trichomanoides* and *Hookeria lucens*, *look* very much like liverworts, but a close examination will show that, although the shoot as a whole is flattened, the leaves are not arranged in two absolutely clear-cut ranks.

Further differences between these two classes of bryophytes are met with in the reproductive characters. Thus, the seta, where it is present in liverworts, lengthens rapidly, after the capsule has reached its full size, into a weak, colourless, semi-transparent structure. In mosses the seta lengthens gradually, becoming at maturity a slender but opaque and comparatively tough 'fruit-stalk' of purplish, red or yellow colour according to the species. The tubular 'perianth' which surrounds the developing capsule in most liverworts is not found outside that group of plants, and the spirally thickened 'elaters' which occur together with spores in the capsules of most liverworts are unknown among mosses. On the other hand, the 'calyptra' which covers the nearly mature moss capsule like a hood is not seen after elongation of the seta in liverworts. The capsule itself, moreover, is very different in typical members of the two classes. Thus, in most liverworts it is ovoid or globose, has no lid, and when ripe splits lengthwise into four 'valves'. In the majority of mosses it is more nearly pear-shaped or cylindrical, with a distinct lid that is shed at maturity to expose the characteristic circle of teeth that surround the mouth of the open capsule. These teeth form the peristome, a structure unknown among liverworts. In some mosses a peristome is lacking, and in the genus *Andreaea* the capsule has no distinct lid but sheds its spores through four longitudinal slits, or 'valves'.

Among liverworts, the genus *Anthoceros* is remarkable in many respects. It has a capsule which is totally unlike that of other liverworts in its long narrow shape, in splitting by two valves only, and in remaining green over a long period. Although this latter character is a feature of resemblance to mosses, *Anthoceros* is a thallose liverwort and could in no circumstance be taken for a moss. In a comparable way, *Sphagnum* and *Andreaea* are remarkable genera among mosses. Reference has already been made to the capsule of *Andreaea*; and the leafy shoot of *Sphagnum* is unlike that of any other moss. But in neither case would there be grounds for confusion with liverworts.

Thus on balance, and taking the sum of the characters, it is usually a simple matter to decide whether a particular plant should be referred to the mosses or the liverworts. This decision having been made, the next step is to identify the species within the group to which it has been referred.

3. IDENTIFICATION OF A MOSS

In a number of cases it is possible, with experience, to identify a moss in the field, and a field key is included in the present work with this object in view. In using this key it is essential to have a lens that magnifies about ten times and to give close attention to all details that can be seen with its aid. Thus, measurements of leaf length, details of leaf shape and character of leaf margin may all be important. Colour, too, is sometimes of value; also the 'texture' of the plant's surface—whether glossy or dull. Often the appearance of the plant when dry affords a useful character; for the leaves in some species become appressed against the stem when dry, in others they become strongly and irregularly curled, whilst in others again they form a close and regular spiral like a corkscrew.

As a rule it will be possible to see, in the field, into which of the two broad groups of mosses a plant falls, and a decision on this point will have to be made at an early stage in working with the key. For, although these two groups are based essentially on the position of archegonia, which are terminal in the acrocarpous mosses and are borne on a side-branch in the pleurocarpous mosses, this distinction is associated in most cases with a readily seen difference of habit. Thus, in the 'acrocarps' one normally sees a number of upright or nearly upright, little-branched stems, with seta and capsule, if present, arising at the apex or tip of each, whilst in the 'pleurocarps' the growth is very commonly prostrate, and wide intricate mats are formed as a result of the extensive branching that takes place. The branching, moreover, will be most commonly of a type known as pinnate, where numerous branches arise at intervals on either side of a main stem, a method of branching never seen in the acrocarps. Seta and capsule, if present, will be lateral in position.

An item of information that should always be noted carefully in the field is the precise surface, or substratum, on which the moss is growing. This should be recorded subsequently on the label if the specimen is being preserved, both for its intrinsic interest and because it may help in making the determination. It will probably also be worth noting, in an acrocarpous species, whether the plant is growing in obvious neat cushions, in wide irregular patches or as scattered individual stems. Use has been made of these features, on occasion, in the field key.

However desirable it may be to make an attempt at identification in the field, the real task of determining an unknown plant will begin subsequently, at home or in the laboratory. It is true that the experienced bryologist can name many species with confidence in the field; but this is because he already knows these plants. It will be a very different matter for the beginner, who would in no instance be wise to rely on a determination made in the field with the aid of the field key, when the moss 'arrived at' would be one with which he was not familiar. Indeed, it is always safest to confirm a determination under the microscope; and where any doubt or difficulty has arisen with hand-lens characters it is essential to call upon the much wider range of precise characters that the microscope makes available.

In making a determination in the laboratory, with all the microscopic characters of the plant available, two courses are open. It may happen that one is able to place the plant with some confidence in its family or even assign it to a genus, when it only remains to read through a few relevant descriptions in order to decide on the species. On the other hand, one may be quite unable to 'place' the plant among related mosses—and this is commonly so with beginners—when it may seem better to attempt the determination with the help of a mechanical aid such as an artificial 'key' to the species. Such a key is provided in this book. It makes use in the main of vegetative characters so that even barren plants should be referable to their species with its aid. After a little experience one should be able to refer a plant fairly quickly to the correct section in the key. This done, it may be 'run down' directly to the species.

Whichever method is to be followed in making the determination, the specimen can be examined in the same way. I give here the routine procedure I have found it useful to adopt. Unless the plant is so fresh as to be still moist it is best to begin by soaking for a few minutes in a dish of water. Usually a small but typical-looking fragment will suffice for this purpose, but at times, as where numerous species have been gathered in a closely intertwined mat and the whole allowed to dry out subsequently, it is best to soak the entire mass in water and, after squeezing out the water, to disentangle the various species. This often reveals fragmentary material of species besides those that were noticed in the field.

The next step is to detach a few leaves, carefully dissecting them off the stem into a drop of water that has already been placed in the middle of a clean glass slide. If the plant be held by a dissecting needle in one hand about half a dozen leaves can generally be pulled off, cleanly and complete, with a pair of forceps in the other hand. When the leaves are very minute or where, as in species of *Drepanocladus*, one is liable by this method to lose the important cells at the basal angles of the leaves, it is better to dissect under a dissecting microscope, using a pair of needles. While a fragment of the plant is being viewed under the dissecting microscope a sharp look-out should be kept for other features of interest, such as 'inflorescences' containing antheridia or archegonia. Ordinarily, however, it will suffice to concentrate on the leaf structure at this stage. Where a much-branched, pleurocarpous moss is being examined it is important to detach leaves from one of the main stems, since those on the finer branches are often quite different in shape. The leaves are arranged in the drop of water and the preparation is covered with a cover-glass. It is then ready for examination under the low power of the compound microscope.

Features which it is important to note in the leaf under examination are: (i) character of leaf margin—whether entire or toothed, and whether plane or rolled inwards or backwards; (ii) surface of leaf—whether smoothly flattened, transversely wavy (undulate), or cast into obvious longitudinal folds (plicate); and, most important of all, (iii) details of cell structure. The presence or absence of a nerve will of course have been noticed already. If there is a white 'hair-point' at the leaf tip it will be important to see whether this is 'smooth' or rough with teeth. The cell structure (often termed by bryologists 'areolation') of the leaf as a whole, however, affords the greatest number of critical characters and requires close study. Under a low power the general shape of the cells may be seen, also whether or not any specialized cells form distinctive patches at the basal angles of the leaf ('angular' or 'alar' cells). A high power will be necessary for further details.

The presence of very fine roughnesses on the surfaces (outer walls) of cells (papillae) is often an important character; and the high power of the compound microscope is usually needed in order to see these. The back of a folded leaf is the best place to examine under the high power for this purpose. Sometimes the outline of the cells in the upper part of a leaf is made obscure by the very numerous papillae. In certain cases it is a great advantage to be able to make precise cell measurements as an aid in identification, and for this a micrometer slide and eyepiece should be available. Once the eyepiece has been calibrated with the aid of the slide it will, of course, give a precise measurement by itself, and I find it an advantage to keep one permanently in the ocular of the microscope. The unit employed is $\frac{1}{1000}$ mm., which is the micron and is commonly expressed by the Greek letter μ. Differences in cell dimensions of the order of a few micra can be used in *Fissidens* and some other genera to separate critical species, and it will be found that use has been made of them in places in the key.

Turning to reproductive characters, it is safe to say that only comparatively rarely will characters of the 'inflorescence' prove of critical importance in making a determination. In certain cases, however, as in the separation of closely allied species in the genera *Bryum*, *Pohlia*, *Brachythecium* and certain others, it becomes important to note the precise distribution of antheridia and archegonia. In describing this bryologists employ four terms of specialized meaning. Thus, when these organs occur on separate plants the species is *dioecious*; when they occur on separate shoots of the same plant, it is *autoecious*; when the two types of organ occur on the same shoot, but not actually intermixed, the plant is said to be *paroecious*; finally, when antheridia and archegonia occur intermixed in a common 'inflorescence' the plant is *synoecious* (see fig. VII, p. 23). All these four conditions occur within the genus *Bryum*. In young 'fruiting' material it is often still possible to determine the arrangement of sex organs; for instance, an autoecious species will usually show the remains of the male branches, with old antheridia still present, and in a synoecious species it will be possible to see old antheridia about the base of the seta. The ellipsoidal ('bomb-shaped') antheridia and flask-shaped archegonia are not difficult structures to recognize.

The characters derived from seta and capsule are of considerable importance in identification. The colour of the seta is often diagnostic, as in separating allied species in the genera *Dicranella* and *Barbula*; in other cases, as in *Brachythecium*, it is important to note whether the surface of the seta is smooth or rough with minute tubercles (papillose). The length of the seta in the mature state is sometimes significant.

The capsule (fig. V, p. 21) affords numerous characters of value in its gross morphology, such as shape, length of 'neck' region (if this is developed), and the angle at which the capsule is held upon the seta. In respect to this latter point capsules are described as erect, horizontal or pendulous, erect capsules on the whole prevailing among acrocarpous mosses, horizontal in many species of *Hypnum*, *Rhytidiadelphus*, etc., and the pendulous condition being especially characteristic of many species of *Bryum*, *Pohlia* and *Mnium*. The amount of contraction that takes place below the mouth of the capsule and the extent of furrowing of the surface of the ripe capsule wall are other characters which may require to be noted. For this purpose it is important to secure thoroughly ripened, but not shrivelled, material. If the calyptra is still present it should be examined. In a few instances it is of such characteristic appearance as to point at once to the genus of the plant, the genera *Encalypta* and *Ulota*, for example, being easily recognized in this way.

The most precise characters furnished by the capsule are those of the peristome, the ring of teeth that surrounds the mouth of the capsule after the lid has fallen (fig. V). In some acrocarpous mosses this comprises a single ring; in many acrocarps, and in all pleurocarpous mosses, however, two rings are present. These are termed the outer and inner peristomes, the inner being much the more delicate structure. It is often essential to see the details of both outer and inner peristomes in order to make a determination (this is especially so in the genus *Bryum*), so that a few words on how best to examine these structures may be of value. In the first place, the general features of the peristome are seen to advantage by examining the mouth of an intact capsule (from which the lid has fallen) by reflected light. The capsule is laid dry upon a slide and examined, against a dark background, under the low power of the compound microscope. To secure an end-on view of the peristome it is sometimes useful to cut off the top of the capsule with scissors and examine this alone. Secondly, for finer detail, it is best to mount as much as possible of the peristome in water on a slide and examine this preparation in the usual way, by transmitted light, using the high power of the microscope. For this purpose a thoroughly soaked capsule should be cut off above the neck, then cut longitudinally in half on a slide in a drop of water. With needles (under a dissecting microscope) the spores may be removed and the delicate inner peristome may often be detached completely from the rim of the capsule mouth. This gives a satisfactory mount consisting of (i) two fragments of capsule wall with outer peristome teeth, (ii) two detached pieces of the inner peristome, which, duly covered, may then be examined under the low and high powers of the compound microscope. Where the inner peristome does not come away readily the two halves of the capsule may require to be mounted and examined without further dissection; then a drop of a wetting agent (e.g. Kodak's) is often useful in getting rid of air bubbles.

It is not always necessary, of course, to assemble *all* the data in this way before attempting to proceed with the determination, although it is often useful to begin by making the necessary microscopic mounts, especially of the leaves of the moss, before proceeding to use the key. Whether one reaches a determination by way of the key, or by means of a more direct approach which entails the ready 'spotting' of the genus of the specimen, it is important to check very carefully each point in the description (and to make use of the figures) before reaching a conclusion. In an introductory book such as this a slight misfit may point to a species that has been excluded from this work; but the affinities of the plant will probably have been gauged, and from this the next step is to refer to a reasonably complete herbarium or a larger work of reference. Even so, some specimens inevitably remain puzzling, and for these the wisest course is to submit them in the last resort to an expert for an opinion.

4. IDENTIFICATION OF A LIVERWORT

The general approach is the same as for a moss, but it will be well to touch briefly on certain features which arise directly out of the structural peculiarities of the hepatics. As with mosses, determination is often possible in the field, at least with the larger liverworts, and the most common and generally prominent of these are accordingly included in the field key. A clear-cut decision can usually be made on whether the plant body is thallose or foliose (leafy). Then, among leafy liverworts, a number of other characters can be satisfactorily seen with a hand lens, whilst it will be possible in thallose forms to make out whether the diamond-shaped or hexagonal markings characteristic of certain genera are present.

Thallose liverworts may be dealt with briefly first, in that they present problems peculiarly their own. In this group, for example, it is often necessary to cut a transverse section of the thallus—at least in order to confirm the determination. The shape of the transverse section thus obtained is highly characteristic in the different species of such genera as *Riccia* and *Riccardia*, and it alone can furnish information about the structure of the cavities that occur within the thallus in such genera as *Marchantia*, *Lunularia* and *Conocephalum*. A fairly rough section may suffice in some cases, but it is well to employ a sharp razor for this purpose. The transverse section will reveal much that cannot be seen from a superficial inspection, including the number of cell layers in the thallus, and, in *Riccia* spp. and some others, sex organs and other details that are not visible on the surface.

The rhizoids, unicellular structures not unlike root-hairs, spring from the surface of the thallus next to the substratum. They are of greater value for identification purposes in liverworts than in mosses. Rhizoids of two kinds, one without and one with tuberculate thickenings of the walls, are characteristic of the Marchantiales, and the distribution and colour of rhizoids are often of importance in naming a leafy liverwort.

The thallose forms, on the whole, should not present great difficulty. Extent of development of midrib, degree and manner of branching, and other easily seen characters are available, and the group as a whole is not a large one. The single large chloroplast in each of the green cells of *Anthoceros* is unlike anything seen elsewhere among hepatics.

By far the majority of the British Hepaticae are 'leafy liverworts' (Jungermanniales Acrogynae), and in making a determination within this group use is made of structural features which do not exist among mosses. One such character—not always easy to grasp at first—is the manner of overlapping of the leaves in the two principal ranks on the stem. This is best understood by reference to the figure (fig. II, p. 18) but may be described briefly here. First, it is important to realize that the leafy shoot of a liverwort normally grows more or less prostrate or ascending, at least when young, and hence is distinctly dorsiventral. This means that there is an upper surface (next the observer) and an under surface (next the substratum), and where underleaves of smaller size and distinct form are present they form a single row along the latter. When, however, the lateral leaves are obliquely inserted on the stem and crowded, so that they overlap, it is plain that they can do this in either of two possible ways. These two types of overlapping of the leaves have necessitated the use of two special terms, 'succubous' and 'incubous', the meanings of which are illustrated in the figure. It will be seen that in the succubous arrangement the observer who views the leafy shoot from above (underleaves, if present, concealed from view beneath the stem) will see the lower edge of each leaf, but the edge nearer the apex of the shoot will be concealed behind the next leaf above; whereas in the incubous arrangement the observer will see the upper edge of each leaf, whilst the lower edge will in this case be concealed behind the next leaf down the shoot. The succubous or incubous leaf arrangement is a constant generic feature and hence is important in classification and useful in making a determination.

Owing to the importance of leaf arrangement and because the leaves of liverworts are often very small and delicate, it is often better in this group to mount whole pieces of the leafy shoot for microscopic examination (in water under a cover-slip) and only attempt to dissect off individual leaves in special cases where this operation seems justified. Fragments of shoot have the advantage that in them a look-out can be kept for male bracts and antheridia, and for female bracts, bracteoles and archegonia, as well as for the generally more conspicuous perianths which surround developing 'fruits'. The distribution of rhizoids, the presence or absence of underleaves too small to be seen readily with a lens, and any special features of these structures may also be noted when mounted shoot fragments are examined under the low power.

The shape of the leaf cells is not often of great value in the identification of a liverwort, although in certain cases it is important, as in the elongated cells that form a vitta or 'false midrib' in the leaf of *Diplophyllum albicans*, or the enlarged cells of nearly square outline forming the leaf margin of *Plectocolea crenulata*. Particular attention, however, must be given to the angles or corners of the cells where small or large thickenings of the walls may often be detected. These are termed 'trigones' by bryologists, but in the present work the expression 'corner thickening' is used. In many genera of leafy liverworts use is made of this character in separating the species. Occasionally the corner thickenings are as large as, or even larger than, the cell cavities themselves.

As with mosses, considerable use is made of cell measurements in identification, and such measurements will commonly be made during the routine examination of the leaf under the high power of the microscope. The cell contents are often of diagnostic value, especially the 'oil-bodies' which are present in the very common leafy liverwort *Nardia scalaris* and a number of others. Again, it is at times important to examine the surface of a folded leaf under the high power for traces of roughness of the cuticle. In the genus *Scapania*, for example, a rough cuticle is characteristic of certain species.

Leaves associated with the organs of reproduction are often markedly different from the ordinary vegetative leaves and are sometimes of special diagnostic value. The antheridia are in many instances borne in the hollow of leaves which are often smaller than the vegetative leaves and are notably concave in form. The leaves associated with the female 'inflorescence' are, however, more strikingly distinct from the ordinary leaves in form and often greatly exceed them in size. Those corresponding in position with the two lateral ranks of leaves are termed bracts, those in the mid-ventral line (corresponding with underleaves) being known as bracteoles (after a rough analogy with flowering plants). In the genus *Frullania*, and elsewhere, the different species are marked by both bracts and bracteoles of distinctive form.

The perianth is a characteristic and conspicuous feature of most leafy liverworts when in the reproductive state (fig. VI, p. 22). It occurs immediately above the bracts and bracteoles just described and is a tube or purse-like structure which serves as a protection for the archegonia and later for the developing sporogonium ('fruit'). It is formed sometimes before, sometimes after, fertilization. It may be pear-shaped, ovoid or narrowly cylindrical in shape, and rounded, compressed or triangular in cross-section. All the structural features of the perianth must be noted with care, since they often form the best means of distinguishing critical species. Sometimes the extremity (mouth) of the perianth is marked by a fringe of fine projections. These are called 'cilia' and consist of slender filaments. At other times the perianth segments are merely 'toothed'.

Although the structure of the different layers of the capsule wall has been extensively employed in the taxonomy of the liverworts, it will not be necessary to take much note of it for the purpose of an elementary work such as this. Nor will the beginner have much occasion to make use of the long, narrow, spirally thickened cells (termed elaters) that are formed among the spores (fig. VI). They are, of course, a characteristic and essential part of the content of the capsule in the majority of liverworts; nevertheless, although they differ considerably in the way they function in spore-dispersal in different groups of hepatics, they do not ordinarily present great differences of form in different species of the same genus or family. Spore dimensions, however, are sometimes of value, whilst in *Fossombronia* and a few other liverwort genera the markings or 'sculpturing' on the surface of the spore are so important as to be practically essential for the determination of the species.

On the whole, owing to the greater diversity of leaf form and leaf arrangement in the foliose liverworts, identification is much less dependent on 'capsule' characters in this group than in the mosses. Moss capsules, moreover, with their wide range of form and peristome structure, offer a far greater number of readily seen structural differences than do the relatively uniform capsules of the liverworts. So the emphasis, as well as the procedure in identification, will tend to be rather different in the two groups.

5. COLLECTION AND PRESERVATION OF BRYOPHYTES

Mosses and hepatics may be collected in small tins, or other small containers which allow individual sods or tufts to be kept separate. Such a method allows the material to be kept fresh for examination, and is especially useful when it is intended to keep a portion of the gathering for subsequent cultivation. Collecting direct into a large container such as a vasculum is not recommended except when material of some species is to be gathered in bulk. I find it useful to employ old envelopes for field collecting. Habitat notes can then be written on the envelope at the time and the different gatherings made in the course of a day can be kept separate. Large envelopes or, better, strong calico bags may be used where an extensive collection is being taken from a single habitat. If, as often happens, much of the material is wet when collected, the first opportunity at the end of the day should be seized to spread out the envelopes and allow their contents to dry thoroughly.

A bryologist may make a successful bryophyte herbarium along the same lines as those followed for flowering plants: lightly pressed specimens can be mounted directly on standard size herbarium sheets, as many as six or eight collections of the smaller species being conveniently preserved on a single sheet. As a refinement on this method, paper or cellophane folders can be employed, the specimens being contained in and protected by the folders, which in turn are mounted on sheets. Many students, however, prefer to maintain their bryophyte collections in the form of a card index. This is a convenient method which ensures adequate protection of the specimens and has the merit of occupying relatively little space. Each gathering, when thoroughly dried (with or without light pressure), is placed in a folder or 'packet' of suitable size; the name, locality, habitat details and date are appended; and the specimen is then ready to be filed away in the collection. The sketch (fig. VIII, p. 24) shows a convenient type of packet, and the way in which such packets may be made out of pieces of plain white paper. It is a good plan to maintain a stock of at least two sizes. Gummed envelopes are quite unsuitable for the purpose.

Wherever the small size or the critical nature of the objects concerned would seem to justify the step, a permanent slide should be made. This is especially useful with the smaller liverworts, and a collection of such preparations will prove of value in the subsequent determination of critical material. Either of the following methods may be employed:

(i) Glycerine jelly preparation. First mount in a mixture of 1 part pure glycerine, 2 parts methylated spirit and 3 parts water. After a few days re-mount in glycerine jelly. A circular cover-slip should be used so that the preparation may subsequently be sealed by 'ringing' with gold size or other suitable medium. Another method which may be found preferable is to proceed as follows. Use two square cover-slips of different sizes and mount the specimen in glycerine jelly, as it were in a sandwich, between the two cover-slips. Then put a large drop of Canada balsam on a slide and lower the sandwich on to it, with the smaller cover-slip downwards. The balsam will then flow out and fill the space below the overlap of the larger cover-slip.

(ii) Gum chloral preparation. Transfer from water direct to gum chloral (Berlese's fluid),* making the drop big enough to spread as a ring round the cover-slip. This serves to seal the preparation and allows for subsequent contraction on drying.

6. DISTRIBUTION AND ECOLOGY OF BRYOPHYTES

Most of the species described in this book are generally distributed throughout Britain. Among those of more limited geographical range, however, may be mentioned numerous species which are largely confined to the west and north. Some of these are found only on or near the west coast and thus show an oceanic type of distribution; others are what may be termed highland plants and are found throughout the

---

* Formula for *gum chloral mounting medium*:

Distilled water	100 c.c.
Gum arabic	40 g.
Glycerine (conc.)	20 c.c.
Chloral hydrate	50 g.

Dissolve the gum arabic in cold water, then add the glycerine and chloral hydrate. Heat gently in a water-bath until the chloral hydrate has dissolved, then filter hot through a no. 31 Whatman filter-paper. The solution of the gum arabic takes up to 48 hours.

more mountainous parts of the country, whilst rare or absent in lowland districts. Certain species, on the other hand, are common in the midlands and south, but become increasingly rare farther north and west.

The subject of bryophyte ecology is one that has attracted considerable attention in recent years, and the importance of bryophytes as 'indicator' plants—revealing something of the potentialities of the land on which they grow—has come to be appreciated to an increasing extent. There is often a very close and precise relation between the acidity or alkalinity of a soil and the flora which it bears. Thus species such as *Tortella tortuosa* and *Neckera crispa* are well known to be strict calcicoles, whilst most species of *Polytrichum* and all *Sphagnum* species are calcifuge. It has been thought inadvisable in an elementary and comparatively non-technical work such as this to introduce the concept of pH as such, but notes on the habitat preferences or requirements of each species are given under a separate head at the end of the description, and an attempt has been made to make this information as precise as possible. In many instances, too, an indication is given in these 'ecological notes' of the distinctive communities of species which exist. Many of these communities are of remarkably constant composition, and every bryologist will readily call to mind certain pairs or groups of species that are habitually associated together; e.g. the *Philonotis fontana–Bryum pseudotriquetrum* association of bog springs on mountains.

At the end of the book are lists of common mosses and hepatics likely to be met with in each of a number of recognizably distinct habitats. These are not intended to represent communities in the strict sense, but it is thought that such lists may prove of help to the beginner. They serve to emphasize the reality of habitat preferences among bryophytes. Moreover, it is exceedingly likely that the first plants which the beginner collects in each of these habitats will be among those included on the list. Thus they may serve at times as 'short cuts' to identification.

GENERAL FIGURES EXPLANATORY OF MATTER
IN THE INTRODUCTION AND GLOSSARY

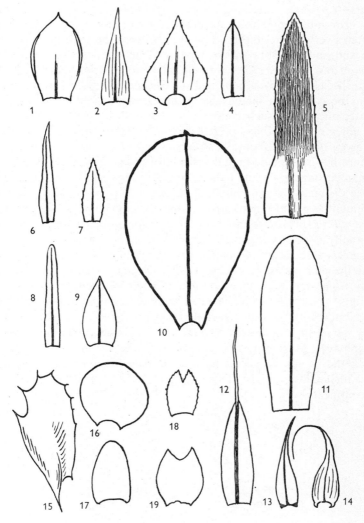

Fig. I. Leaf shapes.

Fig. I. Leaf shapes

1. Apiculate. *Pseudoscleropodium purum.*
2. Lanceolate, acuminate, plicate leaf of *Camptothecium sericeum.*
3. Cordate-triangular, plicate, with well-developed auricles. *Eurhynchium striatum.*
4. Mucronate leaf of *Barbula unguiculata.*
5. Leaf of *Polytrichum aloides,* with hyaline sheath and opaque limb covered by longitudinal lamellae.
6. Linear-lanceolate. *Amphidium mougeotii.*
7. Lanceolate, finely toothed (denticulate). *Aulacomnium androgynum.*
8. Ligulate. *Diphyscium foliosum.*
9. Ovate, acute. *Bryum bicolor.*
10. Obovate, obtuse. *Mnium punctatum.*
11. Lingulate (tongue-shaped). *Encalypta streptocarpa.*
12. Leaf of *Grimmia pulvinata* showing long hyaline hair-point.
13. Falcate leaf of *Blindia acuta* with clearly defined patches of alar cells at base.
14. Falcate leaf of *Drepanocladus uncinatus* with acuminate tip, plicate surface and well-developed auricles at base.
15. Leaf of *Plagiochila spinulosa* showing dentate margin and decurrent base.
16. Orbicular leaf of *Plectocolea crenulata* showing narrow insertion.
17. Rounded ovate leaf of *Jungermannia tristis.*
18. Bifid leaf of *Gymnomitrium crenulatum*; acute sinus.
19. Bifid leaf of *Lophozia ventricosa*; U-shaped sinus.

(1–14, all ×11; 15–19, all × 22.)

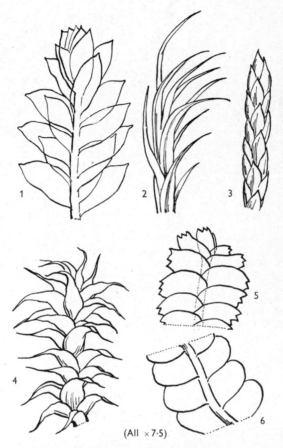

Fig. II. Special types of leaf form or leaf arrangement, producing shoots of characteristic appearance.

1. Complanate. *Neckera complanata.*
2. Falcato-secund. *Dicranella heteromalla.*
3. Imbricate, julaceous. *Bryum argenteum.*
4. Squarrose. *Rhytidiadelphus squarrosus.*
5. Incubous (antical view). *Bazzania trilobata.*
6. Succubous (antical view). *Chiloscyphus polyanthus.*

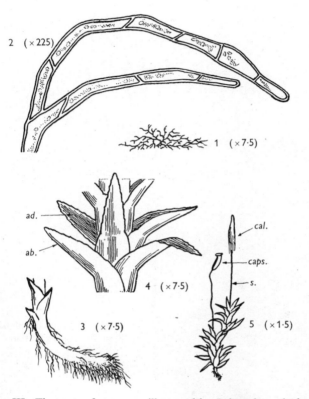

Fig. III. The parts of a moss as illustrated by *Polytrichum aloides*.

1, protonema; 2, fragment of same; 3, base of gametophyte, showing rhizoids and diminutive leaves; 4, part of leafy shoot: *ab.* abaxial (dorsal) surface, *ad.* adaxial (ventral) surface, of leaf; 5, gametophyte surmounted by two sporophytes: *cal.* calyptra, *caps.* old capsule, *s.* seta.

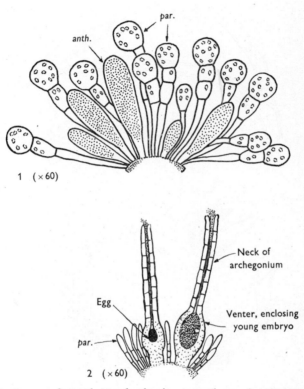

Fig. IV. Organs of sexual reproduction in a moss (somewhat diagrammatic).

1. Part of contents of male 'inflorescence' of *Funaria hygrometrica*: *anth.* antheridium, *par.* paraphyses.

2. Part of contents of female 'inflorescence' of the same species, showing two archegonia and several paraphyses (*par*).

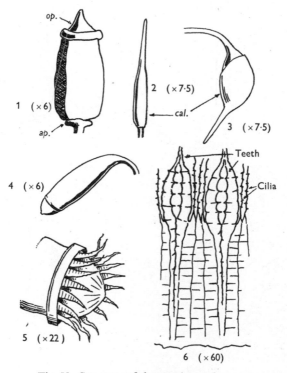

Fig. V. Structure of the moss capsule.

1, erect capsule of *Polytrichum commune*, *ap.* apophysis, *op.* operculum (lid);
2 and 3, two relatively young stages in the development of the capsule of *Funaria hygrometrica*: *cal.* calyptra; 4, inclined capsule of *Bryum capillare*; 5 and 6, peristome of *Bryum capillare* (in 5, the sixteen outer teeth are seen and the dome formed by the inner peristome; 6 shows details of the latter).

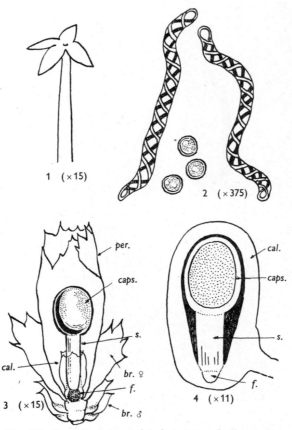

1 (×15)

2 (×375)

3 (×15)

4 (×11)

Fig. VI. Capsule and associated structures in liverworts.

1–3. *Lophocolea heterophylla*: 1, old empty capsule, showing four valves spread crosswise; 2, three spores and two elaters; 3, diagrammatic representation of young sporophyte and related structures (*br. ♂, br. ♀*, male and female bracts, *cal.* calyptra, *caps.* capsule, *f.* foot, *per.* perianth, *s.* seta).

4. *Riccardia pinguis*, young sporophyte and related structures (*cal.* calyptra, *caps.* capsule, *f.* foot, *s.* seta).

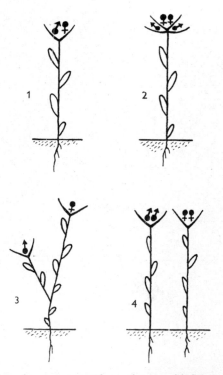

Fig. VII. Diagrams of arrangement of sexual organs ('inflorescence' types) found in different mosses.

1–3. Various monoecious conditions: 1, synoecious; 2, paroecious; 3, autoecious.
4. Dioecious condition. ♂, position of antheridia; ♀, position of archegonia.

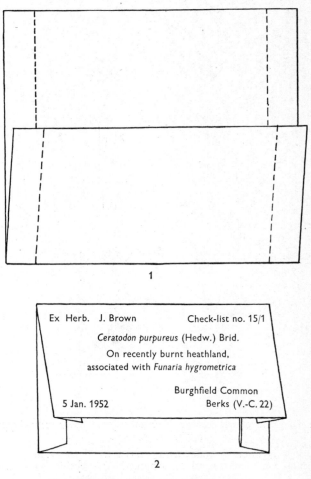

Fig. VIII. A suitable packet for bryophytes:
1, making the packet; 2, completed packet, fully labelled.

GLOSSARY

acidophilous, acid-loving, hence used of plants that grow on substrata which are acid in reaction.

acrocarpous, of mosses; bearing the archegonia, and hence the seta and capsule (fruit), at the *tip* of stem or branch. (N.B. After fertilization the stem may continue to grow so that the fruit appears to be borne laterally, e.g. *Breutelia chrysocoma.*) Acrocarpous mosses form one of the two broad divisions of the true mosses, being distinguished from the pleurocarpous mosses by this character as also by their prevailing growth form or habit. It should be noted, however, that the division into 'Acrocarpi' and 'Pleurocarpi' is not now generally regarded as a natural one, but it is still convenient to retain it.

acuminate, gradually tapering into a relatively long, narrow point; e.g. leaf of *Drepanocladus uncinatus* (fig. I, 14).

acute, sharply pointed, applied to a leaf tip in contrast with one that is bluntly rounded (obtuse); cf. *acuminate*.

alar cells, the cells at the basal angles of a leaf, which often differ from those elsewhere in the leaf, e.g. in *Dicranum, Blindia* (fig. I, 13), *Drepanocladus*, etc.; also known as angular cells. Those parts of the leaf base, in which alar cells occur, sometimes form well-defined clasping auricles (fig. I, 3 and 14).

angular cells, see *alar cells.*

antheridium, the male reproductive organ in bryophytes, ferns, etc.; the antheridia and paraphyses are often surrounded by leaves of special form, the whole then being known variously as the male 'inflorescence' or male 'flower' (fig. IV, 1).

antical, the term applied, in a leafy liverwort, to that part of a stem or leaf that lies in front (or above, if the stem is creeping), hence away from the substratum and nearer to the observer. Cf. especially those liverworts in which each leaf is divided into an antical and a postical lobe (figs. 185–195).

apiculus, a short, abrupt point, e.g. in a leaf tip (fig. I, 1).

apophysis, the more or less solid part of the moss capsule next to the seta and below the spore sac, sometimes forming a distinct swelling of its own, as in *Polytrichum* spp. (fig. V, 1); exceptionally large in *Splachnum.*

appendiculate, the term used of the cilia (q.v.) of the inner peristome when they bear short transverse bars, as in the so-called complete or perfect peristomes of certain species of *Bryum* (fig. V, 6).

appressed, describing leaves that are held more or less erect and closely applied to the stem, e.g. the leaves of most species of *Orthotrichum* when dry (fig. 100).

archegonium, the female reproductive organ, typically a flask-shaped structure, the swollen part of which (the venter) encloses the egg (ovum); a group of archegonia and paraphyses, surrounded by leaves of special form, is known variously as a female 'inflorescence' or female 'flower' (fig. IV, 2).

areolae, the small areas into which a larger surface may be subdivided, by a system of ridges or other means, e.g. on the surface of the spore in certain species of *Fossombronia*; also the surface of the thallus in *Marchantia, Conocephalum,* etc.

areolation, the network of the cells of a leaf, as seen in surface view under the microscope.

auricles, clasping portions of the leaf base, found only in certain mosses, e.g. *Drepanocladus* spp., *Eurhynchium striatum,* etc. (fig. I, 3 and 14).

autoecious, having the male and female organs on the same plant, but in separate inflorescences (fig. VII, 3).

axil, the angle between a leaf and the upward continuation of the stem to which it is attached.

axillary, in the axil, i.e. at the junction of leaf and stem.

base-rich (applied to soils, water of streams, etc.), rich in that class of chemical substance which includes the salts of calcium, magnesium, potassium, etc.; hence basic and the reverse of acid.

bifid, bilobed, cleft into two divisions, e.g. the leaves in many liverworts (fig. I, 18), the peristome teeth of some mosses.

bract, modified leaf of special form associated with the archegonia or antheridia. In many liverworts the precise form of the bracts is of systematic importance, e.g. *Frullania* spp. (figs. 192, 193). Among mosses the bracts are sometimes known as perichaetial leaves. They may form a distinct rosette, resembling a minute flower (figs. 9, 89, 93).

bracteole, the term used in liverworts for modified underleaf of special form associated with reproductive structures (fig. 193).

bulbil, any small bulb-like unit of vegetative propagation, usually produced in a leaf axil; cf. *Pohlia proligera, Bryum erythrocarpum,* etc.

calcicole, favouring rock or soil rich in lime or other bases, the expression 'strict calcicole' implying that a species will grow only in such conditions.

calcifuge, literally 'fleeing from lime', hence in general applied to species that avoid lime-rich rock or soil; a 'strict calcifuge' will grow only in the absence of lime. Since lime-free soils are usually acid, calcifuge plants may also be regarded as acidophilous.

callunetum, vegetation dominated by *Calluna vulgaris* (heather or ling).

calyptra, a protective covering of the young capsule, formed from the enlarged wall of the archegonium; it remains as a cap or hood-like structure on the moss capsule until it is nearly ripe (fig. III, 5 and V, 3) and is sometimes of considerable help in identification. In the liverworts it is conspicuous terminally only in the early stages of sporophyte development; thus it is not carried up as a hood over the capsule, but may be seen at a late stage as a collar around the base of the seta (fig. VI, 3 and 4).

capsule, the part of the sporogonium containing the spores; often loosely known as the fruit (figs. III and V).

carr, the ecologist's term for swampy woodland of alder (*Alnus*), *Salix* spp., *Rhamnus*, etc., in fen country.

chlorophyll, the green pigments of plants, contained in the chloroplasts, and enabling them to perform photosynthesis (synthesizing their own organic substances from inorganic raw materials in the light).

chlorophyllose (chlorophyllous), containing chlorophyll (or chloroplasts).

chloroplasts, specialized units of protoplasm containing the chlorophyll; among bryophytes most commonly taking the form of small green granules, many in each cell; in *Anthoceros* relatively large, and only one or two in each cell.

cilia, any fine, hair-like or thread-like structures, e.g. those between the teeth of the inner peristome in some mosses (cf. *Bryum* spp., fig. V, 6); also chains of cells forming thread-like projections on the margins of leaves or perianth in liverworts (such a leaf or perianth being termed ciliate).

cleistocarpous, capsule opening by irregular splitting, not by a lid or by some other regular method of dehiscence such as slits or valves; e.g. in the moss genera *Pleuridium*, *Phascum*, *Ephemerum*, etc.; also in *Riccia*.

columella, the central, sterile tissue in a capsule.

complanate, with leaves so arranged that the branch as a whole appears to be flattened more or less in one plane (fig. II, 1).

convolute, rolled together, as are the perichaetial bracts of *Barbula convoluta*, which form a tubular sheath about the base of the seta.

cordate, heart-shaped in outline; used of leaves (fig. I, 3).

corner thickenings, the term used here to denote the thickenings of the wall at the angles, or corners, of the cells in many leafy liverworts. Technically known as *trigones*, these thickenings are an important character for the identification of species.

crenulate, used of any edge or margin that bears a number of short, close-set, rounded projections; e.g. leaf of *Gymnomitrium crenulatum*, mouth of perianth in *Plectocolea crenulata*, etc.

crisped (or **crispate**), curled and twisted; e.g. the leaves of many mosses in dry conditions, also the thallus of some liverworts.

cusp, a sharp, rigid point; hence a leaf or shoot tip having this character is termed cuspidate, e.g. shoot of *Acrocladium cuspidatum* (fig. 127).

cuticle, a thin layer of material, chemically different from the rest, covering the external cell walls in some mosses and liverworts. Although the term is in general use there is little definite information as to whether a cuticle, as usually understood, really exists in bryophytes.

cygneous, curved like the neck of a swan, e.g. seta of *Campylopus* spp.

decurrent, used of the base of a leaf when this is continued down the stem as a ridge or wing (fig. I, 15).

dendroid, tree-like; hence used to describe the habit of mosses in which the principal secondary stems resemble miniature trees, e.g. *Climacium, Thamnium.*

dentate, sharply toothed, e.g. leaf margins in a number of mosses, and some liverworts (fig. I, 15).

denticulate, diminutive of dentate, used for leaf margins on which the teeth are relatively minute or obscure.

detritus, any loose gritty or stony material resulting from rock disintegration.

dioecious, bearing male and female sex organs on separate plants (fig. VII, 4).

distichous, refers to leaves that are borne strictly in two ranks, on opposite sides of the stem, e.g. the leaves of *Distichium, Fissidens, Schistostega, Diplophyllum,* etc.

dorsal, the upper surface (of a thallus); the back or abaxial surface (of a leaf), i.e. the surface away from the stem when the leaf lies appressed to it (fig. III, 4).

ecotype, a taxonomic unit within a species; often characterized by minor morphological differences and always by its restriction to a particular set of ecological conditions.

elaters, spirally thickened cells which are intermixed with the spores in the capsules of most liverworts and play a part in spore discharge (figs. 196, 197).

entire, without teeth (e.g. many leaf margins, etc.).

epidermis, the outermost, or covering layer of cells in any organ.

epiphyte, a plant which grows on other plants, but not necessarily parasitically, e.g. mosses or liverworts growing on the twigs, branches or trunks of trees.

eutrophic, used of water relatively rich in mineral nutrients.

excurrent, used of a nerve which runs out, or protrudes, beyond the leaf tip (fig. I, 4 and 12).

exserted, protruding; hence used of any organ that protrudes beyond surrounding organs, e.g. exserted perianth in liverworts, exserted (opposite of immersed) capsule in mosses.

falcate, curved like a sickle, e.g. leaves of *Dicranum* spp., *Drepanocladus* spp., etc. (fig. I, 13 and 14).

falcato-secund, sickle-shaped and uniformly turned to one side of the stem, e.g. leaves in typical states of *Hypnum cupressiforme, Dicranum scoparium, Dicranella heteromalla,* etc. (fig. II, 2).

fibrils, conspicuous spiral thickenings of the walls of the hyaline cells in *Sphagnum* (figs. 1–3).

flagellum (plur. *flagella*), a term used with a special meaning in bryology, i.e. for slender, runner-like branches bearing leaves of greatly reduced size; e.g. in *Bazzania trilobata, Odontoschisma sphagni* (figs. 159, 160, 165).

flexuose, having twists or bends, e.g. the margins of certain leaves.

foliose, leafy; having the plant body differentiated into stem and leaves.

foot, the part of the sporophyte at the base of the seta, its function being to absorb nutrients from the gametophyte.

fusiform, spindle-shaped, i.e. narrow and tapering at both ends; applied to leaf cells of some mosses, also to perianths of some liverworts.

gametophyte, the generation or phase in a plant's life cycle that bears the sexual organs; hence, among bryophytes, the moss or liverwort 'plant' as opposed to the sporogonium, which forms the sporophyte generation.

gemma, a unit of vegetative propagation; varied in shape and characteristically small, they may be unicellular, two-celled or multicellular and bud-like; borne on the leaves (*Ulota phyllantha*), at the tip of the stem, in the leaf axils, etc. (figs. 91, 102, etc.).

glabrous, smooth, implying total absence of hairs.

glaucous, of a green colour that is light in shade and tinged with bluish grey, e.g. leaf of *Polytrichum urnigerum.*

globose, round like a ball or globe, as are the capsules of some mosses and numerous liverworts.

guard cells, two cells of specialized shape and structure that surround the pore of a stoma (cf. fig. 99).

hair-point, a term employed in this work (especially in the key to mosses) in a somewhat special sense, i.e. for a leaf tip that is hair-like in form and different from the rest of the leaf in appearance, being colourless, not green; e.g. leaves of many species of *Tortula, Grimmia, Rhacomitrium* (fig. I, 12).

hyaline, colourless and transparent.

hypnoid, like *Hypnum,* a genus of mosses of which *H. cupressiforme* is a typical and abundant example.

imbricate, used of leaves that (arranged in several rows and attached at different levels) overlap one another like the tiles of a roof (fig. II, 3).

immersed, used of capsules when they are overtopped by the perichaetial leaves (cf. fig. 13); also of stomata when the guard cells are largely concealed beneath the surrounding cells of the capsule wall, e.g. *Orthotrichum anomalum* (fig. 98).

inclined, used of capsules that are held between erect and horizontal or between horizontal and pendulous.

incubous, descriptive of one of the two possible ways in which the leaves of a liverwort may overlap when obliquely inserted on a more or less flattened shoot; it implies that, when this flattened shoot is viewed from above, the forward edge of each leaf rests *upon* the hind edge of the leaf next above it (fig. II, 5). Cf. *succubous*.

inflorescence, a term borrowed from descriptive morphology of higher plants and used among bryophytes to denote the groups of antheridia, or archegonia, and associated structures.

insertion, the attachment of leaves to a stem; thus, among leafy liverworts, it is important to note whether the leaves are transversely, or obliquely, inserted. Thus in *Marsupella* (fig. 181) the insertion is transverse, in *Cephalozia* (fig. 164) it is markedly oblique.

involucre, used for several rather different kinds of enveloping organ in the Hepaticae: e.g. the sheath surrounding each group of archegonia (and later sporogonia) in *Sphaerocarpus*, *Marchantia*, etc.; the flap or tube protecting the archegonia (and later surrounding the base of the calyptra) in *Pellia* spp.; enlarged bracts surrounding the base of the perianth in some leafy liverworts.

isodiametric, used of cells which have an equal diameter in all directions.

julaceous, of branches that are catkin-like, being smoothly cylindrical with the leaves uniformly erect or appressed (fig. II, 3).

lamellae, any thin sheets or plates of tissue; e.g. the longitudinal plates of green tissue on the leaves of *Polytrichum*, *Atrichum*, etc., also the ridges on the spores of some species of *Fossombronia* and certain other liverworts.

lamina, the blade, or expanded part of the leaf, strictly as distinct from the stalk (petiole) or sheathing base; but there being no leaf stalk in bryophytes, the term lamina is commonly used to indicate any expanded part as distinct from the nerve.

lanceolate, shaped like the head of a lance, e.g. leaves of many mosses (fig. I, 7).

ligulate, strap-shaped, proportionally longer and narrower than lingulate, e.g. leaf of *Diphyscium foliosum* (fig. I, 8).

limb, the upper part of the leaf as distinct from its base. Applied to leaves in which the lower part sheathes the stem, e.g. *Polytrichum* (fig. I, 5).

linear, very narrow, so that the edges or sides are nearly parallel; used for both leaves and cells.

lingulate, tongue-shaped, proportionally shorter and wider than ligulate, e.g. leaves of *Atrichum undulatum, Encalypta streptocarpa* (fig. I, 11).

marsupium, a pouch-like structure surrounding the female 'inflorescence' and enlarging considerably after fertilization, found only in certain liverworts, e.g. *Saccogyna, Calypogeia*; also known as the *perigynium.*

micron $(\mu) = \frac{1}{1000}$ mm.

midrib, the vein running down the middle of a leaf, in bryophytes more commonly termed the nerve or costa; it is several cells thick, while the rest of the leaf is only 1–2 cells thick (cf. *vitta* of *Diplophyllum*); a well-defined median rib of thickening in a thallus (sometimes more obvious in a transverse section than in surface view).

monoecious, bearing male and female organs on the same plant. For the different types of monoecious condition among bryophytes see *autoecious, paroecious, synoecious* (cf. fig. VII).

mucro, a very short, abrupt point; thus in the leaves of many mosses (e.g. *Barbula unguiculata, Tortula subulata*) the nerve extends beyond the leaf tip in the form of a mucro (fig. I, 4). Cf. *cusp* and *apiculus.*

mucronate, used of leaves that are abruptly and shortly pointed.

nidus, literally a nest, hence a niche or pocket (of soil or other material) favourable for colonization.

nitrophilous, literally 'nitrogen-loving', hence used of plants which grow in habitats that are rich in nitrogen compounds.

obovate, with an outline like an egg attached by its *narrower* end, e.g. leaf of *Mnium punctatum* (fig. I, 10).

oil bodies, conspicuous inclusions of the leaf cells of certain liverworts, normally colourless and glistening, but varying in shape and chemical composition in different species. Often they somewhat resemble chloroplasts in shape (figs. 167, 173, 179, 194).

orbicular, almost circular in outline, e.g. leaf of *Plectocolea crenulata* (fig. I, 16).

ovate, with an outline like an egg attached by its broader end, e.g. leaf of *Jungermannia tristis* (fig. I, 17).

papillae, small protuberances on the outer wall of any organ or cell; some are just visible with a lens (× 20), such as the papillae on the seta of *Brachythecium rutabulum*; but those on the surfaces of leaf cells are very minute and require the high power of the microscope for study.

papillose, covered with papillae (cf. figs. 4, 38, 133, etc.).

paraphyllia, minute leaf-like or freely branched structures that are borne among the leaves on the stems of some mosses, e.g. *Hylocomium splendens.*

paraphyses, minute threads borne among the antheridia or archegonia in an 'inflorescence'; most commonly multicellular, and club-shaped or hair-like in form (fig. IV).

paroecious, bearing antheridia and archegonia close together, but not mixed, the antheridia being in the axils of bracts below those that surround the archegonia (fig. VII, 2).

pendulous, hanging down, e.g. the ripe capsules of many species of *Bryum.*

perianth, a tube-like or purse-like structure surrounding the archegonia in liverworts and commonly forming the most conspicuous feature of the female 'inflorescence' (fig. VI, 3).

perichaetial, used of the leaves (or bracts) immediately enclosing the archegonia, and hence surrounding the base of the seta, or the capsule in those mosses in which the seta fails to elongate (fig. 13).

perigonial, used of leaves (bracts) surrounding the antheridia, thus forming the outermost part of the male 'flower'. The form of the perigonial leaves (bracts) is diagnostic in certain mosses, e.g. *Philonotis fontana* and *P. calcarea* (fig. 93).

peristome, the ring of teeth that surrounds the mouth of a moss capsule and is visible when the lid of a ripe capsule is removed; it may be single or double (fig. V, 5).

photosynthetic, used of cells or tissues having chlorophyll, and hence able to build up organic carbon compounds from carbon dioxide and water (*photosynthesis*).

pinnate, feather-like; used of lobes or branches that arise in a regular manner on each side of a main axis, e.g. branching of the shoot in *Cratoneuron commutatum* (fig. 116) and of the thallus in *Riccardia multifida* (fig. 200).

plane, used of a leaf margin that is flat, i.e. neither incurved nor recurved.

pleurocarpous, bearing the archegonia, and hence the seta and capsule (fruit), on a short side-branch, and not at the tip of a main stem or branch. Cf. *acrocarpous.*

plicate, used of a leaf that is longitudinally folded or wrinkled, so that 'pleats' or furrows are visible on its surface, e.g. *Eurhynchium striatum*, some species of *Brachythecium*, etc. (fig. I, 2, 3 and 14).

plumose, regularly and closely pinnate like a plume or feather, as are the leafy shoots of *Ptilium crista-castrensis.*

pores, (i) openings, visible with a lens, in the centre of the areolae on the upper surface of the thallus in many Marchantiales (figs. 201–4); (ii) perforations or thin places in a cell wall, e.g. perforations in the hyaline cells of the leaves in *Sphagnum* spp. (figs. 1–3), and thin places in the otherwise thick walls of cells near the base of the leaf in many mosses. (For the latter sense, i.e. pits, see figs. 30, 153, etc.)

porose, having perforations or thin places in the cell wall.

postical, a term applied, in a leafy liverwort, to that part of a stem or leaf that lies behind (or below, if stem is creeping), hence next to the substratum and away from the observer. Cf. *antical*.

processes, a term sometimes used for the teeth of the inner peristome in a moss capsule (fig. V, 6). Cf. *cilia*.

protonema, the young stage of a moss or liverwort produced on the germination of a spore. In the mosses it is usually a system of branched green threads (fig. III, 1 and 2), but in *Sphagnum, Andreaea* and the liverworts it takes various forms, often being a flat expanse of cells. It is normally short-lived (occasionally persistent, e.g. *Polytrichum aloides*), and the adult leafy plant or thallus arises from buds on it.

pseudoperianth, a delicate sheath, usually late to develop, that surrounds the young sporophyte in some liverworts, e.g. in *Fossombronia*.

pseudopodium, a leafless prolongation of the stalk of the gametophyte, e.g. in *Sphagnum, Aulacomnium androgynum*.

quadrate, used of cells that in surface view appear square or nearly so, e.g. those near the apex of the leaf in *Barbula recurvirostra*.

radicles, relatively stout red-brown rhizoids that cover the lower parts of the stems in many mosses and appear like miniature rootlets; in the mass they are often termed the *tomentum*.

receptacles, the various kinds of stalked, often umbrella-like structures on which the sexual organs are borne in *Marchantia* and allied liverworts.

recurved, curved backwards, e.g. leaves or leaf margins in some mosses.

revolute, rolled back, e.g. the leaf margins of *Barbula revoluta*; a more extreme condition than recurved (q.v.).

rhizoids, thread-like outgrowths, almost always unicellular in liverworts but multicellular (septate) in mosses; in the liverworts they are often delicate and colourless, like the root hairs of higher plants; but those of mosses (fig. III, 3) may be relatively large, freely branched and red-brown in colour. See *radicles*.

saprophytic, living on (and obtaining nourishment from) dead organic matter.

secund, turned to one side, e.g. leaves of the moss *Dicranella heteromalla* and the liverwort *Odontoschisma sphagni*.

serrate, closely, regularly and rather sharply toothed, like a saw; applied to leaf margins, etc., but much less common than *dentate* and *denticulate* among bryophytes.

sessile, without a stalk, as are the leaves of all bryophytes; also describes the condition of a capsule which is not borne on a seta.

seta, the stalk on which the capsule (or 'fruit') is normally borne (fig. III, 5).

siliceous, having a high percentage of silica; such rocks, being as a rule relatively poor in bases, bear a flora that contrasts with that of calcareous rocks.

sinus, a depression between two prominences, as on the margin of a leaf or bract; used especially of the bifid leaves of many leafy liverworts, where the precise shape of the sinus is often an important character (fig. I, 18 and 19).

spores, the minute asexual reproductive units of bryophytes (and other cryptogams), formed within the capsule, and ultimately shed, often by special mechanisms; typically unicellular (fig. VI, 2), they are occasionally multicellular (e.g. *Pellia*) or in a few instances retained in clusters of four, termed tetrads, when shed (e.g. *Sphaerocarpus*). They are analogous to the seeds of flowering plants.

sporogonium, the product of the archegonium after fertilization; hence a term used for the sporophyte when, as among bryophytes, that generation never becomes free from the sexual generation.

sporophyte, the whole of the spore-bearing asexual generation or phase in a plant's life cycle; thus in bryophytes it normally includes capsule, seta and foot, and is synonymous with the sporogonium.

squarrose, applied to leaves that spread out at a wide angle with the stem, e.g. those of *Dicranella squarrosa* and *Rhytidiadelphus squarrosus* (fig. II, 4).

stellate, star-like, e.g. the shoot tip of *Campylium stellatum*; also used of the leaf cells in certain liverworts where the corner thickenings are so large that the cell cavity becomes star-like.

stoma (plur. *stomata*), the specialized pores, similar to those on the leaves of higher plants, that occur on the surface of the capsule in *Anthoceros* and in mosses, each consisting of an opening bounded by a pair of 'guard cells' of characteristic shape (figs. 98, 99).

striate, having faint or shallow (usually longitudinal) furrows, e.g. the capsules of certain mosses; also used of leaves.

subinvolucral, just below the involucre.

succubous, descriptive of one of the two possible ways in which the leaves of a liverwort may overlap when obliquely inserted on a more or less flattened shoot; it implies that, when this flattened shoot is viewed from above, the forward edge of each leaf is concealed *beneath* the hind edge of the leaf next above it (fig. II, 6). Cf. *incubous*.

synoecious, having the antheridia and archegonia intermixed in the same 'inflorescence' (fig. VII, 1). Cf. *paroecious, autoecious*.

tetrads, spores in groups of four as they are formed during the meiotic (reduction) division of the spore mother cell, the stage before their final separation into individual spores. In *Sphaerocarpus* and some other liverworts the four spores of a tetrad tend to remain attached, but in all mosses and many liverworts the tetrad soon breaks up into its four constituent spores (cf. fig. 207).

thallose, having the form of a thallus.

thallus, any plant body not differentiated into stem and leaves.

tomentose, covered with a felt of hair. See *tomentum*.

tomentum, a covering of hair, hence used of the dense felt of rhizoids with which the lower parts of the stems are matted together in many mosses, e.g. *Breutelia chrysocoma, Bartramia pomiformis*, etc.

triquetrous, sharply three-angled or three-winged.

tufa, a porous, friable or compact limestone formed by deposition from lime-rich water; found around springs and waterfalls in limestone districts. Mosses play an important part in tufa formation.

underleaves, a row of leaves, usually smaller than the lateral ones, along the underside of the stem in many liverworts; sometimes called amphigastria.

ventral, the under surface (of a thallus or leafy shoot); the front or adaxial surface (of a leaf), i.e. the surface next the stem when the leaf lies appressed to it (fig. III, 4). Cf. *dorsal*.

verrucose, having minute wart-like protuberances; applied to the cuticle of the leaf in some liverworts, e.g. *Mylia taylori* (fig. 171).

vice-counties, divisions, 112 in Great Britain and 40 in Ireland, into which these countries are split for the purpose of botanical records. The boundaries are given in the *Census Catalogues of British Mosses and Hepatics*, and a map is provided in the *Comital Flora* by G. C. Druce.

vitta, a central band of longitudinally elongated cells, as in the leaf of *Diplophyllum albicans*. Cf. *midrib*.

KEY TO MOSSES*

MAKING USE OF MICROSCOPIC CHARACTERS

Key to sections

1 Plants consisting of numerous upright stems on which branches arise strictly in whorls; leaves composed of two types of cell (i) narrow and green, (ii) wide, colourless and strengthened by spiral fibrils. (All are bog or wet-ground plants which look pale when dry, but are vivid green, orange-brown or purple-red when wet, and hold water like a sponge.)   Section 1, p. 39 (*Sphagnum*)
   Plants not as above   2

2 Plant with minute, non-green leafy shoot (no true leaves, but only perichaetial bracts with fringed margins); capsule large, oblique and asymmetrical   *Buxbaumia aphylla* (p. 122)

   Small, dark red-brown to black plants, with small leaves (up to 1 mm.) which lack hair-points; cells in upper part of leaf small (about 10μ), with thick walls and round cavities; capsules opening by 4 longitudinal slits; plants of siliceous rocks on mountains   Section 2, p. 40 (*Andreaea*)
   Plants not as above   *3*

3 Leaves with thin colourless, often sheathing bases, and thick blades rendered opaque by outgrowths of deep green tissue (lamellae) on their upper surfaces   Section 3, p. 40 (*Polytrichum*, etc.)

   Leaves either without such outgrowths or having them limited to very narrow nerve region so that whole leaf blade not thick and opaque   4

4 Leaves *strictly* in 2 ranks on the stems   Section 4, p. 41 (*Fissidens*, etc.)
   (N.B. In a number of mosses the leaves may *appear* to be in 2 ranks, due to the flattening of the shoot as a whole, but closer examination will show that they are not strictly 2-ranked.)
   Leaves not strictly 2-ranked   5

5 Some or all leaves ending in silvery white (hyaline) hair-points, which, when dry, stand out white in contrast with green of leaf
      Section 5, p. 42 (spp. of *Grimmia, Rhacomitrium, Tortula*, etc.)

   Leaves without hair-points which differ in structure and appearance from rest of leaf   6

* Every moss mentioned in this book is included.

6   Moss acrocarpous   7

Moss pleurocarpous   10

(N.B. Beginners may experience difficulty in deciding to which of these groups to refer a moss, and it is impossible to give a single clear-cut criterion on which to base a decision in barren material. If, however, attention be paid to the following three classes of characters a wrong decision should very rarely be made at this point:

(i) *General habit and mode of branching.* Acrocarpous mosses are usually unbranched or sparingly branched, and of *erect or ascending habit*; they are never regularly pinnately branched and the leaves are rarely glossy and 'chaff-like'. Almost all pleurocarpous mosses are *freely branched, often pinnately* so, various in habit but frequently forming dense intricate mats of the *prostrate or ascending*, elaborately branched stems; and the leaves are most often glossy and 'chaff-like' in character.

(ii) *Cell structure and nerves.* Almost all doubtful examples can be settled immediately the leaf is examined under the compound microscope, for the *range* of cell structure is totally different in the two groups. Thus very few acrocarpous mosses have long narrow cells throughout the leaf, whereas this is the cell structure that is most prevalent among the pleurocarps. Again, extremely few pleurocarpous mosses have short (isodiametric) cells in the upper part of the leaf and elongated rectangular cells in the leaf base; but in many acrocarpous families this is the usual type of cell structure. A further useful point: nerveless leaves are very rare among acrocarps; leaves with excurrent nerves are almost equally rare in the pleurocarpous series.

(iii) *Position of archegonia and capsules.* In almost all the acrocarpous series the archegonia—and hence the capsules—arise terminally, i.e. *at the tip* of a stem or branch, whilst in the pleurocarpous mosses they arise, surrounded only by the perichaetial leaves, *on the side* of a stem or branch. Acrocarpous mosses that are particularly likely to be mistaken for pleurocarps are *Cinclidotus fontinaloides*, *Breutelia chrysocoma* and *Mnium affine*.)

7   Leaves proportionately long and narrow, tapering gradually, from base or near it, to long fine points into which the nerve runs

Section 6, p. 44 (*Dicranum, Campylopus*, etc.)

(N.B. In typical leaves of this section the narrow point of the leaf is indeed largely *composed of the nerve*, but all acrocarpous species with narrow finely pointed leaves are included here.)

Leaves lanceolate, whitish green, composed almost entirely of expanded absorbent nerve tissue, several layers of cells thick (plant holds water like a sponge)   *Leucobryum glaucum* (p. 153)

Leaves not as above   8

8   Plants moderately to exceedingly robust (stems 2–10 cm.), leaves
    large (3·5–10 mm. long and 1–3 mm. wide); each leaf *with evident
    border* (formed by very narrow thick-walled cells); cells through-
    out rest of leaf large (15–45 μ across), nearly *isodiametric* except
    in leaf base   Section 7, p. 51 (*Mnium*, etc.)

    Plants slender or robust; leaves not as above (i.e. lacking
    this *combination* of size, border and cell type)   9

9   Leaves proportionately broad (length not more than 2½ times
    breadth), ovate to obovate, usually broadest in mid-leaf,
    narrowing slightly at base, and forming a short acute apex;
    cells either (i) uniformly rhomboid or shortly rectangular
    throughout leaf, or (ii) isodiametric, clearly defined and at least
    16 μ wide in upper part of leaf, and rectangular in leaf base
                        Section 8, p. 52 (*Bryum, Funaria, Pottia*, etc.)

    Leaves in shape as section 8, but with small round cells
    throughout; minute plant (stems 0·5–1·5 cm. tall); bearing
    gemmae in conspicuous terminal cups; capsule with only 4
    peristome teeth   *Tetraphis pellucida* (p. 205)

    Leaves broad, or narrow, but with notably *blunt, rounded or
    'sheared-off' apex*, nerve ceasing in apex or excurrent in short
    point; cells *never* uniformly rhomboid or elongate-rectangular
                        Section 9, p. 56 (*Encalypta*, some spp. of *Tortula*, etc.)

    Leaves proportionately narrow, ovate-lanceolate to narrowly
    lanceolate, not above 4–5 mm. long; always narrowing gradually
    from near base to moderately acute apex, in which the *nerve* is
    almost always lost (thus the nerve is never obviously excurrent
    as in section 6 and some members of sections 8 and 9)
                        Section 10, p. 58 (*Barbula, Orthotrichum*, etc.)

10 Cells in mid-leaf not above 5 times as long as broad, appearing
    square, rounded, shortly rhomboid or elliptical   Section 11, p. 65

    Cells in mid-leaf long and narrow, more than 5 times as long
    as broad   *11*

11 Leaves nerved to mid-leaf or beyond, not curved and turned in
    one direction   Section 12, p. 67

    Leaves nerved to mid-leaf or beyond, showing evident ten-
    dency, at least at tips of branches to be curved and turned in
    one direction   Section 13, p. 71

    Leaves nerveless or with short double nerve only   Section 14, p. 72

SECTION 1

1 Branch leaves broad, very concave, hooded at apex, which is
covered with scales at back; stem leaves not bordered   2
Branch leaves narrower, very rarely hooded at apex, which is
not scaly at back; stem leaves bordered with narrow cells   4

2 Walls of hyaline cells of branch leaves (where these adjoin the
green cells) covered with minute papillae; plant most often
ochraceous   *Sphagnum papillosum* (p. 106)
Walls of hyaline cells of branch leaves quite smooth; plant
usually pale green or reddish   *3*

3 Narrow green cells of branch leaves quite concealed by larger
hyaline cells in surface view on either face of leaf; central
cylinder of stem brown-red or red-purple *Sphagnum magellanicum*
(p. 105)
Narrow green cells exposed in surface view of concave face
of leaf; central cylinder yellowish brown   *S. palustre* (p. 101)

4 Branches stout and obtuse, stem leaves $\frac{1}{3}-\frac{1}{5}$ length of branch
leaves   *Sphagnum compactum* (p. 106)
Branches more slender and tapering, stem leaves proportionately longer   *5*

5 Hyaline cells of branch leaves with many (20-30 per cell) small
pores: plants variable in habit and colour, but never crimson
Sphagnum subsecundum (p. 106)
Hyaline cells of branch leaves with few (up to 8 per cell) large
pores: plants green or purplish crimson   *6*

6 Green cells more exposed abaxially or equally so on both surfaces of the leaf; plants green   *7*
Green cells more exposed adaxially; plants often crimson   *8*

7 Stem leaves longer than broad; margins of branch leaves undulate when dry: aquatic   *Sphagnum cuspidatum* (p. 103)
Stem leaves slightly broader than long; tips of branch leaves
recurved and hook-like when dry; subaquatic
S. recurvum (p. 106)

8 Stem leaves triangular, pointed, their hyaline cells usually
lacking fibrils; some branch leaves usually up to 1·5 mm. long;
plant with notable metallic lustre when dry
Sphagnum plumulosum (p. 104)
Stem leaves tongue-shaped with obtuse apex, their hyaline cells
usually with fibrils at least near leaf tip; branch leaves about
1 mm. long; a slender plant without notable metallic lustre when
dry   *S. rubellum* (p. 106)

SECTION 2

1 Leaves nerved to apex   *Andreaea rothii* (p. 108)
    Leaves nerveless   2

2 Plants 2–7 cm. tall, dark red-brown, leaf cells without papillae
                                      Andreaea alpina (p. 108)
    Plants 1–2 cm. tall, usually appearing blackish, leaf cells
    papillose   *A. rupestris* (p. 107)

    (N.B. If leaves are nerved and plant does not fit *A. rothii*, look for
    short hair-points to some of the leaves; it will probably be
    a species of *Grimmia*, see sections 5 and 10. Other small rock-
    growing acrocarpous mosses that look black at times, but differ
    from *Andreaea* in cell structure, etc., are *Orthotrichum anomalum*
    and *Ulota hutchinsiae*. All are of course entirely different from
    Andreaea in fruit.)

SECTION 3

1 Leaf margin incurved and toothed, lamellae forming wavy band
    of variable width down centre of leaf *Oligotrichum hercynicum* (p.110)
    Leaf margin incurved, without teeth   .   2
    Leaf margin plane, toothed   7

2 Leaf apex terminating in a fine, toothed point, plants usually
    at least 2 cm. tall   3
    Leaf apex obtuse, or fairly acute, without teeth, plants usually
    less than 1·5 cm. tall   5

3 Stems 6–12 cm. tall, the lower part covered with whitish
    tomentum   ·   *Polytrichum alpestre* (p. 120)
    Plant rarely above 5–6 cm. tall, stems not covered with whitish
    tomentum   .   4

4 Leaf point long, whitish, hyaline, plant usually only 2–3 cm. tall
                                Polytrichum piliferum (p. 116)
    Leaf point shorter, reddish, plant usually 4–5 cm. tall
                                      P. juniperinum (p. 115)

5 Leaves acute, not hooded at apex   *Aloina aloides* (p. 164)
    Leaves obtuse, hooded at apex   6

6 Leaves 2–3 mm. long, capsule elongate-cylindrical *Aloina ambigua*
                                              (p. 164)
    Leaves shorter (1·5–2 mm.), capsule narrowly ovoid *A. rigida* (p. 164)

7 Leaves noticeably glaucous (bluish) green   *Polytrichum urnigerum*
                                             (p. 114)
    Leaves deep mid-green   8

8 Plant dwarf (not above 2 cm.), teeth on leaf margin not elongated, small   *9*
Plant usually over 2 cm. tall, often much taller, teeth acute, prominent   *10*

9 Marginal teeth not extending lower than mid-leaf, capsule shortly ovoid, nearly smooth   *Polytrichum nanum* (p. 120)
Leaf margin toothed to base of lamina, capsule cylindrical—rough with papillae   *P. aloides* (p. 113)

10 Each leaf tooth composed of several cells, capsule round in section   *Polytrichum alpinum* (p. 120)
Each leaf tooth composed of 1 cell, capsule prominently angular in section   *11*

11 4–6 rows of clear cells along margin of leaf; cells in sheathing leaf base only 30–60μ long; plant seldom above 6 cm. tall
                                                  Polytrichum gracile (p. 120)
1–3 rows of clear cells along margins of leaf; cells in sheathing leaf base much longer; plants 4–20 cm. tall   *12*

12 Woodland plant, usually 5–10 cm. tall, outermost cells of leaf lamellae conical   *Polytrichum formosum* (p. 117)
Moorland plant, often much above 10 cm. tall, outermost cells of leaf lamellae bifid   *P. commune* (p. 119)

SECTION 4

1 Leaves nerveless, cells very large ($100 \times 25\mu$)   *Schistostega pennata*
                                                                    (p. 203)
Leaves single-nerved, cells small   *2*

2 Leaves narrow, with sheathing bases and very long fine points
                                      Distichium capillaceum (p. 131)
Leaves wider, peculiar in form with additional 'sheathing lamina' at base   *3*

3 Leaf with distinct border formed of clear elongated cells   *4*
Leaf without such border   *7*

4 Cells large, 12–18μ wide, plant aquatic *Fissidens crassipes* (p. 126)
Cells smaller, 8–10μ wide, plant terrestrial   *5*

5 Terminal pair of leaves much longer than the rest; on shaded sandstone and limestone rocks   *Fissidens pusillus* (p. 126)
Terminal pair of leaves not much longer than the rest; on soil   *6*

6   Male 'flower' on short basal branch, capsule curved or
    horizontal   *Fissidens incurvus* (p. 126)
    Male 'flowers' in axils of leaves, capsule erect   *F. bryoides* (p. 123)

7   Plant minute, with only 2–4 pairs of leaves, capsule on terminal
    seta 5–6 mm. long   *Fissidens exilis* (p. 126)
    Plant larger, with more pairs of leaves, capsule terminal or
    lateral   8

8   Nerve of leaf running out into short projecting point, seta arising
    from near base of stem   *Fissidens taxifolius* (p. 124)
    Nerve ceasing just below extreme apex, seta terminal or arising
    laterally, from middle of stem   9

9   Cells of leaf uniform to margin, leaf apex entire or nearly so,
    seta terminal   *Fissidens osmundoides* (p. 127)
    Leaf with distinct margin formed of somewhat larger, more
    clearly defined cells, leaf apex toothed, seta lateral, from near
    middle of stem   10

10 Leaf cells small, 6–10 μ wide, marginal band well defined
                                     Fissidens cristatus (p. 127)
    Leaf cells larger, 10–18 μ wide, marginal band less well defined
                                      F. adianthoides (p. 124)

SECTION 5

1   Leaves nerveless   *Hedwigia ciliata* (p. 250)
    Leaves single-nerved to apex   2

2   Leaves long (about 8 mm.), rigidly straight, nerve $\frac{1}{3}$ width of leaf
    base or more   3
    Leaves shorter, not rigidly straight, nerve much narrower   4

3   Hair-point long, nerve $\frac{1}{2}$ width of leaf base   *Campylopus atrovirens*
                                                           (p. 151)
    Hair-point very short, nerve $\frac{1}{3}$ width of leaf base *C. brevipilus* (p. 152)

4   Cells at base of leaf long and very narrow, with greatly thickened,
    strongly wavy longitudinal walls and thin transverse walls; habit
    usually loose and freely branched   5
    Cells at base of leaf short or long, longitudinal walls seldom
    greatly thickened, not strongly wavy   7
    (N.B. The nearest approach to the *Rhacomitrium* type of leaf-base
        cell structure is found in some species of *Grimmia*, but these
        are always plants of dense, cushion-forming habit.)

5   Hair-point of leaf simply toothed, but teeth not papillose
                                  Rhacomitrium heterostichum (p. 194)
    Hair-point irregularly indented, covered with papillae   6

6   Cells of leaf blade not papillose, plant often very robust, stems 12–25 cm. (very common plant of mountain rocks, moors, etc.)
Rhacomitrium lanuginosum (p. 197)

Cells of leaf blade papillose, stems shorter; plant of gravelly heaths, etc.   *R. canescens* (p. 196)

7   Red-brown to golden green plant forming wide patches on sand dunes; leaf apex tapering gradually into broad base of hyaline hair-point   *Tortula ruraliformis* (p. 158)

Combination of colour, habitat and leaf apex not as above   8

8   Leaf broadening above base, with wide obtuse apex   9

Leaf not broadening above base, tapering to acute apex   14

9   Leaf concave above, highly papillose, bearing spherical gemmae
Tortula papillosa (p. 164)

Leaf concave, margin incurved, lamellae on nerve, capsule almost sessile   *Pterygoneurum ovatum* (p. 164)

Leaf not concave, without gemmae or lamellae, seta long   10

10   Leaf with distinct margin of long narrow cells   *Tortula marginata* (p. 164)

Leaf without margin distinct from rest of leaf   11

11   Hair-point rough with numerous teeth or spines   12

Hair-point smooth or nearly so   13

12   Upper parts of plant usually bright golden green, leaves strongly spreading-recurved when moist   *Tortula ruralis* (p. 157)

Upper parts of plant dull brownish green, leaves nearly straight, not recurved   *T. intermedia* (p. 159)

13   Leaf margin recurved throughout, very common plant on walls and lowland rocks   *Tortula muralis* (p. 162)

Leaf margin recurved in mid-leaf only; plant growing on trees
T. laevipila (p. 160)

14   Upper leaves often curved to one side, bearing gemmae (uncommon plant of siliceous mountain rocks)   *Grimmia hartmanii* (p. 191)

Upper leaves not curved to one side or gemmiferous   15

15   Leaves spirally twisted when dry, giving shoots 'rope-like' character (mountain rock species)   16

Leaves not spirally twisted, nor shoots 'rope-like' when dry   17

16   Plant black below, hair-points on upper leaves often quite long (up to $\frac{1}{2}$ leaf length)   *Grimmia funalis* (p. 190)

Plant brown below, hair-points short, even on upper leaves
G. torquata (p. 190)

17 Hair-point of leaf covered with prominent teeth or spines   *18*
Hair-point smooth, or with a few blunt teeth only   *19*

18 Upper cells of leaf oblong with wavy walls, plant 2–4 cm. tall,
on siliceous rocks   *Grimmia decipiens* (p. 191)
Upper cells of leaf rounded, plant up to 1 cm. tall, usually on
trees or fences   *Orthotrichum diaphanum* (p. 243)

19 Cells in leaf base rarely more than twice as long as broad   *20*
Cells in leaf base 3–6 times as long as broad   *21*

20 Plant usually brownish, hair-points often very short, capsule
immersed, lid and peristome bright red   *Grimmia apocarpa* (p. 185)
Plant greyish green, hair-points very long, capsule on evident
curved seta   *G. pulvinata* (p. 187)

21 Cells in leaf base thick-walled, plant light green, mainly lowland
     Grimmia trichophylla (p. 189)
Cells in leaf base thin-walled, plant grey-green to blackish,
a mountain species   *G. doniana* (p. 186)

SECTION 6

Key to subsections

1 Nerve broad, at least ⅓ total breadth of leaf base   Subsection 1
Nerve narrow, less than ⅓ total breadth of leaf base   *2*

2 Plants small to very small, with all leaves under 2 mm. long
     Subsection 2
Plants of various habit, the longest leaves 2 mm. or more in
length   *3*
(A plant combining tall habit with leaves all under 2 mm. in length
will probably be *Philonotis fontana*.)

3 Leaf margin toothed, at least towards leaf tip   Subsection 3, p. 46
Leaf margin without teeth, being quite entire even near apex
     Subsection 4, p. 49

SUBSECTION 1

1 Plants only a few mm. tall, upper leaves often with lacerated
tips; when fertile bearing large and conspicuous sessile capsules
     Diphyscium foliosum (p. 121)
Plants taller, leaf tips never lacerated; capsules not sessile   *2*

2 Leaf with enlarged (often coloured) cells in basal angles   *3*
Cells in basal angles of leaf not differing markedly from those
just above them   *7*

3 Nerve $\frac{1}{2}$ width of leaf base or even wider, plants robust (commonly 3–10 cm. tall)   *4*
Nerve $\frac{1}{3}$–$\frac{1}{2}$ width of leaf base, plants usually of more slender
habit (1–3 cm. tall)   *5*

4 Upper cells of leaf scarcely longer than broad
                                    Campylopus schwarzii (p. 153)
Upper cells of leaf long and narrow   *C. atrovirens* (forms) (p. 151)

5 $\frac{3}{4}$ of leaf length taken up with long fine channelled point (plant
of mountain woods)   *Dicranodontium denudatum* (p. 153)
Only uppermost $\frac{1}{4}$ of leaf length occupied by fine point   *6*

6 Upper cells elongated-elliptical, extreme leaf tip usually hyaline
                                    Campylopus brevipilus (p. 152)
Upper cells shortly oval, leaf tip never hyaline
                                      C. flexuosus (p. 150)

7 Leaves evidently curved to one side   *Dicranella heteromalla* (p. 139)
Leaves not curved to one side   *8*

8 Leaf narrowed at extreme base, fine point usually occupying
about $\frac{1}{3}$ of leaf length, whole plant commonly of stiff habit and
silky texture   *Campylopus fragilis* (p. 153)
Leaf not narrowed at extreme base, fine point usually $\frac{2}{3}$ of leaf
length, whole plant of weaker habit, leaves less silky and very
deciduous   *C. pyriformis* (p. 149)

SUBSECTION 2

1 Plants of wet fallow fields and pool margins, capsules immersed,
cleistocarpous   *2*
Plants of bare chalk or sandstone rock, capsule on distinct seta,
only 2–6 mm. tall including seta and capsule   *3*
Plants of various habitats, more robust than *Seligeria*, capsules
not immersed   *4*

2 Perichaetial bracts much longer than vegetative leaves, capsule
spherical without protuberance   *Archidium alternifolium* (p. 128)
Perichaetial bracts similar to upper leaves, capsule with short
projecting point   *Pseudephemerum nitidum* (p. 128)

3 Leaf margin toothed just above base, plant on sandstone or
limestone   *Seligeria pusilla* (p. 135)
Leaf margin entire, plant normally on chalk or limestone cliffs
                                      S. calcarea (p. 134)
Leaf margin entire, plant on sandstone, leaf tip longer; seta
strongly curved when moist   *S. recurvata* (p. 135)

4   All leaf cells long and narrow, 3 or 4 times as long as broad   *5*
     Upper cells scarcely or not at all longer than broad   *6*

5   Leaf margin toothed near apex, capsule erect
                          Dicranella rufescens (p. 140)
     Leaf margin almost or quite entire, capsule curved, horizontal
                                 D. varia (p. 138)

6   Leaf margin toothed just above base
                       Eucladium verticillatum (p. 178)
     Leaf margin without teeth just above base   *7*

7   Leaf margin strongly recurved from base to near the notched
     apex; cells all relatively large (10–15μ across) and clearly defined
                     Ceratodon purpureus (small forms) (p. 133)
     Leaf margin lightly recurved below or in mid-leaf only, upper
     cells very small (6–10μ), often somewhat obscure; leaf-apex
     never notched   *8*

8   Plant growing on earth; cells overlying nerve of leaf long and
     narrow   *Barbula fallax* (small forms) (p. 174)
     Plant growing on stone; cells overlying nerve not differing from
     those in rest of leaf blade   *9*

9   Leaves nearly straight, with thick blunt tips; plant often bearing
     gemmae   *Barbula rigidula* (small forms) (p. 176)
     Leaves evidently curved, acute at apex; plant not gemmiferous
                     B. vinealis (small forms) (p. 177)

SUBSECTION 3

1   Plants small to very small (stems up to 1 cm. tall, usually less),
     with spherical or nearly spherical cleistocarpous, immersed
     capsules   *2*
     Plants small to very robust (stems 1–10 cm. or more tall),
     capsules not immersed or cleistocarpous   *5*

2   Capsule spherical, without protuberance, containing only 16–32
     spores, which are very large (150–250μ)
                       Archidium alternifolium (p. 128)
     Capsule nearly spherical, but with small apical protuberance,
     spores not notably few or large   *3*

3   Plant of soft lax habit; perichaetial and vegetative leaves alike
                     Pseudephemerum nitidum (p. 128)
     Plant of more rigid habit; perichaetial leaves much longer than
     others   *4*

4 Antheridia only seen after dissection, in axils of perichaetial bracts; a common plant on sandy heaths, etc.

Pleuridium acuminatum (p. 127)

Antheridia in bud-like clusters, in axils of upper leaves—easily seen without dissection; a much less common plant of damp places   *Pleuridium subulatum* (p. 128)

5 Plants rarely above 4 cm. tall (except *Ditrichum flexicaule*), of slender habit, upper cells of leaf not papillose, patches of distinct alar cells in basal angles of leaf lacking   *6*

Plants of robust habit, 2–10 cm. or more tall, basal angles of leaf composed of very distinct (alar) cells, upper cells of leaf papillose or smooth   *11 (Dicranum,* etc.)

Plants of robust habit, 2–10 cm. or more tall, patches of distinct alar cells lacking; upper cells of leaf papillose   *15 (Bartramiaceae)*

(N.B. A plant combining non-papillose cells with the round capsule of a *Bartramia* will probably be *Plagiopus oederi*.)

6 Leaves evidently curved to one side (falcato-secund), nerve broad ($\frac{1}{5}$–$\frac{1}{3}$ width of leaf at base)   *7*

Leaves variously spreading, not falcato-secund, nerve narrower   *8*

7 Leaf point long and fine, capsule without swelling at neck

Dicranella heteromalla (p. 139)

Leaf point shorter, leaves not usually so strongly falcato-secund, capsule with knob-like swelling in neck   *D. cerviculata* (p. 140)

8 Leaf apex almost entirely occupied by nerve, upper cells scarcely longer than broad   *9*

Leaf apex wider, basal cells of leaf long and very narrow, with thick longitudinal walls   *Ptychomitrium polyphyllum* (p. 234)

Leaf apex variable, often long and fine but not wholly occupied by nerve; all cells of leaf long and narrow, uniformly thin-walled   *10*

9 Plant under 1 cm. tall, leaves spreading at right angles to stem, woodland rides, uncommon   *Ditrichum cylindricum* (p. 131)

Plant tall, 2–7 cm. (occasionally taller), leaves flexuose, calcareous grassy places, common   *D. flexicaule* (p. 130)

10 Plant of sandstone rocks, capsule narrowly club-shaped to cylindrical (2·5 mm. long) with tapering neck

Orthodontium gracile (p. 207)

Plant of rotten wood and peaty banks, abundant in some districts, capsule shorter and broader (up to 2 mm.) often somewhat curved and asymmetrical   *O. lineare* (p. 207)

Plant of cinder heaps and greenhouses, capsule broadly pearshaped (about 2 mm. long)   *Leptobryum pyriforme* (p. 206)

11 Cells in upper part of leaf scarcely longer than broad, not porose   *12*
    Cells in upper part of leaf long, narrow, with thick walls inter-
    rupted by evident thin places (porose)   *13*

12 Cell cavities round, walls much thickened at corners; leaf apex
    only moderately acute; plant of marshes and bogs
                                   Aulacomnium palustre (p. 227)
    Cell cavities angular, not much thickened at corners; leaf
    tapering to a very long fine point; on mountains
                                     Dicranum fuscescens (p. 143)

13 Leaves 10–14 mm. long, very strongly and uniformly curved to
    one side   *Dicranum majus* (p. 144)
    Leaves shorter, straight or moderately curved to one side   *14*

14 Leaves usually transversely undulate, leaf point broad, back of
    nerve smooth or nearly so   *Dicranum bonjeani* (p. 146)
    Leaves very rarely transversely undulate, leaf point narrow,
    back of nerve with several toothed ridges near leaf apex
                                        D. scoparium (p. 147)

15 Leaves long, 4 mm. or more, their apices long and narrow   *16*
    Leaves shorter, 1–3 mm. broad-based and with much shorter
    apices   *19*

16 Leaves rigidly spreading at right angles to stem, deeply plicate
                             Breutelia chrysocoma (p. 233)
    Leaves not rigid, variously spreading or flexuose, not plicate   *17*

17 Upper cells of leaf much longer than broad, leaf base con-
    spicuously sheathing   *Bartramia ithyphylla* (p. 231)
    Upper cells of leaf scarcely longer than broad, leaf base not
    conspicuously sheathing   *18*

18 Capsules nearly spherical, looking like miniature green apples
    when young, seta about 2 cm. long; plant common on mountain
    ledges and on heathy banks   *Bartramia pomiformis* (p. 230)
    Capsules somewhat longer, not raised clear above leaves; rather
    rare plant of mountain ledges   *B. halleriana* (p. 231)

19 Plants bright yellowish or glaucous green, leaves short (1–2 mm.),
    cells in mid-leaf narrow, 6–10μ wide   *Philonotis fontana* (p. 231)
    Plants usually vivid emerald green, leaves larger and longer
    (2–3 mm.), cells in mid-leaf 10–15μ wide   *P. calcarea* (p. 232)

1 Plants only a few mm. tall, capsules almost sessile among leaves   *2*

   Plants taller, capsules not almost sessile among leaves   *3*

2 Capsule large (looking like a grain of wheat), with conspicuous whitish peristome, leaves flat; on peaty banks

                          Diphyscium foliosum (p. 121)

   Capsule smaller, without peristome, leaves with inrolled margins; on calcareous ground   *Weissia crispa* (p. 183)

3 Leaves markedly curved to one side, cells of basal angles not distinct   *Ditrichum heteromallum* (p. 129)

   Leaves slightly or markedly curved to one side, cells of basal angles (alar cells) forming distinct (often coloured) patches   *4*

   Leaves not at all curved to one side, without distinct alar cells   *7*

4 Upper cells of leaf several times as long as broad

                              Blindia acuta (p. 135)

   Upper cells of leaf not, or scarcely, longer than broad   *5*

5 Leaves very strongly curved to one side, capsule with swelling in neck   *Dicranum falcatum* (p. 148)

   Leaves at most lightly turned to one side, capsule without neck swelling   *6*

6 Leaves much crisped and curled when dry (2–2·5 mm. long)

                     Dicranoweisia crispula (p. 143)

   Leaves only lightly twisted or flexuose when dry (4 mm. long)

                     Dicranum scottianum (p. 148)

7 Upper part of leaf thick and opaque (cells in several layers), plant strictly maritime   *Grimmia maritima* (p. 184)

   Upper part of leaf not thickened as above, plants not strictly maritime   *8*

8 Leaf margin strongly incurved   *9*

   Leaf margin not incurved, being plane or locally recurved   *10*

9 Capsule with rudimentary peristome

                       Weissia controversa (p. 182)

   Capsule with no peristome and very narrow mouth

                       W. microstoma (p. 183)

10 Plants strongly tinged with reddish brown below

                     Barbula recurvirostra (p. 175)

   Plants lacking marked reddish brown tinge below   *11*

11 Leaves very long (the longest reaching 4–7 mm.) with sharp contrast between clear cells of leaf base and opaque green cells above   *12*
     Leaves shorter, without sharp contrast between clear basal and green upper cells   *13*

12 Line between clear and opaque cells well defined, oblique; common plant in calcareous places   *Tortella tortuosa* (p. 179)
     Line between clear and opaque cells less well defined, not oblique; plant not a calcicole   *Trichostomum tenuirostre* (p. 182)

13 Leaves long (the longest reaching 2·5–4 mm.), tapering to fine points   *14*
     Leaves shorter (1·5–2 mm.), points less fine   *17*

14 Cells long and narrow throughout leaf   *15*
     Cells in upper part of leaf short, isodiametric   *16*

15 Rather rare plant of sandstone rocks, capsule narrowly club-shaped to cylindrical (2·5 mm. long), with tapering neck
                                 Orthodontium gracile (p. 207)
     Plant of rotten wood and peaty banks, abundant in some districts, capsule shorter and broader   *O. lineare* (p. 207)

16 Leaves narrow throughout length, upper cells clearly defined, plant forming dense cushions on wet rocks in mountains
                                 Amphidium mougeotii (p. 235)
     Leaves tapering from broad base, upper cells obscure and densely papillose, plant growing in loose patches, chiefly on soil   *Barbula cylindrica* (p. 177)

17 Leaf margin strongly recurved from base to near the notched apex; cells all relatively large, often square in outline and clearly defined; abundant species on many kinds of substratum
                                 Ceratodon purpureus (p. 133)
     Leaf margin lightly recurved at base or in mid-leaf only, upper cells very small (6–10 μ), often somewhat obscure; leaf apex never notched; not normally mountain ledge species, nor forming deep cushions   *18*

18 Plant growing on earth; cells overlying nerve of leaf long and narrow   *Barbula fallax* (p. 174)
     Plant growing on stone; cells overlying nerve not differing from those in rest of leaf blade   *19*

19 Leaves nearly straight, with thick blunt tips; plant often bearing gemmae   *Barbula rigidula* (p. 176)
     Leaves evidently curved, acute at apex; plant not gemmiferous
                                       B. vinealis (p. 177)

SECTION 7

1   Nerve of leaf with outgrowths consisting of longitudinal plates
of green cells (lamellae)   2
Nerve without such outgrowths   3

2   Leaves undulate, cells rather small, about 20μ wide
Atrichum undulatum (p. 109)
Leaves not undulate, cells larger, reaching $35-45\mu$ wide
A. crispum (p. 110)

3   Leaf margin without teeth   4
Leaf margin toothed   6

4   Nerve ceasing well below apex, leaf border unthickened, capsule
nearly globose   *Mnium pseudopunctatum* (p. 227)
Nerve reaching apex or nearly so, leaf border thickened or not,
capsule ovoid or ellipsoidal   5

5   Plant dark green, leaf border several cells thick; common plant
of wet rock ledges, etc.   *Mnium punctatum* (p. 226)
Plant reddish to purple-black, leaf border only 1 cell thick; rare
plant of deep bogs   *Cinclidium stygium* (p. 227)

6   Longest leaves 8–12 mm. long, transversely undulate
Mnium undulatum (p. 224)
Leaves shorter, without obvious transverse wrinkles or undula-
tions   7

7   Teeth along leaf margins consisting of double structures   8
Teeth along leaf margins single structures   9

8   Plant reddish below, nerve not spiny at back, uncommon species
of mountain rock ledges   *Mnium marginatum* (p. 227)
Plant seldom reddish below, nerve spiny at back, abundant
species in many habitats   *M. hornum* (p. 223)

9   Leaves rounded at apex, nerve projecting in short entire point,
lid of capsule long-beaked   *Mnium longirostrum* (p. 227)
Leaves tapering somewhat at apex, nerve projecting in longer,
toothed point, lid of capsule bluntly conical   10

10   Plant of slender habit (stems 1–3 cm. long), pale green; cells
of leaf small, about 20μ wide   *Mnium cuspidatum* (p. 227)
Plant of more robust habit, dark green, cells of leaf $25-40\mu$ wide   11

11   Plant prostrate or nearly so, habitat damp woods chiefly
Mnium affine (p. 227)
Plant erect, 5–10 cm. tall, habitat marshes   *M. seligeri* (p. 227)

SECTION 8

Key to subsections

1 Minute cleistocarpous mosses with globose or nearly globose capsules immersed among leaves, sessile or on very short setae. (All have leafy shoots only a few millimetres tall)   Subsection 1
Mosses not cleistocarpous; usually taller, capsules not globose in form or sessile in position   *2*

2 Basal cells of leaf somewhat elongated, upper rounded or quadrate. (All are small plants, with stems 5–20 mm. tall)
Subsection 2 (*Pottia*)
Leaf cells uniformly isodiametric (20–30μ); robust pale green plant (stems 2–3 cm. tall) occurring locally on calcareous rock ledges   *Mnium stellare* (p. 227)
Cells throughout leaf uniform and elongated to some extent— being either hexagonal to rectangular (*Funaria* type) or rhomboid (*Bryum* type)   *3*

3 Nerve of leaf ceasing below extreme apex   Subsection 3
Nerve of leaf extending to extreme apex or projecting as a point of varying length   Subsection 4

SUBSECTION 1

1 Leaf margin toothed near apex   *2*
Leaf margin entire   *4*

2 Leaves nerveless   *Ephemerum serratum* (p. 201)
Leaves nerved   *3*

3 Nerve lost just below apex, cells elongated (40–80 × 20μ); capsule nearly spherical, but with small apical protuberance
Physcomitrella patens (p. 201)
Nerve extending beyond apex into short point, upper cells scarcely elongated (20–40 × 15–20μ); capsule perfectly spherical
Acaulon muticum (p. 170)

4 Plant bright or deep green; a very common species on fallow soil
Phascum cuspidatum (p. 168)
Plant brownish green or reddish brown; uncommon species   *5*

5 Capsule fully immersed among leaves, plant reddish brown
Phascum floerkeanum (p. 170)
Capsule raised just clear of leaves, plant brownish green   *6*

6 Seta straight   *Pottia recta* (p. 168)
Seta strongly curved downwards   *Phascum curvicollum* (p. 170)

SUBSECTION 2

1   Leaf margin distinctly toothed near apex; chiefly maritime, the most robust species of the group (leafy shoots about 8 mm., seta 8–14 mm.)   *Pottia heimii* (p. 168)
Leaf margin entire or merely rendered rough by papillae, or projecting cell walls; more minute plants (seta not above 8–9 mm.)   2

2   Capsule elongated, narrowly ovoid to ellipsoidal   3
Capsule shorter, urn-shaped   4

3   Ripe capsule red-purple, peristome well developed
         Pottia lanceolata (p. 165)
Ripe capsule light brown, peristome rudimentary or absent.
         P. intermedia (p. 168)

4   Leaf margin recurved, upper cells strongly papillose; plant very minute (leafy shoot, seta and capsule together 3–6 mm. tall)   *Pottia davalliana* (p. 167)
Leaf margin not recurved, upper cells scarcely papillose; plant taller (leafy shoot, seta and capsule together 6–10 mm. tall)
         P. truncata (p. 166)

SUBSECTION 3

1   Leaf margin distinctly toothed near apex   2
Leaf margin entire, or (at most) indistinctly notched (*Funaria* spp.)   *11*

2   Densely gregarious; conspicuous owing to massed orange-red setae and flask-shaped capsules; on animal excrement in mountain country   *Splachnum ampullaceum* (p. 202)
Plant less densely gregarious as a rule, growing on some other substratum; capsules never flask-shaped   3

3   Upper leaves large, 8–12 mm. long and 3–4 mm. wide, forming striking rosette   *Rhodobryum roseum* (p. 221)
All leaves much smaller, not forming rosettes   4

4   Leaves all broadly ovate or obovate; capsules erect, shortly and broadly pear-shaped   *Physcomitrium pyriforme* (p. 200)
Some or all leaves narrower, approaching lanceolate; capsules horizontal or pendulous, narrowly pear-shaped (exc. *Pohlia delicatula*)   5

(N.B. This group of species is included here and in section 10, the leaves often showing considerable range of variation in shape on a single stem.)

5   Leaf cells up to 20 μ wide   6
    Leaf cells less than 15 μ wide   7

6   Usually tall (1–8 cm.), conspicuously pale glaucous green; seta 2–2·5 cm. long   *Pohlia albicans* (p. 210)
    Usually dwarf (about 1 cm.), without marked glaucous hue, often reddish; seta 0·8–1·2 cm. long   *P. delicatula* (p. 209)

7   Leaves glaucous green with bright metallic lustre
                                 Pohlia cruda (p. 210)
    Leaves without glaucous hue or metallic lustre   8

8   Plant usually bearing bulbils, capsules very rare   9
    Plant without bulbils, commonly fertile, with abundant capsules   10

9   Bulbils roundly ovoid or narrowing to a point at one end
                                 Pohlia annotina (p. 210)
    Bulbils long and irregular in shape, suggestive of a glove
                                 *P. proligera** (p. 210)

10   Neck of capsule very long, as long as rest of capsule, the whole 4–6 mm. in length; plant chiefly on mountain ledges
                                 Pohlia elongata (p. 210)
    Neck of capsule very short (whole capsule 3 mm. long); an abundant plant on heaths, moors, etc.
                                 P. nutans (p. 207)

11   Plants densely gregarious, on decaying animal matter in hill country; usually very fertile with massed setae and erect flask-shaped capsules   12
    Plants less densely massed, on some other substratum; capsules never flask-shaped, inclined or pendulous when ripe   13

12   Leaves with short points, seta weak and pale, apophysis of capsule ovoid-globose   *Splachnum ovatum* (p. 203)
    Leaves with long points, seta firm and red, apophysis narrowly pear-shaped   *Tetraplodon mnioides* (p. 203)

13   Plant silvery grey in upper parts   14
    Plant green, yellow-green or reddish in upper parts   15

14   Plant notably pinkish in lower parts; uncommon species, of base-rich mountain ledges   *Plagiobryum zierii* (p. 210)
    Plant lacking pinkish colour below; abundant lowland species
                                 Bryum argenteum (p. 216)

15   Leaf with well-defined border of distinct narrow cells
                                 Funaria obtusa (p. 200)
    Leaf without well-defined border   16

      * See footnote on p. 210 for the modern concept of *P. proligera*.

16 Cells of leaf with square ends, 30μ wide or more   *17*
     Cells of leaf narrow, about 15μ wide, with pointed ends   *18*

17 Plant very small, capsule erect and symmetrical (on seta 5–10 mm. long); upland banks   *Funaria attenuata* (p. 200)
     Plant larger, capsule oblique and asymmetrical (on seta 2–5 cm. long); abundant species on burnt ground, etc.
                             F. hygrometrica (p. 198)

18 Shoots narrowly cylindrical (leaves closely overlapping and appressed when dry), nerve ceasing well below rounded apex of very short, rounded concave leaf, whole plant glossy yellow-green   *Anomobryum filiforme* (p. 210)
     Shoots wider, nerve extending almost to extreme apex of longer, more pointed leaf   *19*

19 Dwarf green plant (about 5 mm. tall), not glossy, commonly fertile, chiefly lowland   *Bryum bicolor* (p. 217)
     Taller plant (1–5 cm.), very glossy and often variegated with red or orange, rarely fertile, chiefly upland and western
                             B. alpinum (p. 218)

SUBSECTION 4

1 Leaf border more than 1 cell thick, plant commonly pinkish in colour   *Bryum pallens* (p. 213)
    Leaf border, if present, not thickened, colour of plant various   *2*

2 Leaves broadest near apex, spirally curled when dry, nerve protruding in long point   *Bryum capillare* (p. 219)
    Leaves broadest near base, only lightly twisted when dry, protruding points short or absent   *3*

3 Plant in rigid tufts, with notable metallic lustre, often tinged deep crimson   *Bryum alpinum* (p. 218)
    Plant of softer habit, with little or no metallic lustre, green or purple-brown   *4*

4 Leafy shoots 3–10 cm. long; growing in bogs and other very wet places   *Bryum pseudotriquetrum* (p. 214)
    Leafy shoots rarely above 1–2·5 cm. long; not bog species   *5*

5 Plant short, seta 0·8–2 cm. long, ripe capsules deep red   *6*
    Plant more robust, seta 1·5–3 cm. long, ripe capsules brownish   *7*

6 Neck of capsule very short; a common plant on waste ground, etc.   *Bryum bicolor* (p. 217)
    Neck of capsule longer, lower leaves often with red gemmae; common   *B. erythrocarpum* (p. 221)

Neck of capsule as last, not gemmiferous, uncommon plant of walls *B. murale* (p. 221)

7 Appendiculate cilia present on inner peristome of capsule 8
Cilia of inner peristome absent or rudimentary 9

8 Common plant of wet ground, spores large 20–25 μ
Bryum intermedium (p. 221)
Common plant of dry ground, spores small, 8–14 μ
B. caespiticium (p. 215)

9 Transverse bars on outer peristome connected by vertical or oblique lines *Bryum pendulum* (p. 221)
Transverse bars on outer peristome without such connecting lines *B. inclinatum* (p. 221)

SECTION 9

1 Cells in leaf base with very thick wavy longitudinal and thin transverse walls 2
Cells in leaf base lacking this character 3

· 2 Leaf apex very broad; dark green plant of rocks in and by streams *Rhacomitrium aciculare* (p. 191)
Leaf apex narrow, though blunt; yellow-green plant of wet rock ledges *R. aquaticum* (p. 192)

3 Leaf margin toothed near apex 4
Leaf margin without teeth 5

4 Plant 1–5 cm. tall, leaf tapering from rather broad base
Dichodontium pellucidum (p. 141)
Plant 1 cm. or less, leaf tongue-shaped *Rhabdoweisia denticulata* (p. 140)

5 Leaves very widely spreading (squarrose), upper cells elongated; plant of wet detritus by mountain streams
Dicranella squarrosa (p. 136)
Leaves not very widely spreading, upper cells not longer than broad; plants of other habitats 6

6 Leaves rather long (3–5 mm.), with evident border (at least in lower part of leaf), due to special cells or thickening 7
Leaves short or long without evident border 8

7 Plant prostrate with long stems (5–18 cm.); aquatic
Cinclidotus fontinaloides (p. 170)
Plant erect, stems short (1·5–4 cm.); on rocks and tree stumps by rivers *C. mucronatus* (p. 171)
Plant erect, stems even shorter (5 mm.) but with cylindrical capsule on seta 2–3 cm. long; terrestrial *Tortula subulata* (p. 161)

8   Nerve ceasing below apex   *9*
     Nerve extending to apex or beyond in short point   *13*

9   Leaves very large (up to 5 mm. long) upper cells with large
     'stud-like' papillae   *Encalypta streptocarpa* (p. 155)
     Leaves smaller (2·5 mm. or less) upper cells with minute
     papillae or smooth   *10*

10 Plant semi-aquatic, upper cells of leaf clearly defined, round,
     thick-walled   *Orthotrichum rivulare* (p. 244)
     Plant terrestrial, upper cells of leaf obscure, angular, thin-walled   *11*

11 Leaves strongly spirally twisted when dry; common lowland
     species   *Barbula convoluta* (p. 171)
     Leaves only lightly twisted when dry   *12*

12 Nerve reddish, leaf margin not revolute
                               Gymnostomum aeruginosum (p. 177)
     Nerve not reddish, leaf margin revolute   *Barbula tophacea* (p. 177)

13 Leaf apex very broad and flattened (truncate); uncommon
     plant, on tree roots by water   *Tortula latifolia* (p. 164)
     Leaf apex not truncate, but tapering somewhat; habitat
     different   *14*

14 Leaves large (3–4 mm. long), upper cells with large 'stud-like'
     papillae; calyptra of capsule extinguisher-shaped   *15*
     Leaves shorter and narrower; papillae, if present, minute;
     calyptra of capsule not extinguisher-shaped   *16*

15 Edge of calyptra fringed; uncommon plant of mountain ledges
                                  Encalypta ciliata (p. 156)
     Edge of calyptra not fringed; common lowland species on
     limestone   *E. vulgaris* (p. 154)

16 Sharp oblique line of demarcation between clear cells of leaf
     base and green cells above   *Tortella flavovirens* (p. 181)
     No such sharp oblique line of demarcation   *17*

17 Leaf apex hood-like in form   *Trichostomum crispulum* (p. 182)
     Leaf apex not hood-like in form   *18*

18 Leaf margin broadly recurved, almost to nerve *Barbula revoluta* (p.176)
     Leaf margin plane or only narrowly recurved   *19*
     Leaf margin incurved   *22*

19 Leaves about 2·5–3 mm. long, margins not recurved
                           Trichostomum brachydontium (p. 181)
     Leaves shorter, margins recurved to some extent   *20*

20 Plants in low tufts a few millimetres tall, seta about 5 mm. long;
uncommon plant of soil-capped walls, chiefly near the sea
Desmatodon convolutus (p. 164)
Plants taller, seta much longer, very common plants of soil,
paths, etc.　　　　　　　　　　　　　　　　　　　　　　　*21*

21 Perichaetial leaves convolute, seta yellow　*Barbula convoluta* (p. 171)
Perichaetial leaves not convolute, seta red　*B. unguiculata* (p. 172)

22 Capsule nearly globose, almost sessile among perichaetial
leaves　　　　　　　　　　　　　　　　*Weissia crispa* (p. 183)
Capsule ovoid-ellipsoidal, on evident seta　　*W. tortilis* (p. 183)

SECTION 10

Key to subsections

1 Leaf margin toothed to some extent　　　　　　　　　　　*2*
Leaf margin quite entire　　　　　　　　　　　　　　　　*3*

2 Cells of leaf uniformly elongated (narrowly rhomboid), thin-
walled, not papillose　　　　　　　Subsection 1 (*Pohlia* spp.)
Cells in upper part of leaf isodiametric or nearly so, papillose
or smooth　　　　　　　　　　　　　　　　　Subsection 2

3 Cells at base of leaf long and very narrow, with greatly
thickened, strongly wavy longitudinal walls and thin transverse
walls; habit usually loose and freely branched; on siliceous
rocks　　　　　　　Subsection 3, p. 61 (*Rhacomitrium* spp.)
This combination of characters lacking　　　　　　　　　*4*

　(N.B. Species of *Ulota* and some species of *Orthotrichum* have
　　long narrow cells in the leaf base with longitudinal walls much
　　thickened, but these walls are never strongly and regularly wavy
　　as in *Rhacomitrium*, and the plants form neat cushions on trees
　　or rocks.)

4 Outlines of cells in upper part of leaf obscure owing to thin
walls and dense covering of papillae, these cells always small
and isodiametric (6–10μ across)　　　　　Subsection 4, p. 61
Outlines of these cells moderately clear to very well defined
(6–18μ across)　　　　　　　　　　　Subsection 5, p. 62

　(N.B. Many species in this subsection have papillose leaf cells, but
　　the papillae are not such as to obscure the cell wall outline. Species
　　in which the condition of the cells in question varies from obscure
　　to moderately well-defined are included in both subsections.)

SUBSECTION 1

1 Leaf cells up to 20 μ wide   *2*
    Leaf cells less than 15 μ wide   *3*

2 Usually tall (1–8 cm.), conspicuously pale glaucous green
                                            Pohlia albicans (p. 210)
    Usually dwarf (about 1 cm.), glaucous hue seldom well developed;
    often reddish below   *P. delicatula* (p. 209)

3 Leaves glaucous green, with bright metallic lustre or opalescence;
    uncommon plant of mountain ledges   *Pohlia cruda* (p. 210)
    Leaves without glaucous hue or metallic lustre

4 Plant usually bearing bulbils, capsules very rare
    Plant without bulbils, commonly fertile, with abundant capsules

5 Bulbils roundly ovoid or narrowing to a point at one end
                                            Pohlia annotina (p. 210)
    Bulbils elongated and irregular in shape, suggestive of a glove
                                            *P. proligera** (p. 210)

6 Neck of capsule very long (as long as body of capsule) the
    whole 4–6 mm. long   *Pohlia elongata* (p. 210)
    Neck of capsule short, whole capsule about 3 mm. long; abun-
    dant species   *P. nutans* (p. 207)

  (N.B. In all these species of *Pohlia* the nerve ceases in or just below
    the leaf tip. A plant appearing to 'key out' here with nerve
    excurrent will probably be a species of *Bryum*. Most species of
    Bryum, however, having a broader form of leaf, are dealt with in
    section 8, subsection 4 of this key.)

SUBSECTION 2

1 Some leaves up to 3–5 mm. in length   *2*
    Leaves all 2·25 mm. or shorter   *6*

2 Leaves plicate, cells in leaf base with very thick longitudinal
    walls; forms cushions on siliceous rock; usually very fertile
                                            Ptychomitrium polyphyllum (p. 234)
    Leaves not plicate; cells in leaf base without greatly thickened
    longitudinal walls   *3*

3 Leaf margin at most bluntly notched; cells in upper part of leaf
    very small (6–10 μ)   *4*
    Leaf margin sharply toothed; cells in upper part of leaf larger
    (10–18 μ)   *5*

  * See footnote on p. 210 for the modern concept of *P. proligera*.

4 Cells in leaf base very long (up to $80 \times 12\mu$), thin-walled; the
   outlines of those in upper part of leaf obscured by papillae;
   on acid ledges   *Trichostomum tenuirostre* (p. 182)
   Cells in leaf base shorter (up to about $50 \times 12\mu$); outlines of
   upper cells not greatly obscured by papillae; in calcareous
   habitats (lowland)   *T. sinuosum* (p. 182)

5 Leaves fairly uniform in size (about 3 mm.); many of the
   upper cells shortly rectangular; gemmae absent but nearly
   spherical capsules commonly present   *Plagiopus oederi* (p. 231)
   Leaves very diverse in size (often ranging from 1·5 to 4 mm. on
   single plant); upper cells with nearly round cavities and walls
   with thickened corners; gemmae almost always present in
   greenish stalked clusters (like pin-heads)
                                  Aulacomnium androgynum (p. 229)

6 A few teeth on leaf margin near base only; small pale green
   plant of wet calcareous habitats   *Eucladium verticillatum* (p. 178)
   A few blunt notches at extreme leaf tip only; common plants
   on earth, soil covered walls, etc.   7
   Leaves conspicuously notched or toothed for some distance
   back from apex   8

7 Cells in upper part of leaf about 10μ wide, their outlines somewhat obscured by papillae; common on calcareous soil and
   usually easy to recognize by rusty red colour of older parts
                                  Barbula recurvirostra (p. 175)
   Cells in upper part of leaf larger ($10-15\mu$), their outlines nearly
   square and well-defined; abundant in many habitats
                                  Ceratodon purpureus (p. 133)

8 Leaves with broad clasping bases and narrower, more or less
   squarrose tips   *Dichodontium pellucidum* (p. 141)
   Leaves not as above, lanceolate or nearly strap-shaped   9

9 Leaves only 1–1·5 mm. long, cells towards leaf tip studded with
   conspicuous papillae   *Leptodontium flexifolium* (p. 184)
   Leaves longer; conspicuous papillae lacking   10

10 Plant less than 1 cm. tall, leaf nearly parallel-sided, the apical
   part with edges rendered jagged by numerous projecting cells
                                  Rhabdoweisia denticulata (p. 140)
   Plant several centimetres tall; leaf lanceolate, tapering; margin
   near leaf apex with a few widely spaced notches
                                  Cynodontium bruntonii (p. 140)

SUBSECTION 3

1 Stems 1–2·5 cm., habit more or less erect and rigid; cells in upper part of leaf in 2 layers   *Rhacomitrium ellipticum* (p. 198)
Stems longer, habit looser; cells in upper part of leaf not in 2 layers   *2*

2 Stems 4–12 cm. long, with few lateral branches; on wet rock ledges by waterfalls, etc.   *Rhacomitrium aquaticum* (p. 192)
Stems with numerous lateral branches; forming wide cushions or mats on boulders   *3*

3 Plant light yellowish green to tawny; hair-point completely lacking   *Rhacomitrium fasciculare* (p. 193)
Plant deep olive or blackish green above, dark brown to black below; a few short hair-points often present on upper leaves
      R. heterostichum var. *gracilescens* (p. 194)

SUBSECTION 4

1 Leaf margins incurved   *2*
Leaf margins plane or recurved to some extent   *4*

2 Nerve very wide, about 70μ at base of leaf; uncommon plant, only on calcareous banks, walls, etc.   *Weissia tortilis* (p. 183)
Nerve narrower; common plants on soil, mud-capped walls, etc.   *3*

3 Capsule with very narrow mouth; peristome teeth lacking
      Weissia microstoma (p. 183)
Capsule with wider mouth; rudimentary peristome teeth present
      W. controversa (p. 182)

4 Longest leaves 4–5 mm. long   *5*
No leaves as long as this   *7*

5 Cells in leaf base only slightly elongated (up to 30–40μ)
      Barbula cylindrica (p. 177)
Cells in leaf base greatly elongated (80–100 × 10–15μ)   *6*

6 Hyaline cells of leaf base extending up margins; calcicole species
      Tortella tortuosa (p. 179)
Hyaline cells not extending up margins; plant of moist acid rock ledges   *Trichostomum tenuirostre* (p. 182)

7 Plants forming deep, dense cushions on mountain rock ledges and crevices   *8*
Plants forming loose tufts or low cushions mainly in lime-rich lowland habitats   *10*

8   Plant 2–6 cm. tall; longest leaves 2–3 mm. long; cells in upper part of leaf varying from well-defined to moderately obscure, very rarely fertile   *Amphidium mougeotii* (p. 235)

     Plant 1–3 cm. tall; longest leaves 2 mm. long; upper leaf cells as last, commonly fertile   *A. lapponicum* (p. 236)

     All leaves shorter (1–1·5 mm.); all upper cells usually very obscure due to covering of papillae   9

9   Leaf apex acute, leaf channelled; shoot tips vivid green, stems matted together with red-brown tomentum below

                                Anoectangium compactum (p. 179)

     Leaf apex obtuse, leaf flat or nearly so; shoot tips dull green, red-brown tomentum lacking   *Gymnostomum aeruginosum* (p. 177)

10   Older parts of plants strongly tinged with rusty red

                                Barbula recurvirostra (p. 175)

     Rusty red colour lacking   11

11   Cells overlying nerve of leaf elongated; other cells of upper part of leaf with rounded cavities and walls thickened at corners   *Barbula fallax** (p. 174)

     Cells overlying nerve of leaf not elongated; other leaf cells tending to be square in outline, with walls not much thickened at corners   12

12   Longest leaves exceeding 2·5 mm. in length, those at shoot tips markedly flexuose; habitat most commonly soil

                                Barbula cylindrica (p. 177)

     Longest leaves scarcely 2 mm. long, curved but not markedly flexuose; habitat almost always rock   *B. vinealis* (p. 177)

SUBSECTION 5

1   Leaves all short, the longest scarcely up to 2 mm.   2

     Leaves longer, the longest on any typical stem being 2·25–4 mm.   13

2   On trees   3

     On rocks or soil   5

3   Cells in upper part of leaf quadrate, not papillose; seta purple, 1–2 cm. long   *Ceratodon purpureus* (p. 133)

     Cells in upper part of leaf rounded-hexagonal, papillose; seta shorter, 6–10 mm. long   4

4   Cells in upper part of leaf 8–12μ across; peristome rudimentary or absent; common species   *Zygodon viridissimus* (p. 236)

     Cells in upper part of leaf 10–15μ across; peristome well developed; rather rare species   *Z. conoideus* (p. 238)

   * If the leaf shape has been misinterpreted in the key to sections the very common *Barbula convoluta* may also key out here.

5 Stems long (4–10 cm.), trailing; on wet rocks in streams
                            Grimmia alpicola var. *rivularis* (p. 190)
   Habit and habitat different   6

6 Minute hyaline tips visible on some of upper leaves   7
   All leaves without traces of hyaline tips   8

7 Leaves spirally twisted so that shoots have rope-like character
   when dry; uncommon plant, on mountain rock-ledges
                                      Grimmia torquata (p. 190)
   Rope-like character lacking in dry state; common species on
   rocks and walls generally   *G. apocarpa* (p. 185)

8 Margins recurved from base to near apex of leaf; upper cells
   rather large (10–15μ across), mostly quadrate and very clearly
   defined   *Ceratodon purpureus* (p. 133)
   Margins of leaf at most irregularly and locally recurved; upper
   cells usually smaller (6–12μ across)   9

9 Leaves only about twice as long as broad, with notably blunt
   apices, appressed in dry state   *Barbula trifaria* (p. 176)
   Leaves longer in proportion to breadth, with moderately to
   very acute apices; curled and twisted when dry   10

10 In deep dense cushions (stems 4–10 cm. tall), on calcareous rock
   ledges on mountains; cells in upper part of leaf very clearly
   defined, often slightly elongated
                          Gymnostomum recurvirostrum (p. 178)
   (Cf. also *Amphidium mougeotii* and *A. lapponicum* which usually
      have longer leaves.)
   Stems shorter; in loose tufts or low cushions, chiefly in calcareous lowland habitats; upper leaf cells isodiametric   11

11 Cells overlying nerve of leaf elongated; upper leaf cells with
   rounded cavities and walls much thickened at corners; plant
   on soil   *Barbula fallax* (p. 174)
   Cells overlying nerve not elongated; upper leaf cells not much
   thickened at corners; plant on rock   12

12 Leaves curved, with acute apices and cells tending to be square
   in outline   *Barbula vinealis* (p. 177)
   Leaves straight, with relatively blunt apices, cells regularly
   hexagonal; plant often with gemmae   *B. rigidula* (p. 176)

13 On trees   14
   On rocks or soil   22

14 Longitudinal walls of cells in leaf base without special thickenings; upper leaf cells quadrate to shortly rectangular; seta about 1 cm. long *Dicranoweisia cirrata* (p. 142)
Longitudinal walls of cells in leaf base either very thick or with irregular thickenings; upper cells with round cavities and walls thickened at corners; seta not above 4 mm. in length *15*

15 Leaves much curled and twisted when dry *16*
Leaves nearly straight and appressed when dry *19*

16 Bearing clusters of multicellular gemmae at tips of leaves; capsules extremely rare *Ulota phyllantha* (p. 244)
Leaf tips without gemmae; capsules commonly present *17*

17 Cells in leaf base with only a few irregular lumps of thickening in their longitudinal walls; peristome orange-red
Orthotrichum pulchellum (p. 244)
Cells in middle of leaf base very long and narrow, with uniformly thick longitudinal walls; peristome brown *18*

18 Leaves up to 2·5 mm. long, upper leaf cells 8–12μ across; capsule and seta together about 4 mm. long *Ulota crispa* (p. 245)
Leaves up to 3–4 mm. long, upper leaf cells 12–16μ across; capsule and seta together about 8 mm. long *U. bruchii* (p. 247)

19 Leaf margin lightly recurved near base only; upper surface of leaf bearing numerous long, multicellular gemmae
Orthotrichum lyellii (p. 241)
Leaf margin widely recurved almost to apex; upper surface of leaf not gemmiferous or only sparsely so *20*

20 Ripe capsule without ridges and furrows
Orthotrichum striatum (p. 244)
Ripe capsule marked with longitudinal ridges and furrows *21*

21 Leaf cells highly papillose; stomata of capsule superficial
Orthotrichum affine (p. 240)
Leaf cells at most only faintly papillose; stomata of capsule immersed *O. tenellum* (p. 244)

22 Leaves narrow throughout (only about 0·2–0·3 mm. wide in widest part) *23*
Leaves wider (0·4–0·8 mm. wide in widest part) *25*

23 Upper leaf cells without papillae; seta about 1 cm. long; on lowland walls, etc. *Dicranoweisia cirrata* (p. 142)
Upper leaf cells papillose to some extent; seta only a few millimetres long; plants forming dense cushions on mountain rock ledges *24*

24 Plant 2–6 cm. tall, longest leaves 2–3 mm.; very rarely fertile
Amphidium mougeotii (p. 235)
Plant 1–3 cm. tall, longest leaves about 2 mm., commonly fertile   *A. lapponicum* (p. 236)

25 On siliceous rocks   26
On calcareous rocks or moderately lime-rich soil   29

26 Leaf with back of nerve 2-winged and cells in 2–3 layers towards apex   *Grimmia patens* (p. 191)
Back of nerve not winged, leaf cells not in 2–3 layers   27

27 Cells in extreme base of leaf long and very narrow, with very thick longitudinal walls; plant dark brown to blackish
Ulota hutchinsiae (p. 247)
Cells in leaf base wider, with walls only irregularly thickened; plant dark green   28

28 Stems 2–5 cm. long; stomata on capsule superficial
Orthotrichum rupestre (p. 244)
Stems 1–2 cm. long; stomata on capsule immersed
O. anomalum (typical form) (p. 239)

29 Leaf apex relatively broad; cells in upper part of leaf 10–15μ wide, with round cavities; plants forming dense rounded cushions on limestone rock   30
Leaf tapering to narrow apex; upper leaf cells 6–10μ wide, nearly square in outline; forming loose tufts on rock or soil   31

30 Capsule broadly ovoid, quite immersed among perichaetial leaves; calyptra almost without hairs
Orthotrichum cupulatum (p. 244)
Capsule narrowly ovoid-cylindrical, on short but evident seta; calyptra hairy   *O. anomalum* var. *saxatile* (p. 239)

31 Plants strongly tinged rusty red below; paroecious or synoecious
Barbula recurvirostra (p. 175)
Plants lacking rusty red tinge; dioecious   *B. cylindrica* (p. 177)

SECTION 11

1 Shoots twice or thrice pinnate   2
Shoots simply pinnate or irregularly branched   4

2 Apical cell of branch leaf acute, undivided; abundant species
Thuidium tamariscinum (p. 260)
Apical cell of branch leaf obtuse, crowned with 2 or 3 papillae   3

3 Tips of stem leaves long and fine, recurved when moist
Thuidium philiberti (p. 262)
　　Tips of stem leaves shorter, not recurved when moist
T. delicatulum (p. 262)

4 Shoots evidently flattened in one plant (complanate), the leaves
　　thus having the appearance of being arranged more or less in
　　2 ranks　　　　　　　　　　　　　　　　　　　　　　　　5
　　Shoots not flattened in one plane　　　　　　　　　　　8

5 Leaves strongly transversely undulate　　　*Neckera crispa* (p. 252)
　　Leaves not undulate　　　　　　　　　　　　　　　　6

6 Leaves large (6 mm. long), cells very large ($300 \times 50\mu$)
Hookeria lucens (p. 257)
　　Leaves smaller, cells much smaller ($40\text{–}70 \times 6\text{–}10\mu$)　　7

7 Nerve single to beyond mid-leaf, plant commonly fertile
Homalia trichomanoides (p. 254)
　　Nerve very short, double; plant rarely fertile　*Neckera complanata*
(p. 253)

8 Plants with 'tassel-like' or 'miniature tree' habit
Thamnium alopecurum (p. 255)

　　(Cf. also *Climacium dendroides*, which, however, has rather longer
　　　　leaf cells and is therefore not included in this section.)
　　Plants without this characteristic habit　　　　　　　9

9 Leaf apex very obtuse, rounded; robust and common plant on
　　calcareous banks, etc.　　　　*Anomodon viticulosus* (p. 259)
　　Leaf apex acute　　　　　　　　　　　　　　　　10

10 Leaf apex having form of grapnel; robust, rather uncommon
　　plant of western Britain　　　*Antitrichia curtipendula* (p. 251)
　　Leaf apex without this character　　　　　　　　　*11*

11 Leaves nerveless　　　　　　　　　　　　　　　　*12*
　　Leaves single-nerved to mid-leaf or above　　　　　*14*

12 Leaves strongly plicate, margins entire　*Leucodon sciuroides* (p. 251)
　　Leaves not plicate, margins toothed　　　　　　　*13*

13 Plant of very slender habit (shoots thread-like), leaves up to
　　0·75 mm. long, not concave　*Heterocladium heteropterum* (p. 259)
　　Plant of rather robust habit, leaves up to 1·5 mm. long; concave
Pterogonium gracile (p. 251)

14 Leaf margin regularly toothed　　　　　　　　　*15*
　　Leaf margin entire or only very faintly indented　　*16*

15 Plant robust, with rigid, regularly pinnate branches, upper cells
of leaf round; in relatively dry, grassy calcareous places
Thuidium abietinum (p. 261)
Plant of much weaker, more slender habit, upper cells rhomboid; in wet calcareous places *Cratoneuron filicinum* (p. 262)

16 Upper cells of leaf rounded, scarcely longer than broad *17*
Upper cells of leaf rhomboid, 2–5 times as long as broad *18*

17 Plants epiphytic on trees, seta less than 1 mm. long
Cryphaea heteromalla (p. 251)
Plants at bases of trees, seta about 1 cm. long
Leskea polycarpa (p. 258)

18 Plants of very slender habit, nerve of leaf thin; terrestrial or on
trees *19*
Plants more robust, nerve of leaf stout; aquatic *20*

19 Leaves about 0·5 mm. long, not widely spreading; nerve to about
mid-leaf; common *Amblystegium serpens* (p. 268)
Leaves about 1 mm. long, widely spreading; nerve extending
above mid-leaf; uncommon *A. varium* (p. 269)

20 Nerve disappearing in acute leaf-apex, margin faintly indented
near tip of leaf *Hygroamblystegium tenax* (p. 269)
Nerve extending to extreme point of blunt leaf-apex, margin
entire *H. fluviatile* (p. 270)

SECTION 12

1 Plants of dendroid or sub-dendroid habit, principal secondary
stems bare below, with numerous crowded branches near their
summits *2*
Plants lacking this characteristic habit *4*

2 Principal secondary stems upright with crown of branches at
summits, giving striking resemblance to 'miniature trees'
Climacium dendroides (p. 249)
Habit sub-dendroid only, plants forming dense cushions or mats
on boulders or tree bases *3*

3 Branch leaves narrowly ovate-lanceolate, finely drawn-out at
apex; their margins toothed all round *Isothecium myosuroides*
(p. 281)
Branch leaves mostly wider, concave, apex not finely drawn out,
margins lightly toothed near leaf tip only *I. myurum* (p. 280)

4 Leaves obtuse or apiculate (not acute), with clearly defined
   patches of specialized (alar) cells at basal angles; marsh plants 5
   Leaves various, if obtuse or apiculate without clearly defined
   patches of alar cells; habitat various   8

5 Leaf with apiculate tip, plant with strong tinge of purplish red
                          Acrocladium sarmentosum (p. 279)
   Leaf with broadly rounded apex, plant green or tinged with
   yellow or orange   6

6 Leaves narrowly oblong, held erect against stem so that habit
   appears very slender   *Acrocladium stramineum* (p. 279)
   Leaves widely heart-shaped to ovate, spreading and so giving
   shoots more robust appearance   7

7 Plant distantly and irregularly branched, alar cells not forming
   very well-defined auricles   *Acrocladium cordifolium* (p. 279)
   Plant densely branched—giving bushy habit, alar cells of leaf
   forming very well-defined auricles   *A. giganteum* (p. 279)

8 Leaf surface scored by obvious longitudinal folds or furrows
   (plicate)   9
   Leaf surface not plicate or only lightly so   13

9 Stem leaves broadly heart-shaped, triangular (almost as broad
   as long), their margins closely toothed all round
                          Eurhynchium striatum (p. 293)
   Stem leaves ovate-lanceolate, their margins entire or only faintly
   and irregularly indented   10

10 Leaves widening just above base, then suddenly contracted near
   the rather long fine tip, leaf cells shorter and wider towards base 11
   Leaves gradually narrowing from extreme base to apex, leaf
   cells very long and narrow except at extreme base   12

11 Plant of slender habit, leaf margin usually quite entire; common
   on acid stony ground   *Brachythecium albicans* (p. 284)
   Plant more robust, leaf margin faintly indented near apex;
   uncommon plant, chiefly of calcareous habitats *B. glareosum* (p. 291)

12 Plant bright green, habit creeping; on rocks, walls and tree
   trunks   *Camptothecium sericeum* (p. 282)
   Plant yellowish, habit tufted with ascending branches; on calcareous ground   *C. lutescens* (p. 283)

13 Plant forming intricate prostrate patches of growth, stems
   regularly pinnate or bi-pinnate, stem leaves almost as broad as
   long, suddenly contracted into long narrow apex, branch leaves
   much smaller, ovate-lanceolate *Eurhynchium praelongum* (p. 294)

Stems not regularly pinnate or bi-pinnate, stem- and branch-leaves not as above   *14*

14 Nerve very stout in lower part of leaf, suddenly narrowing
Cirriphyllum crassinervium (p. 292)
Nerve of leaf not showing this character   *15*

15 Leaves markedly concave, obtuse or apiculate   *16*
Leaves not concave, or, if slightly so, tapering to acute apices   *18*

16 Plant robust, leaves 2 mm. long or more, shoots pale whitish green   *Pseudoscleropodium purum* (p. 299)
Plant more slender, leaves under 2 mm. long, shoots not whitish green   *17*

17 Plant not markedly glossy, growing on ground in south-west Britain; seta rough with papillae, capsule with short lid
Scleropodium illecebrum (p. 291)
Plant very glossy, on calcareous rock ledges and walls; seta smooth, lid of capsule long-beaked   *Eurhynchium murale* (p. 299)

18 Plants large, with their longest leaves 2 mm. long or over   *19*
Plants small to very small (even if forming extensive patches, the shoots slender), leaves not attaining 2 mm. in length   *26*

19 Leaf suddenly contracted near apex into long fine point
Cirriphyllum piliferum (p. 291)
Leaf without sudden contraction into hair-like point   *20*

20 Leaf margin entire   *21*
Leaf margin toothed, at least near apex   *23*

21 Well-defined auricles lacking   *Leptodictyum riparium* (p. 267)
Well-defined auricles present at base of leaf   *22*

22 Auricles decurrent, base of leaf hollowed, plant dioecious
Drepanocladus aduncus (p. 270)
Auricles not decurrent, base of leaf not hollowed, plant autoecious and commonly fertile   *Campylium polygamum* (p. 267)

23 Leaves several times as long as broad, drawn out into long fine points   *Drepanocladus fluitans* (forms) (p. 271)
Leaves scarcely twice as long as broad, points acute but not attenuated   *24*

24 Stems rather rigid, with long, little-branched divisions, deep green; seta smooth, plant strictly of aquatic habitats
Eurhynchium riparioides (p. 296)
Plant usually of much softer habit, yellowish to mid-green; stems with shorter, more branched divisions, seta rough; plants of terrestrial or sub-aquatic habitats   *25*

25 Cells at basal angles of leaf inflated, forming well-defined
   auricles, plant dioecious, habitat commonly sub-aquatic
                              Brachythecium rivulare (p. 287)
   Well-defined auricles not formed, plant autoecious; habitat
   terrestrial, very abundant   *B. rutabulum* (p. 286)

26 Longest leaves only 0·5–0·75 mm. long, narrowly ovate; habit
   extremely slender; leaf cells about 6–7 times as long as wide; an
   uncommon plant. (Cf. the much commoner species *Amblystegium serpens*, with more finely pointed leaves and cells 3–5
   times as long as wide)   *Rhynchostegiella pallidirostra* (p. 299)
   Leaves longer, the longest on any shoot reaching 1–1·75 mm.   27

27 Leaves ovate, not more than twice as long as broad, not tapering
   to fine points, their margins toothed   28
   Leaves lanceolate or ovate-lanceolate, more than twice as long
   as broad, tapering to fine points   30

28 Plant aquatic or sub-aquatic, dark green
                    Eurhynchium riparioides (small forms) (p. 296)
   Plant terrestrial   29

29 Leaves concave and held more or less erect and overlapping on
   shoots, their margins lightly toothed; plant with numerous
   glossy, yellowish green ascending branches
                              Scleropodium caespitosum (p. 291)
   Leaves not concave, widely spreading, their margins strongly
   toothed; plant dull yellow-green to tawny, more or less prostrate
   and irregularly branched   *Eurhynchium swartzii* (p. 295)

30 Leaf extremely long and narrow, 6–8 times as long as broad;
   bright yellowish green plant of calcareous places
                              Rhynchostegiella tenella (p. 299)
   Leaf wider, 3–5 times as long as broad   31

31 Nerve extending into apex of leaf *Brachythecium populeum* (p. 291)
   Nerve only to mid-leaf or just above   32

32 Leaves widely spreading, with long, fine entire points; yellowish
   plant of calcareous places   *Campylium chrysophyllum* (p. 266)
   Leaves not widely spreading, their points less fine and margins
   toothed   33

33 Leaves narrowly ovate-lanceolate, tapering to fairly narrow
   acute points, very silky; seta rough, lid of capsule short
                              Brachythecium velutinum (p. 288)
   Leaves wider, shorter in the point, less silky; seta smooth, lid of
   capsule long-beaked   *Eurhynchium confertum* (p. 298)

SECTION 13

1 Plants very robust (branches 2–3 mm. wide, leaves 3–4 mm. long); rather rare species. *2*
  Plants slender to moderately robust only *3*

2 Leaves transversely undulate wet and dry, spiny papillae at back of leaf, plant of dry calcareous hillsides
                    Rhytidium rugosum (p. 318)
  Leaves wrinkled when dry only, no spiny papillae at back of leaf, plant of marshes   *Drepanocladus lycopodioides* (p. 275)

3 Leaves somewhat concave, their apices not drawn out to long fine points *4*
  Leaves not concave, apices drawn out into long fine points *5*

4 Leaf margin incurved near tip, entire *Hygrohypnum luridum* (p. 275)
  Leaf margin not incurved, toothed *Brachythecium plumosum* (p. 290)

5 No auricles at base of leaf *Drepanocladus revolvens* (p. 273)
  Auricles, composed of distinct (alar) cells, present at base of leaf *6*

6 Leaf margin entire throughout *Drepanocladus aduncus* (p. 270)
  Leaf margin finely toothed, at least near apex *7*

7 Leaves, when moist, scored with distinct longitudinal folds or furrows (plicate) *8*
  Leaves, when moist, not plicate or only very faintly so *9*

8 Nerve of very strongly curled, finely pointed leaf very narrow at base (30–35μ); plant autoecious; habitat commonly terrestrial (rocks, walls, etc.)   *Drepanocladus uncinatus* (p. 274)
  Nerve of less strongly curved, more shortly pointed leaf wide at base (60–100μ); plant dioecious; in calcareous bogs and by waterfalls   *Cratoneuron commutatum* (p. 263)

9 Only upper leaves curved to one side, plant of soft texture, green to brown in colour, autoecious *Drepanocladus fluitans* (p. 271)
  Leaves more uniformly curved to one side, plant of more rigid texture, commonly purplish in colour, dioecious, almost confined to mountain bogs   *D. exannulatus* (p. 275)

SECTION 14

Key to subsections

1   Leaves curved to one side, at least at tips of branches   Subsection 1
    Leaves straight throughout   *2*

2   Branches flattened in one plane, leaves thus appearing to be
    arranged more or less in 2 irregular ranks   Subsection 2
    Branches not flattened, leaves arranged evenly all round stems
                                                Subsection 3

SUBSECTION 1

1   Stems rigid, red; robust plant, common and conspicuous in hill
    districts   *Rhytidiadelphus loreus* (p. 315)
    Stems less rigid, rarely with any tinge of red   *2*

2   Leaf margin distinctly toothed   *3*
    Leaf margin entire (or with a few minute indentations near
    apex)   *5*

3   Stem leaves with broad, oblong plicate base, shoots regularly
    pinnate and plume-like; rather rare moss of highland woods
                               Ptilium crista-castrensis (p. 311)
    Stem leaves heart-shaped, with clasping auriculate base, shoots
    pinnate or irregularly branched   *4*

4   Habit close and tufted, pinnate character of branching usually
    well-marked, leaves strongly curled to one side
                                 Ctenidium molluscum (p. 310)
    Habit loose, branching irregular, leaves at most only lightly
    curved to one side   *Hyocomium flagellare* (p. 311)

5   Habitat rocks in streams or mountain bogs, plants of soft, nearly
    prostrate growth   *6*
    Habitat various, but not aquatic, habit commonly stiffer and less
    prostrate (except in some forms of *Hypnum cupressiforme*)   *9*

6   Plant very robust (branches 2·5–5 mm. wide, leaves 2–4 mm.
    long), habitat bogs and wet rocks   *Scorpidium scorpioides* (p. 277)
    Plant more slender (branches to 1·5 mm. wide, leaves smaller),
    habitat almost always wet rocks in streams   *7*

7   Cells at basal angles of leaf large, hyaline and thin-walled
                                Hygrohypnum ochraceum (p. 276)
    Cells at basal angles of leaf smaller, greenish or orange coloured   *8*

8   Leaf margin much incurved above to form almost hooded apex;
    alar cells ill defined, greenish   *Hygrohypnum luridum* (p. 275)
    Leaf margin lightly incurved only, apex acute, not hooded; alar
    cells well defined, orange in old leaves   *H. eugyrium* (p. 276)

9   Cells at basal angles of leaf well defined, enlarged, hyaline and
    thin-walled   *Hypnum patientiae* (p. 309)
    Cells at basal angles of leaf not as above   10

10 Habit always slender, leaves scarcely curved to one side, cells
    6–10 times as long as broad, with rather ill-defined patches of
    quadrate cells in basal angles; uncommon autoecious species
                                    Pylaisia polyantha (p. 309)
    Habit slender to robust, leaves much curved to one side or nearly
    straight, cells 10–15 times as long as broad, with very well-
    defined patches of short cells in basal angles; very common
    dioecious species
            Hypnum cupressiforme (with numerous varieties) (p. 307)
    Habit very slender, leaves rather regularly but lightly curved
    or turned to one side, cells 15–20 times as long as broad,
    without distinct patches in basal angles; autoecious
                                    Isopterygium pulchellum (p. 303)

SUBSECTION 2

1   Plant robust, with long branches 3–5 mm. wide, whitish green
    in colour, leaves transversely undulate
                                    Plagiothecium undulatum (p. 305)
    Plant slender to moderately robust, bright or olive green, leaves
    not transversely undulate   2

2   Cells somewhat enlarged in basal angles of leaf; plants
    moderately robust (leaves about 1·5–2 mm. long)   3
    Cells not enlarged in basal angles of leaf; plants slender (leaves
    not above 0·75–1 mm. long)   4

3   Plant bright shining green, cells about 12μ wide; autoecious
                                    Plagiothecium denticulatum (p. 304)
    Plant usually dull olive green, cells about 16μ wide; dioecious
                                    P. silvaticum (p. 306)

4   Leaves forming short acute points only, cells 7–10μ wide
                                    Isopterygium depressum (p. 303)
    Leaves drawn out into long fine points, cells 4–8μ wide   5

5   Leaf margin lightly toothed towards apex
                                    Isopterygium elegans (p. 302)
    Leaf margin quite entire   *I. pulchellum* (forms) (p. 303)

SUBSECTION 3

1   Plants exclusively aquatic (submerged); stems very long (15 cm.–1 m.)   *2*
Plants not aquatic, rarely growing submerged   *3*

2   Lower leaves folded and keeled at back
             Fontinalis antipyretica (p. 247)
Leaves all concave and rounded, not keeled   *F. squamosa* (p. 249)

3   Leaf margin distinctly toothed   *4*
Leaf margin entire or only faintly and bluntly indented   *8*

4   Stem greenish, plant of prostrate or pendulous habit and soft texture   *Hyocomium flagellare* (p. 311)
Stem red, plant of ascending or erect habit and rather rigid texture   *5*

5   Branching regularly twice pinnate, the shoots somewhat flattened, frond-like   *Hylocomium splendens* (p. 317)
Branching simply pinnate or irregular, never twice pinnate   *6*

6   Plant of very robust bushy habit, bright green in colour; leaf 3–5 mm. long   *Rhytidiadelphus triquetrus* (p. 313)
Plant only moderately robust, lacking bushy habit, dull green, leaf 2–3 mm. long   *7*

7   Minute paraphyllia present among leaves on stem, leaves plicate
             Hylocomium brevirostre (p. 318)
Paraphyllia absent, leaves not plicate
             Rhytidiadelphus squarrosus (p. 314)

8   Leaves bluntly rounded at apex   *9*
Leaves drawn out to long fine points   *11*

9   Cells at basal angles of leaf not enlarged, nor forming well-defined auricles; habit rather compact; uncommon plant of calcareous dunes, etc.   *Entodon orthocarpus* (p. 302)
Alar cells swollen, forming well-defined auricles; habit loose; very common plants of heaths, marshes, etc.   *10*

10   Shoot tips rendered round (in section) and sharply pointed by the rolling of the apical leaves (cuspidate), stems green to reddish, auricles decurrent   *Acrocladium cuspidatum* (p. 278)
Shoot tips scarcely cuspidate, stems always red, auricles not decurrent   *Pleurozium schreberi* (p. 300)

11   Plants very robust, with numerous ascending branches, leaves concave, densely crowded, each leaf narrowing rather abruptly above the middle to form a long fine point   *12*

Plants slender to moderately robust, leaves not concave, nor
conspicuously crowded, tapering gradually from base to apex   *13*

12 Leafy shoot 4–5 mm. wide, leaf 3 mm. long; very robust species
almost confined to Outer Hebrides   *Myurium hebridarum* (p. 252)
Leafy shoot 2–3 mm. wide, leaf 2 mm. long; widespread plants
generally   Robust vars. of *Hypnum cupressiforme* (p. 309)

13 Leaves at shoot tips spreading in star-like manner, well-defined
auricles composed of swollen cells at leaf base; in bogs and
marshes   *Campylium stellatum* (p. 265)
Leaves not spreading as above, well-defined auricles lacking;
habitat different   *14*

14 Evident pink or reddish tinge present; uncommon plants of
base-rich mountain ledges   *15*
Pink or reddish tinge lacking   *16*

15 Leaves plicate, 3 mm. long   *Orthothecium rufescens* (p. 302)
Leaves not plicate, 1 mm. long   *O. intricatum* (p. 302)

16 Cells of leaf 6–10 times as long as broad; patches of alar cells
rather ill-defined; uncommon autoecious species of tree trunks
                                         Pylaisia polyantha (p. 309)
Cells of leaf up to 10–15 times as long as broad; patches of alar
cells well-defined; common dioecious plant of trees and rocks
                          Hypnum cupressiforme var. *resupinatum* (p. 308)

KEY TO LIVERWORTS*

MAKING USE OF MICROSCOPIC CHARACTERS

Key to sections

1   Plant a thallus (without differentiation into stem and leaves)
Section 1

     Plant foliose (differentiated into stem and leaves)
2

2   Shoot radially symmetrical, leaves being in 3 ranks, all of nearly equal size
Section 2, p. 80

     Shoot dorsiventral (i.e. presenting obvious upper, or antical, and lower, or postical, surfaces); leaves in 2 principal ranks, with or without an additional rank of smaller underleaves
3

3   Leaves simple, or divided into 2, 3 or 4 almost equal lobes
4

     Leaves deeply divided into 2 portions, one distinctly smaller than the other and either bent back over the larger one or forming a small, often sac-like appendage
8

4   Leaves simple, with entire or minutely toothed margins, but no division of leaf into small number of obvious lobes
5

     Leaves notched or more deeply divided into 2, 3 or 4 almost equal lobes
6

5   Plant with underleaves, which are readily seen under microscope
Section 3, p. 81

     Plant without underleaves, or underleaves few and difficult to detect even under microscope
Section 4, p. 82

6   Plant with underleaves, which are readily seen under microscope
Section 5, p. 84

     Plant without underleaves, or underleaves few and difficult to detect even under microscope
7

7   Leaves transversely inserted on stem
Section 6, p. 85

     Leaves obliquely inserted on stem
Section 7, p. 87

8   Underleaves lacking, the deeply and unequally divided leaves thus giving impression of a shoot with leaves in 4 ranks
Section 8, p. 89

     Underleaves present, thus giving impression of shoot with leaves or appendages in 5 distinct ranks
Section 9, p. 91

* Every liverwort mentioned in this book is included.

SECTION 1

1 Plant forming minute, nearly circular patches or rosettes, less than 2 cm. in diameter   2
   Plant consisting of a number of variously branched, more or less elongate thallus segments, not forming compact nearly circular rosettes   *11*

2 Thallus mostly only 1 cell thick, commonly almost covered by the conspicuous inflated involucres   *3*
   Thallus thicker, without inflated involucres   *4*

3 Involucres broadly pear-shaped, spore clusters (tetrads) about 100 μ in diameter   *Sphaerocarpus michelii* (p. 396)
   Involucres narrowly club-shaped and pointed, spore tetrads about 135 μ in diameter   *S. texanus* (p. 396)

4 Thallus spongy in texture, transverse section showing at least the uppermost $\frac{2}{3}$ to be composed of a lattice-work of air chambers   *5*
   Thallus not showing this texture or structure, chambers if present being scattered throughout the tissue of the thallus   *7*

5 Pores in upper surface giving direct entry into uppermost air chambers   *Riccia crystallina* (p. 396)
   No pores in upper surface   *6*

6 Plant growing on mud at margins of ponds; ventral scales inconspicuous   *Riccia fluitans* (terrestrial form) (p. 393)
   Plant usually floating, tinged with purple, occasionally growing on mud at pond margins, thallus fringed with tongue-shaped, toothed ventral scales   *Ricciocarpus natans* (p. 392)

7 Thallus segments each with median groove near apex; numerous small chloroplasts in each cell; tuberculate rhizoids present   *8*
   Thallus segments without median groove; 1 large chloroplast in each cell, tuberculate rhizoids lacking   *9*

8 Groove wide and shallow   *Riccia glauca* (p. 395)
   Groove narrow and deep, section of thallus thus usually V-shaped   *R. sorocarpa* (p. 396)

9 Thallus usually smooth and nearly flat, no mucilage cavities; spores yellow   *Anthoceros laevis* (p. 398)
   Thallus irregularly lobed and crisped, with large mucilage cavities scattered throughout tissues; spores black   *10*

10 Thallus annual, forming rosettes 0·5–1·5 cm. wide, section usually only 8–12 cells thick in middle; capsule 1–3 cm. long
Anthoceros punctatus (p. 397)
Thallus perennial, forming larger rosettes, section usually 12–22 cells thick in middle; capsule 4–6 cm. long (or even longer)
A. husnoti (p. 399)

11 Plant robust, segments of mature thallus 6–10 mm. wide   *12*
Plant small, segments of mature thallus 5 mm. wide or less   *20*

12 Thallus differentiated into upper green and lower colourless layers, tuberculate rhizoids present   *13*
Thallus of uniform tissue, tuberculate rhizoids absent   *17*

13 Plant bearing gemmae   *14*
Plant without gemmae   *15*

14 Gemmae borne in depressions flanked by crescent-shaped ridges of tissue   *Lunularia cruciata* (p. 391)
Gemmae in goblet-shaped cups   *Marchantia polymorpha* (p. 387)

15 Plant very robust, thallus covered with conspicuous hexagonal markings, in the centre of each a pore easily visible to naked eye
Conocephalum conicum (p. 389)
Plants less robust, hexagonal markings inconspicuous, pores scarcely visible to naked eye   *16*

16 Thallus section honeycombed with air-chambers
Reboulia hemisphaerica (p. 392)
Thallus section showing simple series of air-chambers, in 1 layer only   *Preissia quadrata* (p. 388)

17 Thallus of brittle fleshy texture, no midrib present
Riccardia pinguis (p. 383)
Thallus of thinner, translucent texture, ill-defined mid-rib present   *18*

18 Thallus section without bands of thickening, plant calcicole
Pellia fabbroniana (p. 381)
Thallus section usually with brown bands of thickening between some of the cells, plants of neutral to acid soils   *19*

19 Plant usually strongly tinged with dark red; involucre a short cylinder; chiefly a plant of north and west Britain
Pellia neesiana (p. 381)
Plant green, or with faint tinge of red, involucre a scale-like flap; very common species in most parts of Britain   *P. epiphylla* (p. 380)

20 Thallus section differentiated into upper green and lower colourless regions, pores in upper surface; plant blackish beneath; uncommon species. (Cf. small states of *Preissia*)
Targionia hypophylla (p. 392)
Thallus not thus differentiated, without pores *21*

21 Midrib clearly defined *22*
Midrib very ill-defined or lacking *26*

22 Thallus segments with lobed and wavy margins, midrib prominent on underside only; uncommon plant of moist sand dunes
Moerckia flotowiana (p. 382)
Margins of thallus not lobed and wavy, midrib equally evident on upper- and undersides of thallus *23*

23 Thallus pinnately branched, covered on both surfaces with short single hairs; uncommon calcicole species *Metzgeria pubescens* (p. 383)
Thallus with fork-branching, hairs on underside of midrib and at margins only *24*

24 Hairs strongly curved, hook-like; a rather rare western species
Metzgeria hamata (p. 383)
Hairs straight or nearly so *25*

25 Hairs mostly in pairs, cells of thallus about 50μ wide; plant occurring chiefly on rocks *Metzgeria conjugata* (p. 383)
Hairs single, cells of thallus about 35μ wide; very common plant of trees and rocks *M. furcata* (p. 382)

26 Thallus divided into very narrow (linear) segments, which are only 0·5–1 mm. broad *27*
Thallus lobed or branched, the divisions broader, never linear *28*

27 Branching of forked type, plant aquatic
Riccia fluitans (floating forms) (p. 393)
Branching pinnate, plant of moist terrestrial habitats
Riccardia multifida (p. 385)
(Cf. also slender states of *R. sinuata*.)

28 Thallus showing forked type of branching *29*
Thallus pinnately branched or lobed *32*

29 Plant bearing crescent-shaped ridges, with gemmae
Lunularia cruciata (small states) (p. 391)
Plant without gemmae *30*

30 Thallus irregularly crisped or fringed at margins; each cell with only one large chloroplast
Anthoceros spp. (looser-growing states, see also *9*)
Plant lacking these characters *31*

31 Thallus rather pale green, with obvious leaf-like lateral lobes or appendages, flask-shaped gemma receptacles commonly present   *Blasia pusilla* (p. 381)

Thallus dark green, lacking leaf-like appendages and flask-shaped gemma receptacles   *Pellia* spp. (small states),* pp. 380–81.

32 Thallus broad (2–5 mm.), lateral lobes few, short and irregular, texture fleshy and brittle   *Riccardia pinguis* (p. 383)

Thallus narrow (0·5–1·5 mm.) lateral lobes numerous, longer

     R. sinuata (p. 386)

SECTION 2

1   Leaves divided into 2 segments only   *2*

Leaves divided into 3 or more segments   *5*

2   Plant pale blue-grey when dry, shoots slender, leaves very small (0·3 mm. or less)   *3*

Plant red-brown to dark brownish green, robust, leaves longer (over 1 mm.)   *4*

3   Stems several centimetres long, perianth extending well beyond bracts; plant forming conspicuous grey tufts or cushions on wet mountain rocks, etc.   *Anthelia julacea* (p. 321)

Stems a few millimetres long, perianth only just extending beyond bracts, plant inconspicuous and rather rare, usually on soil near tops of mountains   *A. juratzkana* (p. 322)

4   Plants tall (to 12 cm.), forming conspicuous red-brown tufts on mountain ledges; with the habit of a tall slender species of *Dicranum*   *Herberta hutchinsiae* (p. 322)

Plants shorter (3–7 cm.), dull olive or brownish green, not suggestive of a species of *Dicranum*   *H. adunca* (p. 322)

5   Each leaf segment fringed with long thread-like cilia   *6*

Leaf segments 3 to 4, not further subdivided   *8*

6   Plant strikingly pale green, with superficial resemblance to a *Thuidium*, leaf cells elongate rectangular

     Trichocolea tomentella (p. 325)

Plant olive green to reddish brown, leaf cells isodiametric, rounded hexagonal with thickened corners   *7*

7   Broadest leaf segment about 20 cells wide at base, leaves divided only to mid-leaf; on rocks, soil and trees

     Ptilidium ciliare (p. 322)

---

* Terrestrial species of *Riccia*, with rosettes imperfectly formed, will 'key out' here too.

Broadest leaf segment about 10 cells wide at base, leaves divided to ¾ leaf length; mainly on trees   *P. pulcherrimum* (p. 323)

8 Leaf segments only 1 cell wide to the base, shoot thus suggestive of branched filamentous alga *Blepharostoma trichophyllum* (p. 323)
Leaf segments 2 or more cells wide towards base   9

9 Plant forming large swollen cushions, leaf cuticle smooth or nearly so, perianth fringed with short teeth only
Lepidozia trichoclados (p. 329)
Plant forming small dark green patches, leaf cuticle very rough, perianth fringed with long cilia   *L. setacea* (p. 328)

(Cf. also *L. reptans*, in which the leaves are lobed rather than divided into narrow segments. Owing to the distinctly smaller size of one of the rows of leaves, neither *L. setacea* nor *L. reptans* should come strictly into this section, and this is true to some extent of *Ptilidium* spp. also.)

SECTION 3

1 Leaves ovate-oblong or oblong with broad (truncate) apices, underleaves large, broad or bilobed, cells without corner thickenings   2
Leaves nearly orbicular, sometimes broader than long, underleaves small and lance-shaped, cells with corner thickenings   6

2 Leaves opposite or nearly so, underleaves orbicular, toothed
Saccogyna viticulosa (p. 345)
Leaves distinctly alternate, underleaves not as above   3

3 Underleaves divided to ⅓ into broad rounded lobes, leaves ovate-oblong, distinctly longer than broad *Calypogeia trichomanis* (p. 330)
Underleaves divided to near base into 2 narrow tooth-like divisions, leaves oblong with truncate apices, only slightly longer than broad   4

4 Plant dull green to brownish, leaf cells 25–35μ across; lobes of perianth without obvious teeth *Chiloscyphus polyanthus* (p. 341)
Plant bright or pale green, leaf cells larger; lobes of perianth sharply toothed   5

5 Habit more or less prostrate, leaf cells 35 × 45–60μ; in ditches and on moist banks and decaying logs
Chiloscyphus pallescens (typical form) (p. 343)
Habit nearly erect, leaf cells somewhat smaller; in mountain springs and among grass in very wet places
C. pallescens var. *fragilis* (p. 343)

6   W M

6   Leaf cells very large, 45–55μ across   7
    Leaf cells smaller, 20–35μ across   8
7   Plant forming dense yellowish-brown to red-purple tufts on
    mountain ledges, etc., leaf cuticle rough, gemmae scarce or
    absent   *Mylia taylori* (p. 343)
    Plant not forming dense tufts, green to yellow-brown, in bogs;
    cuticle smooth, gemmae almost always abundant at tips of
    narrower, ovate upper leaves   *M. anomala* (p. 344)
8   Plant often reddish, with long stems (3–12 cm.), growing in
    mountain streams   *Nardia compressa* (p. 355)
    Plant never reddish, with short stems (1–3 cm.), terrestrial or
    bog species   9
9   Glistening oil-bodies in cells, 'flagella' absent; very common
    plant on soil of banks, roadsides, etc.   *Nardia scalaris* (p. 354)
    Oil-bodies absent, 'flagella' growing from undersides of stems   10
10 Stems 1–2 cm. long; cells with corner thickenings almost as wide
    as cell cavities; gemmae usually present; on peaty banks,
    decaying stumps and sandstone rocks   *Odontoschisma denudatum*
                                                                      (p. 335)
    Stems 2–8 cm. long; corner thickenings of cells smaller; gemmae
    absent; in *Sphagnum* bogs   *O. sphagni* (p. 334)

SECTION 4

1   Leaf margin toothed   2
    Leaf margin entire   4
2   Leaf margin beset with 20–40 small teeth   *Plagiochila asplenioides*
                                                                      (p. 359)
    Leaf margin with 5–10 large teeth   3
3   Stems 3–12 cm. long; leaf base decurrent; corner thickenings
    of cells distinct but small   *Plagiochila spinulosa* (p. 361)
    Stems 1–3 cm. long; leaf base not decurrent; corner thickenings
    of cells very large (some almost equalling cell cavities)
                                                      P. punctata (p. 362)
4   Leaf cells very large, 45–55μ across   5
    Leaf cells smaller, 20–35μ across (occasionally wider in *Plecto-
    colea hyalina*)   6
5   Plant forming dense yellowish brown to red-purple tufts on
    mountain ledges, etc.; leaf cuticle rough, gemmae scarce or
    absent   *Mylia taylori** (p. 343)

* Plants thus marked should not strictly come into this section but may accidentally 'key out' here since the underleaves are small and easily overlooked.

Plant not forming dense tufts, green to yellow-brown; in bogs; cuticle smooth, gemmae almost always abundant at tips of narrower, ovate upper leaves   *M. anomala** (p. 344)

6 Plant aquatic, growing partly or wholly submerged in mountain streams or lakes; stems commonly 6–12 cm. long   7
   Plant terrestrial; stems much shorter   9

7 Leaf kidney-shaped to orbicular   *Nardia compressa** (p. 355)
   Leaf heart-shaped to broadly ovate   8

8 Cell walls dark brown, without corner thickenings
                                    Jungermannia cordifolia (p. 352)
   Cell walls colourless, minute corner thickenings present
                                    J. tristis (aquatic forms) (p. 353)

9 Plant of peat bogs; 'flagella' growing from underside of stems
                                    Odontoschisma sphagni (p. 334)
   Plant of other habitats; 'flagella' normally absent   10

10 Marginal row of leaf cells 2–3 times as large as those internal to them, forming evident border   *Plectocolea crenulata* (p. 355)
   Marginal row of leaf cells scarcely different from the rest   11

11 1–3 conspicuous glistening oil bodies in each cell of leaf
                                    Nardia scalaris (p. 354)
   Oil bodies lacking   12

12 Rhizoids violet or red   13
   Rhizoids colourless or brownish   14

13 Rhizoids red, whole plant pale and glistening, mainly lowland species   *Plectocolea hyalina* (p. 356)
   Rhizoids violet-coloured, plant dark green to red-brown; mountain species   *P. obovata* (p. 356)

14 Plant minute, stems 0·5–1 cm. long; narrow pointed perianths nearly always present   *Jungermannia pumila* (p. 353)
   Plant larger, stems 1–3 cm., perianths obtusely club- or pear-shaped   15

15 Leaves round (orbicular)   *Jungermannia sphaerocarpa* (p. 353)
   Leaves longer than broad (ovate to oblong)   *J. tristis* (p. 353)

* Plants thus marked should not strictly come into this section but may accidentally 'key out' here since the underleaves are small and easily overlooked.

SECTION 5

1 Leaves 2-lobed or divided to near base into 2 segments   *2*

   Leaves 3- or 4-lobed, or divided to near base into 3 or 4 segments   *11*

2 Plants minute, forming dark green or brownish patches, leaves less than 0·3 mm. long, leaf cells 10–18 μ across
                              Cephaloziella spp., pp. 335–38

   Plants larger, leaves larger, 0·3 mm. long or more, leaf cells above 18 μ across   *3*

3 Underleaves conspicuous, deeply divided into 2 or more lobes or teeth   *4*

   Underleaves inconspicuous, lance-shaped, entire or merely notched at apex   *9*

4 Leaves succubously arranged   *5*

   Leaves incubously arranged   *8*

5 Leaves near shoot tips usually entire *Lophocolea heterophylla* (p. 340)

   Leaves all alike, 2-pronged   *6*

6 Plant dioecious; perianths generally absent; growing in wet grassy places   *Lophocolea bidentata* (p. 338)

   Plant monoecious; perianths generally present   *7*

7 Plant pale green, leaf cells about 40 μ across; chiefly on rotting logs and stumps; common   *Lophocolea cuspidata* (p. 339)

   Plant dark green, leaf cells up to 50 μ across; on moist banks and walls; uncommon   *L. alata* (p. 341)

8 Stems 2–4 cm. long, leaf apices narrow, notched; underleaves divided into 2 broad irregular lobes; common   *Calypogeia fissa* (p. 331)

   Stems 1–2 cm. long, leaf apices broadly truncate, with 2 divergent points; underleaves divided into thread-like segments; uncommon, chiefly western   *C. arguta* (p. 332)

9 Plants without 'flagella'; on calcium-rich mountain ledges
                              Leiocolea muelleri (p. 348)

   Plants bearing leafless or small-leaved shoots ('flagella') besides ordinary leafy shoots   *10*

10 Leaves divided to $\frac{1}{6}$ or $\frac{1}{5}$ only; uncommon plant of moist peaty banks   *Cladopodiella francisci* (p. 334)

   Leaves divided to $\frac{1}{4}$ or $\frac{1}{3}$; plant of bogs or moorland pools
                              C. fluitans (p. 334)

11 Underleaves closely resembling lateral leaves in form, but slightly smaller   *12*

Underleaves of different form from lateral leaves, and less than $\frac{1}{4}$ their size   *16*

12 Leaves divided to $\frac{1}{3}$ or $\frac{1}{2}$ leaf length only   *13*
Leaves divided to base into narrow segments   *15*

13 Leaf segments acute; plant forming dense yellowish or whitish green cushions; chiefly near west coast   *Lepidozia pinnata* (p. 329)
Leaf segments obtuse; habit not as above   *14*

14 Common species forming loose green patches on banks, etc.; monoecious   *Lepidozia reptans* (p. 327)
Chiefly western species occurring as scattered stems in *Sphagnum*, etc.; dioecious   *L. pearsoni* (p. 329)

15 Plant forming large, dark yellow-green cushions, tips of bracts and mouth of perianth fringed with shortly projecting cells
     Lepidozia trichoclados (p. 329)
Plant forming small, dark green compact patches, tips of bracts and mouth of perianth fringed with long multicellular threads (cilia)   *L. setacea* (p. 328)

16 Leaves incubous, narrowed at 3-toothed apex, underleaves approximately kidney-shaped, irregularly lobed   *17*
Leaves succubous, widest near 3-lobed apex, underleaves lance-shaped or divided into 2 narrow segments   *18*

17 Plant forming large light green to whitish patches; very robust, leafy shoot 3–4 mm. across   *Bazzania trilobata* (p. 326)
Plant forming yellowish to red-brown tufts; leafy shoots 0·5–2 mm. across   *B. tricrenata* (p. 327)

18 Special slender shoots (with appressed, nearly entire leaves) arising from some stem apices; leaf cells small, about 18μ
     Orthocaulis attenuatus (p. 350)
Special slender shoots lacking; leaf cells larger, about 25μ
     O. floerkii (p. 349)

SECTION 6

1 Leaf length 0·5 mm. or less   *2*
Leaf length (at least of largest leaves on shoot) 1 mm. or more   *9*

2 Shoots cylindrical and catkin-like due to closely appressed leaves, 0·5 mm. or less in diameter; plants of mountain rocks   *3*
Shoots not as above; plants of various habitats   *5*

3   Leaf margin with elongated, strongly projecting cells; plant often red-brown, looking like fine copper wire
$$Gymnomitrium\ crenulatum\ \text{(p. 358)}$$
Leaf margin without elongated projecting cells, colour brownish yellow to whitish green   4

4   Leaf lobes obtuse, margin crenulate due to convex outer walls of cells, colour of plant usually pale green
$$Gymnomitrium\ obtusum\ \text{(p. 359)}$$
Leaf lobes acute, cells of leaf margin lacking convexity of outer walls, colour usually brownish yellow.   *G. concinnatum* (p. 359)

5   Leaf lobes drawn out to fine points; plant usually tinged rose-red, growing on rotting wood   *Nowellia curvifolia* (p. 334)

    (N.B. Plants lacking characteristic sac-like swelling of *Nowellia* on one side of leaf base and 'keying out' here will probably be *Cephalozia* spp. or *Cephaloziella* spp. in which the oblique leaf insertion was not very obvious.)

    Leaf lobes not drawn out to fine points, plants of rock ledges or banks on mountains   6

6   Leaf cells evenly thickened, gemmae commonly present
$$Sphenolobus\ minutus\ \text{(p. 352)}$$
Leaf cells with evident corner thickenings, gemmae lacking   7

7   Leaves widely spreading; plant of moderate altitudes
$$Marsupella\ funckii\ \text{(p. 358)}$$
Leaves incurved to semi-appressed; plants found chiefly at tops of mountains   8

8   Leaves much narrowed at notched apex, 0·3 mm. long in middle of shoot; cells 10–14μ wide; chiefly on rock
$$Gymnomitrium\ adustum\ \text{(p. 359)}$$
Leaves scarcely narrowed at broadly bilobed apex, 0·5 mm. long in middle of shoot; cells 14–18μ wide; chiefly on soil   *G. varians*
(p. 359)

9   Leaves bilobed for $\frac{1}{8}$ to $\frac{1}{5}$ leaf length, not channelled or keel-like below, perianth united to inner bracts for part of its length; very common mountain plant   *Marsupella emarginata* (p. 356)
Leaves divided for $\frac{1}{3}$ to $\frac{1}{2}$ leaf length, channelled or keel-like below, perianth free   10

10   Divisions of leaf unequal in lower leaves, leaf cuticle rough with papillae   *Scapania aequiloba* (p. 367)

    (Cf. other species of *Scapania* with leaves deeply and unequally divided into 2 lobes, and hence included in section 8 of this key.)

    Divisions of leaf nearly or quite equal throughout the shoot   *11*

11 Plant compact, 0·5–2 cm. tall; on rocks and rocky banks
Scapania compacta (p. 367)
Plant taller, 3 cm. or more; chiefly by streams *S. subalpina* (p. 367)

SECTION 7

1 Leaves irregularly and variously lobed or toothed   2
Leaves regularly 2-, 3- or 4-lobed   6

2 Stem 0·5–1·5 cm. long, rhizoids violet, very numerous, leaf cells
large (about 50 μ)   3
Stem longer, rhizoids not violet, scarce, leaf cells smaller
(16–30 μ)   5

3 Surface of spore with raised ridges that form a regular network
(areolate); on peat   *Fossombronia dumortieri* (p. 379)
Surface of spore not areolate, but with ridges which project
as spines around margin; on mineral soil   4

4 16–24 spines around margin of spore *Fossombronia pusilla* (p. 378)
28–36 spines around margin of spore   *F. wondraczeki* (p. 379)

5 Stem 3–12 cm. long, leaves oblong, with strongly decurrent
antical margins   *Plagiochila spinulosa* (p. 361)
Stem 1–2·5 cm. long, leaves round-ovate, antical margin not
markedly decurrent   *P. punctata* (p. 362)

6 Leaves 3–4-lobed   7
Leaves 2-lobed   9

7 Leaf lobes markedly unequal so that whole leaf asymmetrical
Tritomaria quinquedentata (p. 352)
Leaf lobes equal or nearly so   8

8 Leaves mostly 4-lobed   *Barbilophozia barbata* (p. 352)
Leaves 3-lobed; shoots with small-leaved catkin-like tips
Orthocaulis attenuatus (p. 350)

9 Minute plants, stems 0·2–1 cm., leaves about 0·1–0·3 mm. long,
leaf cells 10–18 μ   10
Larger plants, stems usually longer, leaves 0·5 mm. or longer
(on well-developed shoots), leaf cells above 18 μ, usually much
larger   12

10 Underleaves present throughout; plant dioecious
Cephaloziella byssacea (p. 336)
Underleaves absent, or present only near inflorescences; plant
monoecious   *11*

11   Plant dark brown to reddish; each leaf lobe 4–5 cells wide at its
base; cells about 12μ across   *Cephaloziella rubella* (p. 338)
Plant green to brownish; each leaf lobe 4–10 cells wide at its
base; cells 11–18μ   *C. hampeana* (p. 337)

12   Leaf cells very large, 35–60μ, lacking corner thickenings   *13*
Leaf cells smaller, 18–35μ, corner thickenings rarely lacking
altogether (in the few cases where they are lacking the smaller
cells will effectively separate these plants)   *15*

13   Leaf lobes rounded; on calcareous rocks and soil
                           Leiocolea turbinata (p. 347)
Leaf lobes acute; on non-calcareous soils   *14*

14   Bracts 2-lobed, with toothed margins; plant with small-leaved
'flagella' branches   *Cephalozia bicuspidata* (p. 332)
Bracts palmately lobed, with entire segments; plant without
'flagella'   *C. connivens* (p. 333)

15   Margins of leaf lobes toothed   *Lophozia incisa* (p. 347)
Margins of leaf lobes entire   *16*

16   Leaves lobed to ⅓ leaf length, the notch or 'sinus' so formed
being V-shaped   *17*
Leaves merely emarginate or lobed to less than ⅓ leaf length,
the 'sinus' U-shaped. (In cases where lobing extends to nearly
⅓ leaf length the widely U-shaped sinus will distinguish plants
of this group)   *18*

17   Medium-sized blackish green plant (1–2 cm. tall), leaf lobes
rounded, cell cavities angular; gemmae lacking
                           Gymnocolea inflata (p. 348)
Small reddish brown plant (0·5 cm. long), leaf lobes acute, cell
cavities rounded; reddish gemmae commonly present
                           Isopaches bicrenatus (p. 351)
(Cf. also *Lophozia excisa* which sometimes has angular 'sinus'.)

18   Stems long, 4–7 cm., western mountain plant with habit of
a *Bazzania*   *Anastrepta orcadensis* (p. 352)
Stems short, not above 3 cm.   *19*

19   Leaves very shallowly notched, many of them almost or quite
entire   *Nardia geoscyphus* (p. 355)
Leaves all regularly bilobed ·   *20*

20   Leaf lobes incurved and hence pincer-like, forming horseshoe-
shaped 'sinus'   *Cephalozia media* (p. 334)
Leaf lobes not incurved, 'sinus' not horseshoe-shaped   *21*

21 Leaves lobed only to $\frac{1}{7}$ leaf length, cells small (19–24μ)

Lophozia alpestris (p. 347)

Leaves more deeply lobed, cells larger (24–35μ)   *22*

22 Plant very small (0·5–1 cm.); paroecious, antheridia just below female inflorescence; purple-red gemmae commonly present

Lophozia excisa (p. 347)

Plant larger (1–3 cm.); dioecious; gemmae greenish   *23*

23 Plant green, leaf cells with minute corner thickenings; very common species .   *Lophozia ventricosa* (p. 346)

Plant tinged reddish, leaf cells with large corner thickenings; uncommon   *L. porphyroleuca* (p. 347)

SECTION 8

1 Each leaf deeply divided into 2 segments or lobes, the larger expanded, the smaller sac-like   *2*

Each leaf deeply divided into 2 segments or lobes, both of which are expanded   *3*

2 Minute yellowish green plant (stems less than 1 cm. long) of calcareous places   *Cololejeunea calcarea* (p. 378)

Medium-sized red-brown plant (3–5 cm.) of trees and shaded rocks in west Britain   *Radula aquilegia* (p. 370)

Robust purplish or orange-red plant (stems 4–16 cm. long) of moorland in west Scotland and Ireland   *Pleurozia purpurea* (p. 371)

3 Leaf lobes acute, the larger bifid at tip; clusters of red gemmae commonly present at shoot tips   *Tritomaria exsectiformis* (p. 352)

Leaf lobes narrowly oblong-ovate, a band of elongated narrow cells forming pseudo-midrib down centre of each lobe

Diplophyllum albicans (p. 362)

Leaf lobes wider, rounded ovate, without pseudo-midrib of narrow cells   *4*

4 Leaves incubous, the postical lobe smaller than the antical, rhizoids attached to postical lobes of leaves   *5*

Leaves succubous, the antical lobe smaller than the postical or the 2 lobes sub-equal; rhizoids never attached to leaf lobes   *7*

5 Plant paroecious, conspicuous tubular perianth nearly always present   *Radula complanata* (p. 369)

Plants dioecious, perianth seldom present   *6*

6   Plant reddish brown, leaf lobes markedly convex, the smaller one inflated; leaf cells 15–21 μ, rather thick walled
                                  Radula aquilegia (p. 370)
    Plant dark green, leaf lobes nearly flat; leaf cells 20–25 μ, thin-walled   *R. lindbergiana* (p. 370)

7   Postical leaf lobe less than twice the size of antical lobe   *8*
    Postical leaf lobe 2–3 times the size of the antical lobe   *12*

8   Leaf lobes almost equal in size   *9*
    Postical lobe $\frac{1}{3}$–$\frac{1}{2}$ larger than antical   *10*

9   Stem 1–2 cm. long, leaves not sharply keeled; inflorescence usually paroecious; plant of rocky banks *Scapania compacta* (p. 367)
    Stem 3 cm. or longer, leaves sharply keeled; dioecious; plant of gravelly detritus by streams   *S. subalpina* (p. 367)

10   Plant usually of non-calcareous banks and walls, margin of antical leaf lobe reflexed, cuticle smooth *Scapania gracilis* (p. 366)
    Plants of calcareous places, margin of antical leaf lobe not reflexed, cuticle rough   *11*

11   Slender plant, stems 2–4 cm., lobes of upper leaves nearly equal in size   *Scapania aequiloba* (p. 367)
    Robust plant, stems 3–10 cm., lobes of all leaves unequal in size
                                     S. aspera (p. 367)

12   Leaf lobes with entire margins   *13*
    Leaf lobes toothed   *15*

13   Plant small (stems 1–2 cm.), antical leaf lobe narrowly ovate, pointed; on loamy banks   *Scapania curta* (p. 367)
    Plants larger (2–8 cm.), antical leaf lobe heart-shaped to nearly circular; in wet places   *14*

14   Plant of moderate size (2–5 cm.), antical leaf lobe pointed, rhizoids extending along stem   *Scapania irrigua* (p. 367)
    Plant often very robust, antical leaf lobe round, rhizoids scarce, not extending along stem   *S. undulata* (p. 365)

15   Green plant of woodland banks, etc., leaf teeth 2–3 cells long and acute; dark red-brown gemmae often present
                               Scapania nemorosa (p. 363)
    Commonly purple-red (sometimes dark green) plant of streams and wet ledges, leaf teeth 1–2 cells long and blunt; gemmae greenish   *S. undulata* (p. 365)
    (N.B. Dentate forms = *S. dentata* Dum., here included under *S. undulata*.)

SECTION 9

1 Plant microscopically small (stems 2–4 mm. long), antical leaf
   lobe sac-like and ending in a long beak; rare plant of moist
   shaded rocks in the west   *Colura calyptrifolia* (p. 378)
   Plant larger (stems 5 mm. or longer); beaked sacs lacking   *2*

2 Plants robust (stems 3–10 cm.), green, postical leaf lobe not
   inflated   *3*
   Plants mostly slender or small (0·5–3 cm.), postical leaf lobe
   inflated, sac-like   *5*
   Plant usually robust, reddish brown to coppery purple (except
   Jubula which is dark green), postical leaf lobe helmet-shaped *9*

3 Taste biting, underleaves conspicuously toothed
                                          Porella laevigata (p. 369)
   Taste mild, underleaves entire or with few teeth at base only   *4*

4 Underleaves entire, with scarcely decurrent bases, close-set;
   postical lobes of lateral leaves ½ as wide as underleaves
                                        Porella platyphylla (p. 368)
   Underleaves with a few teeth at longly decurrent bases, distant;
   postical lobes of lateral leaves less than ½ as wide as underleaves
                                              P. cordeana (p. 369)

5 Minute plant (stems 4–8 mm., leaf 0·3 mm. long), sac-like
   postical lobe nearly as large as expanded antical lobe
                                        Microlejeunea ulicina (p. 376)
   Larger plants (leaves about 0·8 mm. long), sac-like postical
   lobe ⅓ size of antical lobe or less   *6*

6 Underleaves nearly circular in outline, rather rare western
   species   *Marchesinia mackaii* (p. 378)
   Underleaves deeply bilobed   *7*

7 Postical lobes of lateral leaves twice as large as underleaves
                                            Lejeunea patens (p. 376)
   Postical lobes of lateral leaves smaller than underleaves   *8*

8 Underleaves close-set, much larger than postical lobes of lateral
   leaves   *Lejeunea cavifolia* (p. 375)
   Underleaves more distant, only 1½ times as large as postical
   lobes of lateral leaves   *L. lamacerina* (p. 376)

9 Plant dark green, expanded antical leaf lobe strongly toothed,
   uncommon plant of wet rocks in west *Jubula hutchinsiae* (p. 375)
   Plants reddish brown to coppery purple, antical leaf lobe entire *10*

10 Some enlarged cells present among ordinary cells of antical
  leaf lobe *11*
  Enlarged cells lacking *12*

11 Enlarged cells forming interrupted line or 'false midrib'; underleaves with recurved margins; common and widespread species
  Frullania tamarisci (p. 372)

  Enlarged cells scattered; underleaves with plane margins; small
  western species marked by leaves which break off readily
  F. fragilifolia (p. 375)

12 Postical (helmet-like) leaf lobes larger than underleaves;
  perianth tubercled; common plant throughout Britain
  Frullania dilatata (p. 373)
  Postical (helmet-like) leaf lobes smaller than underleaves;
  perianth smooth; confined to the west *F. germana* (p. 375)

SYNOPSIS OF CLASSIFICATION

MUSCI (*Mosses*)

Synopsis of super-orders, orders, families and genera* in the British Flora

SPHAGNALES
SPHAGNACEAE
Sphagnum

ANDREAEALES
ANDREAEACEAE
Andreaea

BRYALES†

POLYTRICHALES
POLYTRICHACEAE
Atrichum
Oligotrichum
Polytrichum

BUXBAUMIALES
BUXBAUMIACEAE
Buxbaumia
Diphyscium

FISSIDENTALES
FISSIDENTACEAE
Fissidens
Octodiceras

DICRANALES
ARCHIDIACEAE
Archidium
DICRANACEAE
Pleuridium
Ditrichum
Distichium
Saelania
Ceratodon
Brachydontium

Seligeria
Trochobryum
Blindia
Trematodon
Pseudephemerum
Dicranella
Rhabdoweisia
Cynodontium
Oncophorus
Dichodontium
Dicranoweisia
Arctoa
Dicranum
Dicranodontium
Campylopus
Paraleucobryum
Leucobryum

POTTIALES
ENCALYPTACEAE
Encalypta
POTTIACEAE
Pottieae (subfam.)
Tortula
Aloina

* Genera not considered in this work are in italic.
† The series from here to the end of the Orthotrichaceae constitute the old group 'Acrocarpous Bryales', use of which is made in the keys in this work.

Pottieae (subfam.) (*cont.*)
  Desmatodon
  Pterygoneurum
  Stegonia
  Pottia
  Phascum
  Acaulon
Cinclidoteae (subfam.)
  Cinclidotus
Trichostomeae (subfam.)
  Barbula
  Gymnostomum
  Gyroweissia
  Eucladium
  Anoectangium
  Tortella
  Pleurochaete
  Trichostomum
  Weissia
  Leptodontium

GRIMMIALES
  GRIMMIACEAE
    Coscinodon
    Grimmia
    Rhacomitrium

FUNARIALES
  DISCELIACEAE
    Discelium
  FUNARIACEAE
    Funaria
    Physcomitrium
    Physcomitrella
  EPHEMERACEAE
    Nanomitrium
    Ephemerum
  OEDIPODIACEAE
    Oedipodium
  SPLACHNACEAE
    Tayloria
    Tetraplodon
    Haplodon
    Splachnum

SCHISTOSTEGALES
  SCHISTOSTEGACEAE
    Schistostega

TETRAPHIDALES
  TETRAPHIDACEAE
    Tetraphis

EUBRYALES
  BRYACEAE
    Mielichhoferia
    Orthodontium
    Leptobryum
    Pohlia
    Epipterygium
    Plagiobryum
    Anomobryum
    Bryum
    Rhodobryum
  MNIACEAE
    Mnium
    Cinclidium
  AULACOMNIACEAE
    Aulacomnium
  MEESIACEAE
    Paludella
    Meesia
    Amblyodon
  CATOSCOPIACEAE
    Catoscopium
  BARTRAMIACEAE
    Plagiopus
    Bartramia
    Conostomum
    Bartramidula
    Philonotis
    Breutelia
  TIMMIACEAE
    Timmia

ISOBRYALES
  PTYCHOMITRIACEAE
    Campylostelium
    Ptychomitrium
    Glyphomitrium

ORTHOTRICHACEAE
  Amphidium
  Zygodon
  Orthotrichum
  Ulota
FONTINALACEAE*
  Fontinalis
CLIMACIACEAE
  Climacium
HEDWIGIACEAE
  Hedwigia
CRYPHAEACEAE
  Cryphaea
LEUCODONTACEAE
  Leucodon
  Antitrichia
  Pterogonium
MYURIACEAE
  Myurium
NECKERACEAE
  Leptodon
  Neckera
  Homalia
  Thamnium

HOOKERIALES
HOOKERIACEAE
  Daltonia
  Hookeria
  Cyclodictyon

HYPNOBRYALES
THELIACEAE
  Myurella
FABRONIACEAE
  Myrinia
  Habrodon
LESKEACEAE
  Leskea
  Lescuraea
  Pseudoleskea

THUIDIACEAE
  Heterocladium
  Anomodon
  Thuidium
  Helodium
HYPNACEAE
  Amblystegieae (subfam.)
    Cratoneuron
    Campylium
    Leptodictyum
    Hygroamblystegium
    Amblystegium
    Amblystegiella
    Drepanocladus
    Hygrohypnum
    Scorpidium
    Acrocladium
  Brachytheciae (subfam.)
    Isothecium
    Scorpiurium
    Camptothecium
    Brachythecium
    Scleropodium
    Cirriphyllum
    Eurhynchium
    Rhynchostegiella
  Entodonteae (subfam.)
    Pterygynandrum
    Orthothecium
    Entodon
    Pseudoscleropodium
    Pleurozium
  Plagiotheciae (subfam.)
    Isopterygium
    Plagiothecium
  Sematophylleae (subfam.)
    Sematophyllum
  Hypneae (subfam.)
    Pylaisia
    Homomallium
    Hypnum

* The series from here to the end, with the exception of *Hedwigia*, constitute the old group 'Pleurocarpous Bryales' use of which is made in the keys in this work.

Hypneae (subfam.) *(cont.)*
  Ptilium
  Ctenidium
  Hyocomium
  Ptychodium

Hylocomiae (subfam.)
  Rhytidium
  Rhytidiadelphus
  Hylocomium

HEPATICAE (*Liverworts*)

Synopsis of orders, families, and genera* in the British Flora

JUNGERMANNIALES
HAPLOMITRINEAE (suborder)
HAPLOMITRIACEAE
Haplomitrium

JUNGERMANNINEAE (sub-order)
(JUNGERMANNIALES ACROGYNAE)

PTILIDIACEAE
  Herberta
  Anthelia
  Mastigophora
  Hygrobiella
  Pleuroclada
  Ptilidium
  Blepharostoma
  Trichocolea
LEPIDOZIACEAE
  Bazzania
  Lepidozia
  Telaranea
CALYPOGEIACEAE
  Calypogeia
CEPHALOZIACEAE
  Cephalozia
  Nowellia
  Cladopodiella
  Odontoschisma
  Eremonotus
  Adelanthus

CEPHALOZIELLACEAE
  Cephaloziella
  Prionolobus
  Dichiton
HARPANTHACEAE
  Lophocolea
  Chiloscyphus
  Pedinophyllum
  Mylia
  Harpanthus
  Geocalyx
  Saccogyna
JUNGERMANNIACEAE
  Lophozia
  Leiocolea
  Isopaches
  Sphenolobus
  Anastrophyllum
  Gymnocolea
  Anastrepta
  Tritomaria
  Orthocaulis
  Chandonanthus
  Barbilophozia
  Jungermannia
  Jamesoniella
  Nardia
  Plectocolea
  Southbya
  Acrobolbus
  Gongylanthus

* Genera not considered in this work are in italic.

MARSUPELLACEAE
  Marsupella
  Gymnomitrium
PLAGIOCHILACEAE
  Plagiochila
SCAPANIACEAE
  Diplophyllum
  Douinia
  Scapania
PORELLACEAE
  Porella
RADULACEAE
  Radula
PLEUROZIACEAE
  Pleurozia
FRULLANIACEAE
  Frullania
  Jubula
LEJEUNEACEAE
  Marchesinia
  Harpalejeunea
  Drepanolejeunea
  Lejeunea
  Microlejeunea
  Colura
  Cololejeunea

METZGERINEAE (sub-order)
**(JUNGERMANNIALES
  ANACROGYNAE)**
  FOSSOMBRONIACEAE
    Fossombronia
    Petalophyllum

PELLIACEAE
  Pellia
BLASIACEAE
  Blasia
PALLAVICINIACEAE
  Pallavicinia
  Moerckia
METZGERIACEAE
  Metzgeria
RICCARDIACEAE
  Riccardia
  Cryptothallus

MARCHANTIALES
MARCHANTIACEAE
  Marchantia
  Preissia
  Conocephalum
  Lunularia
  Dumortiera
REBOULIACEAE
  Reboulia
TARGIONIACEAE
  Targionia
RICCIACEAE
  Ricciocarpus
  Riccia

SPHAEROCARPALES
SPHAEROCARPACEAE
  Sphaerocarpus

ANTHOCEROTALES
ANTHOCEROTACEAE
  Anthoceros

SPHAGNALES

SPHAGNUM

General notes on the genus. The genus *Sphagnum* is readily known as such by the unique branching system (in whorls), the distinct stem and branch leaves, and the peculiar differentiation of the tissues of the leaf to give a regular pattern of alternating hyaline and green cells. Moreover, the conspicuous spongy (water-holding) masses or cushions formed by *Sphagnum* in bogs and other wet places cannot readily be mistaken, even in the field, for any other genus (cf. however, *Leucobryum*). Nevertheless, the recognition of the species of *Sphagnum* is a difficult matter, although the difficulty appears less if one neglects the rarer species and all the minor varieties and forms that have been described.

Not only are the leafy shoots of *Sphagnum* unique in the characters referred to above, but the capsules too are quite unlike those of all other mosses. Thus the ripe capsule is raised on a short stalk (pseudopodium) which forms part of the gametophyte; it is invariably widely ovoid or globose before dehiscence, and lacks a peristome. Eventually there is a lateral shrinkage of the wall with the result that an internal air pressure is set up and the lid is finally blown off and the spores released explosively.*

In making determinations of *Sphagnum* spp. it is most satisfactory to begin by carefully dissecting off (i) several leaves from a main stem and (ii) several branch leaves, and mounting each of these series in a drop of water. It is also useful to cut a rough transverse section of the stem. These three preparations will furnish all the information that is normally required for the purpose, capsule characters being of no value in the separation of the different species. In making the mounts of leaves it is important that some should have the adaxial, others the abaxial, surface upwards, as the narrow green cells will often be more fully exposed to view on one surface than on the other, and this constitutes a useful character. For the critical study of this and other details, sections of the leaves are very desirable.

Close attention must be paid, not only to the precise shape of the leaves, but also to all details of the cell structure in each type of leaf. It may be helpful to list relevant points: (i) presence or absence of a border composed of extremely narrow cells; (ii) character of leaf

---

* See the description in *Spore Discharge in Land Plants*, by C. T. Ingold (Clarendon Press, Oxford, 1939).

apex, e.g. hood-like or flattened, toothed or not, scaly at back or smooth; (iii) green cells more exposed on abaxial or on adaxial surface of leaf; (iv) distribution of fibrils and pores in hyaline cells of stem leaves; (v) number, size and distribution of pores in hyaline cells of branch leaves. With a little practice these five points are not difficult to observe and, taken in conjunction with information about the general habit of the plant and the leaf shape, should make for ready identification of the species. The stem section may be used as a confirmatory character. Its principal use is to show the number of layers of wide hyaline cells surrounding the inner thick-walled tissues. These outer cells may form three or four well-defined layers (as in *S. palustre*, *S. plumulosum* and others), several relatively ill-defined layers (as in *S. recurvum*), or a single well-defined layer (as in *S. subsecundum*). This character is especially useful in separating *S. subsecundum* from allied plants.

Further, the different species of *Sphagnum* have their own habitat preferences, and a knowledge of these will often help in the determination of specimens.* Finally, both habit and colouring may provide valuable information; for some species (e.g. *S. compactum*) have a characteristic growth form, whilst each species has a definite colour range within which it keeps. Thus red colouring is never developed in certain species (e.g. *S. palustre*, *S. subsecundum*); but even in species where it is usual the red colour is intensified by sunny and very acid conditions.

Many of the species are extraordinarily variable in habit and are well known to produce, under very wet and rather dry conditions respectively, forms that differ entirely from one another. Many of the aquatic states of *S. subsecundum* and *S. cuspidatum* are especially puzzling. *Sphagnum* is indeed a notably difficult genus systematically, and no unanimity exists on the question of the correct delimitation of the species. A thorough grasp of the genus can be gained only after close study and experience, and all that can usefully be done in a work of this scope is to indicate the best lines along which the beginner may proceed.

1. SPHAGNUM PALUSTRE L. (*S. cymbifolium* Ehrh.)

S. palustre is one of the commonest and most easily known species of *Sphagnum*. Robust in habit, it is usually marked by a pale, whitish green colour, frequently tinged with brighter green and more rarely with delicate salmon pink; but it is never bright purplish red. More characteristic even than the colour are the swollen, blunt-tipped branches,

---

* See descriptions of species; also habitat lists, pp. 407, 410.

due to the crowded, concave branch leaves with their blunt hooded apices. The stem leaves are rather large, normally expanding from a wide base to an even wider apex.

Certain microscopic characters will be found useful: (i) the outermost three layers of cells in the stem are large, hyaline and display conspicuous spiral thickenings; (ii) the stem leaves lack a border of narrow cells; and (iii) the rounded tips of the branch leaves are rough with

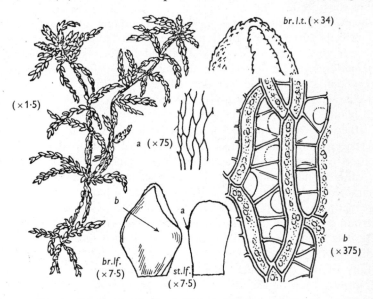

Fig. 1. *Sphagnum palustre*: *br.lf.* branch leaf, *br.l.t.* tip of same, *st.lf.* stem leaf.

scaly projections at the back. This combination of characters will distinguish *S. palustre* from all other common species except *S. papillosum* and *S. magellanicum*, both of which differ in prevailing colour and other details.*

S. palustre is dioecious; the capsule is not very frequent but is more so than in most species of the genus.

Ecology. It is very widespread, but, on the whole, avoids the wettest places, although it will grow at times with *S. cuspidatum* in wet moorland hollows. *S. palustre*, however, has no floating forms. It is somewhat intolerant of shade, but has been found, on occasion, in the 'field layer' of highland pine forest. It tolerates a wide range of acidity and is said to have a low mineral requirement.

* See notes on additional species.

2. SPHAGNUM CUSPIDATUM Ehrh. ex Hoffm. emend.

S. cuspidatum is a well-marked species, differing from most others in its more aquatic habitat, very long stems (12–40 cm.) and long leaves. In colour it is usually light green or yellowish brown—never red. Numerous varieties and forms have been named, some of which grow habitually submerged in pools; the leafy shoots are then remarkable

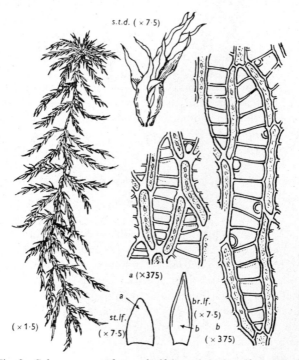

Fig. 2. *Sphagnum cuspidatum*: *br.lf.* branch leaf, *st.lf.* stem leaf, *s.t.d.* tip of branch in a dry state.

for their beautiful feather-like form, and the whole plant bears a superficial resemblance to an alga.

The stem of *S. cuspidatum* is green or brown. The stem leaves are broadly triangular in shape, longer than broad, and have borders of narrow cells—much as in *S. plumulosum*. They differ from the stem leaves of that species, however, in the better development of fibres and pores in the cells towards the apex. In *S. recurvum*, which is closely related to *S. cuspidatum*, the stem leaves are shortly triangular, broader than long, their cells mostly without fibres or pores.* The branch leaves

* See notes on additional species.

are the most distinctive feature of *S. cuspidatum*. They are very long (1·5–3 mm.), narrow in outline and usually markedly wavy at the margins when dry. They have little or none of the metallic lustre (when dry) of *S. plumulosum*. The leaves are sometimes curved and slightly turned to one side, as in *S. subsecundum*, but that plant has not the long narrow branch leaf of this species.

In the cell structure of the branch leaves the following two features should be noted: the green cells are exposed on both surfaces of the leaf, and the colourless cells are long and rather narrow (cells in mid-leaf: $140–220 \times 15–20\mu$), with few pores.

Ecology. The characteristic habitat of *S. cuspidatum* is in little pools in moorland; thus in the 'hollow–hummock' succession of bog it occupies the hollows which mark the first phase of the cycle. It tolerates a wide range of acidity and its mineral requirement is low. Dr W. Watson noted that in Somerset it was frequently associated with *Drepanocladus fluitans*, *D. exannulatus*, *Cladopodiella fluitans* and *Gymnocolea inflata*.

3. SPHAGNUM PLUMULOSUM Röll. (*S. acutifolium* Ehrh. var. *subnitens* (Russ. & Warnst.) Dixon)

This is one of the commonest of the 'acute-leaved' species of *Sphagnum*. With branches tapering to relatively fine points and leaves that—with naked eye or lens at least—seem acute, it has an appearance totally different from *S. palustre*. It is smaller in all its dimensions and is much more like certain other species, such as *S. cuspidatum*, so that careful attention to microscopic detail will sometimes be necessary for differentiation. Usually, however, *S. plumulosum* is marked by a strong tinge of red in its general colouring (a feature also seen in *S. rubellum*), and by the bright metallic lustre which the shoot tips exhibit when dry. No other species shows such evident metallic lustre, though several may show some trace of it. The stems are typically slender and the whole plant is of notably soft texture.

The stem is green or red, never brown; the outer layers of cells are without the fibrous thickenings seen in the stem of *S. palustre*. The stem leaf is triangular, pointed and usually lacks fibrils. It possesses a strong border of very narrow cells, this border sometimes becoming 6–8 cells wide at the leaf base. The branch leaves are narrow, and nearly oval in general outline, with margins incurved towards the tip. Under the microscope the leaf apex is seen to be not strictly acute, but sheared off (truncate), and marked by several prominent teeth; this form of leaf tip is common to many species of *Sphagnum*.

In the branch leaf the narrow green cells are exposed only on the upper (morphologically, ventral) surface of the leaf. The colourless cells are large and rather wide in proportion to their length (cells in mid-leaf measure $130-200 \times 20-40\mu$), with few large pores in each cell.

It is worth noting that the catkin-like male branches are always red in *S. plumulosum*.

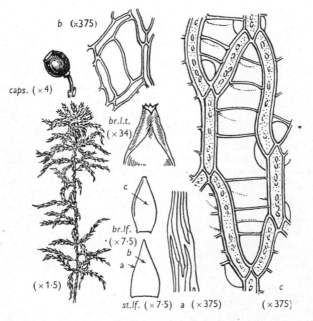

Fig. 3. *Sphagnum plumulosum*: *br.lf.* branch leaf, *br.l.t.* tip of same, *st.lf.* stem leaf, *caps.* capsule. The cells at *c* represent the adaxial surface of the leaf.

Ecology. It is the commonest species of *Sphagnum* over vast areas of moorland and blanket bog in Wales, west and north England, and Scotland. It is less common on the wet heaths of southern England. In the 'hollow–hummock' succession it has a place between the *S. cuspidatum* of water-filled hollows and *S. rubellum* of elevated hummocks; it is, indeed, one of the chief hummock-builders. It differs from *S. palustre* in being intolerant of the less markedly acid habitats.

ADDITIONAL SPECIES

(a) Species related to S. palustre

S. magellanicum Brid. (*S. medium* Limpr.) is not a very common species. It differs from *S. palustre* in that the woody cylinder of the stem ranges in colour from red-brown to deep purplish red (yellowish brown in *S. palustre*), and the plant as a whole is nearly always strongly tinged with red or purple. Also, the outer cells of the stem have few fibrils and the branch leaves are very distinct in that the narrow green cells are not exposed on either surface of the leaf, being completely enclosed by the hyaline cells. *S. magellanicum* occurs in the drier and more markedly acid habitats of bog country and is conspicuous at times as a hummock-former.

S. papillosum Lindb. is a rather common plant of the comparatively dry parts of moors, and has been noted as an important hummock-former. The branch leaves differ from those of *S. palustre* chiefly in the presence of minute papillae (visible under the high power) on the walls of the hyaline cells where these adjoin the green cells. The plant as a whole is usually strongly tinged with yellowish brown; the branches are blunter and more swollen, and the habit usually more compact than in *S. palustre*.

(*b*) *Species related to* S. cuspidatum

S. recurvum P. Beauv. (*S. intermedium* Hoffm.), which usually grows submerged in pools and deep ditches, is in some districts a commoner plant than *S. cuspidatum*. It resembles *S. cuspidatum* closely in general structure, but is usually a deeper green and has minute triangular stem leaves which are usually broader than long. Moreover, the tips of the branch leaves tend when dry to become strongly bent outwards or 'recurved' in the form of hooks.

S. subsecundum Nees. In the broad sense in which it is used here the name *S. subsecundum* is taken to include a number of plants that are treated by some authorities as separate species, e.g. *S. auriculatum* Schimp., *S. crassicladum* Warnst. and *S. inundatum* Russ. Although extremely variable in habit, *S. subsecundum* may be distinguished from all other species by the combination of two readily seen characters: (i) a single well-defined layer of large hyaline cells forming the outermost layer of the stem, and (ii) hyaline cells of the branch leaves having numerous small pores arranged in regular rows along the sides of each cell. The colour of the plant may be red-brown or orange, but is most commonly green or yellow-green, and is never rose-red. *S. subsecundum* is said to have a relatively high mineral requirement and occurs at times in fens (though only in the more acid patches). Elsewhere it is widely distributed and generally common.

(*c*) *Species related to* S. plumulosum

S. rubellum Wils. (*S. acutifolium* Ehrh. var. *rubellum* (Wils.) Russ.) is a very common plant that is not as a rule hard to identify. It is very near *S. plumulosum*, but differs in the following characters: (i) stem leaves tongue-shaped or at most only obtusely pointed, usually with some fibrils in the upper cells; (ii) branch leaves on the whole shorter (about 1 mm. against 1·5 mm. in *S. plumulosum*); (iii) lack of metallic lustre when dry. Other features are the slender habit, the prevailing red colour implied in the name, and a tendency for the leaves often to be turned slightly to one side when dry.

(*d*) *S. compactum* DC. (*S. rigidum* Schp.), which is abundant on many moors in the north and west, and by no means absent from the southern counties, can often be recognized in the field by its small, compact tufts and short stems. It can be confused only with certain forms of *S. palustre* and *S. papillosum*, and from both it may be known at once by the appearance of the leaves under the microscope; thus in *S. compactum* the minute stem leaf has a border of very narrow cells, and the narrow tip of the branch leaf is not hooded but sheared off (truncate) with 5–7 teeth.

ANDREAEALES

4. ANDREAEA RUPESTRIS Hedw. (*A. petrophila* Ehrh.)

The species of *Andreaea* have a characteristic appearance, by which the genus can be readily recognized. They form widely scattered blackish or dark red-brown patches on rocks in mountainous country, and it is this blackness which—together with the small size of the plants and the restricted habitat of mountain boulders—points at once to *Andreaea*; moreover, the capsule (described below) is unique among British genera.

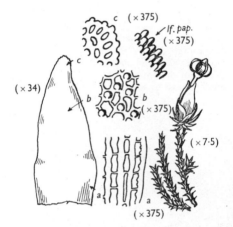

Fig. 4. *Andreaea rupestris.* On the right are a few stems as they appear in dry conditions, with the 4-valved capsule wide open; *lf.pap.* papillose back of leaf.

Certain species of *Grimmia* (also blackish and growing on rocks) might be confused with *Andreaea*, but they are mostly larger plants with some, at least, of the leaves whitish (hyaline) at their tips. *A. rupestris* is one of three species of *Andreaea* that are common in Britain.

The stems in this species are usually only 1–2 cm. in length, repeatedly branched, and bear close-set leaves of variable shape. The absence of a nerve (distinguishing this species from *A. rothii*) can be seen with a lens, and is at once evident under the microscope. The leaves are commonly ovate-lanceolate, but vary from obtuse to acute at their tips and from nearly straight to strongly curved in form. Although the leaf appears black when dry, when moistened and viewed by transmitted light it is reddish brown or dark olive green.

Further microscopic characters of this nerveless leaf are the very thick-walled cells, elongated in the leaf base, short and rounded above, and the strongly papillose nature of the back of the leaf; the latter distinguishes it from *A. alpina*.

Antheridia and archegonia occur on the same plant, but on separate branches. The capsule in *A. rupestris*, as in all species of *Andreaea*, opens by four longitudinal slits, which gape widely apart in the ripe state. It is exceedingly small (0·5 mm. long) and, although raised distinctly above the enlarged perichaetial leaves, is not conspicuous.

Ecology. It is characteristic of hard siliceous rocks in an early stage of erosion. It is restricted in the main to ground above about 1200 ft.

ADDITIONAL SPECIES

A. alpina Sm. is a taller-growing species (2–7 cm.) of a dull purplish brown colour. The leaves are nerveless as in *A. rupestris*, but the cells are not papillose.

A. rothii Web. and Mohr is a plant in habit much like *A. rupestris*, but differs in the leaves, which are nerved to the apex and usually more strongly curved.

Both these species are not uncommon on mountains, and *A. rothii*, especially in the north-west, descends almost to sea-level.

BRYALES

POLYTRICHALES

5. ATRICHUM UNDULATUM (Hedw.) P. Beauv. (*Catharinea undulata* Web. & Mohr)

This is a common and generally distributed species in woods, and is also met with on heath and waste land in the open. Combining the habit of a *Polytrichum* with the frail-textured leaves of a *Mnium*, it is usually an easy species to recognize. In its narrow, undulate leaves with toothed margins it resembles *M. undulatum*, but in its upright un-branched stems *Atrichum* is quite different. Most often it is found in fruit; it is then distinct from all other British mosses.

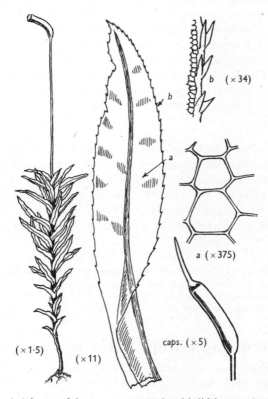

Fig. 5. *Atrichum undulatum*: *caps.* capsule with lid intact; that in the plant on the left has lost the lid exposing the epiphragm.

Beneath the soil is a creeping root-like structure. From this arise the numerous simple erect stems, 2–5 cm. tall and bearing leaves which are small near the ground, but rapidly increase in size upwards. The upper leaves, which are 6–9 mm. long, are deep or yellowish green, translucent and tongue-shaped. When moist they are widely spreading, with well-marked transverse undulations; when dry they become shrivelled and curled.

Under the microscope it can be seen that each tooth of the leaf margin is a double structure, while most of the leaf is composed of rounded cells about 20μ in diameter. In the leaf base the cells are elongate-rectangular—a distinction from *Mnium undulatum*, which has short cells in the leaf base and the marginal teeth single. That plant, of course, also lacks the series of thin upright plates or lamellae that run along the upper surface of the nerve in *Atrichum undulatum*. These lamellae usually number 4–7, and are 3–5 cells high. Similar, but more numerous, lamellae are found in *Oligotrichum* and *Polytrichum*. At the back of the nerve, near where it vanishes in the leaf apex, and on the adjoining leaf surface, are further scattered teeth.

The male inflorescence is terminal, the female arising the following year on the same stem. The red seta is 2–4 cm. long; the capsule is narrowly cylindrical and curved. The lid has a beak almost as long as the capsule. There are 32 peristome teeth, and, as in *Polytrichum*, the mouth of the capsule (after the fall of the lid) is closed by a delicate membrane.

Ecology. *Atrichum undulatum* has been regarded as an indicator of soil with good humus decomposition, but is, in fact, found in almost all types of British woodland. Both highly calcareous and extremely acid conditions, however, are avoided. Clearings in woods are a favourite habitat; on heaths and waste land, where it is frequent, the plants are often stunted and sometimes of a lighter, yellowish green.

ADDITIONAL SPECIES

A. crispum (James) Sull. (*Catharinea crispa* James) is known from the west and north of England and from Wales. It grows by rocky streams, and is distinguished from *Atrichum undulatum* by the larger cells of the leaf ($25–45\mu$) and lamellae only 1–3 cells high. Moreover, the leaves are scarcely undulate.

6. OLIGOTRICHUM HERCYNICUM (Hedw.) Lam. & DC.

O. hercynicum is a small moss, 1–2 cm. high as a rule, and in consequence rather likely to be overlooked. It is often plentiful on screes and other stony places on mountains, but is absent from almost all the southern and eastern counties.

This little moss is of interest in showing a rather solid leaf texture of the *Polytrichum* type, and a capsule which is structurally akin to that of *Atrichum*. Thus in a sense it is intermediate between these two genera; but when first found it is hardly likely to suggest either *A. undulatum* or one of the large and familiar species of *Polytrichum*. When moist it appears deep or vivid green, sometimes tinged with reddish, and most resembles a very small species of *Polytrichum*. In dry conditions, however, the colour appears dull, and the leaves curl up, with their margins strongly incurved. Then it might easily be passed over for some quite unrelated moss.

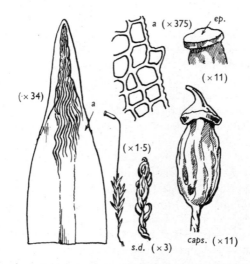

Fig. 6. *Oligotrichum hercynicum*: *caps.* capsule with lid (dry, not fully ripe), *ep.* epiphragm, *s.d.* part of leafy shoot in a dry state.

The short upright stems, appearing scattered, are connected below with a well-developed rhizoid system, as in *Atrichum undulatum*. Each leaf is awl-shaped, with wide sheathing base and acute tip. About 12 lamellae run a wavy course the length of the broad midrib and give the leaf its rather solid texture. A few smaller lamellae occur at the back of the nerve, at the leaf tip. Widely spaced teeth are found on the leaf margin and at the back of the leaf tip.

The plant is dioecious, the male plants being smaller than the female. In its main features the capsule agrees with that of *Atrichum*. The seta, however, is only about 2 cm. long, the capsule is not curved, the lid has only a short beak, and the calyptra is beset with a few long hairs. In this combination of characters the capsule is unique among British mosses.

Ecology. It grows almost always on loose soil. Thus it is found in those parts of a non-calcareous scree where fine detritus has accumulated or something like true soil has been formed. It is sometimes abundant in the moss–lichen community of mountain-top detritus, and on loose gravelly soils flanking moorland, at lower altitudes. *Oligotrichum hercynicum* is probably confined to acid conditions.

POLYTRICHUM

General notes on the genus. All species of *Polytrichum* are alike in that their apparently scattered and little-branched erect stems spring in the first instance from a complicated underground rhizoid system, with the result that the plants tend to grow in extensive patches that may perhaps be likened to miniature forests. All agree too in having leaves which show an obvious differentiation into a broad colourless sheathing base and a narrow, green or brownish expanded limb. The upper or adaxial surface of the limb of each leaf is largely covered by a series of longitudinally attached plates of green tissue termed lamellae. In a transverse section of the limb of the leaf the lamellae appear as a series of erect chains of cells, the terminal cell of each chain differing from the rest and furnishing a very reliable character for the separation of some of the species. Thus it forms the best means of distinguishing between *P. commune*, *P. formosum* and *P. alpinum*, and is a useful confirmatory character in *P. urnigerum*. The necessary transverse section is not difficult to prepare if a leafy shoot is grasped in the thumb and forefinger of the left hand, care being taken to ensure that numerous leaves are held appressed to the stem. If a number of swift cuts then be made, using a sharp razor, many leaf sections will be secured, and among them should be some fragments that are thin enough to show the structure well. Alternatively, the lamellae may be scraped off the adaxial surface of the leaf with a sharp scalpel, mounted in water and examined; these will reveal the distinguishing characters of most of the species, but it must be admitted that the transverse section is much more satisfactory for showing the clear-cut difference in lamellar structure between *P. commune* and *P. formosum*.

All the European species of *Polytrichum* are dioecious. Male 'flowers' are often reddish or orange and are indeed more flower-like than most structures so called among bryophytes. The capsule has several unusual features. Thus, after the fall of the lid there is a conspicuous whitish membrane (epiphragm) closing the mouth, and there are 32 or 64 peristome teeth. The tips of the teeth are united with the epiphragm, and the spores are shed through the small holes that separate each tooth from the next. Apart from *Atrichum* and *Oligotrichum*, this type of capsule and spore dispersal is not found elsewhere among British mosses.

The name *Polytrichum* refers to the exceedingly hairy calyptra, which is a constant feature throughout the genus.

7. **POLYTRICHUM ALOIDES** Hedw.

This moss takes its specific name from a not entirely fanciful resemblance to a miniature aloe plant. This impression is due to the short stem with its rosette-like tuft of stiff, dark green, awl-shaped leaves,

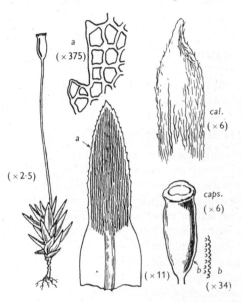

Fig. 7. *Polytrichum aloides*: *cal.* calyptra, *caps.* capsule without lid.

from which arises, in a fertile female plant, the scarlet seta that supports the ovoid-cylindrical capsule. The plants form short turfs (1–2 cm. tall), the erect stems developing from a persistent protonema. *P. aloides* should be looked for on the soil of steep, shaded banks in hilly districts, where it is usually abundant; but it is rare in many parts of south and south-east England. Sometimes one of the most conspicuous features of the plant is the delicate whitish epiphragm that covers the mouth of the ripe capsule after the lid has fallen. These discs of opaque white catch the eye from a distance.

To one who has first become familiar with some of the larger species, such as *P. commune*, this small plant will scarcely suggest a *Polytrichum*.

Under the microscope, however, the parallel lamellae on the leaf surface are seen—it is these which impart to the leaf its solid texture—and the translucent margin, 2–3 cells wide, confirms the relationship. The margin is sharply toothed, several cells usually composing each tooth.

The young capsule is enveloped in a shaggy brown calyptra which, together with the scarlet seta, makes the plant conspicuous at certain seasons. The circular (not polygonal) cross-section of the capsule separates this and the next species from the other, mainly taller-growing species described here (only *P. alpinum* in Britain combines taller habit with rounded capsule). The best character for separating *P. aloides* from the less common *P. nanum* is the papillose surface of the capsule.

Ecology. It is common on steep banks in woodland cuttings at moderate altitudes, and on disturbed soil, as by roads, in moorland country; it also occurs in quarries. In all these habitats it seems able to act as a pioneer on bare, loose soil. It ascends to the alpine zone, growing in soil-filled rock clefts or on soil-capped ledges. *P. aloides* is probably restricted to soils that are at least moderately acid.

8. **POLYTRICHUM URNIGERUM** Hedw.

The best field character for recognizing this species is the pale glaucous green colour of the leaves, in contrast with the prevailing deep dull green of most species of *Polytrichum*. In structural detail it resembles *P. aloides*, but is generally taller, the upright stems varying from 1·5 to 6 cm. in height. On loose soil in mountain country the two species often grow together; then the almost blue-green colour of *P. urnigerum* will separate even stunted plants of that species from *P. aloides*. When moist, the plants are conspicuous with their red stems crowned by widely spreading leaves; but in dry conditions the leaves are held straight or incurved, appressed to the stem.

The leaf is lanceolate, acute and up to 1 cm. in length. It has the usual *Polytrichum* characters of a broadly sheathing base and longitudinal lines of lamellae on the broad midrib of the limb. It is, however, longer, more sharply pointed and with more prominent marginal teeth than that of *P. aloides*.

Under the microscope the nerve is seen to be sharply toothed at the back, whilst each tooth of the leaf margin is composed of several cells. A sure diagnostic feature is provided by the leaf section. This will reveal that the end cell of each row forming a lamella is enlarged and covered with papillae.

This species is dioecious, as are all the European species of *Polytrichum*. The seta is 2–3 cm. long. The capsule tends to be slightly longer and more narrowly cylindrical than in *P. aloides*, with a wider mouth.

Ecology. It is more restricted to hilly and mountainous districts than is any other species described here. It grows on loose soil by moorland roads or on the finer detritus at the edges of screes. It also occurs on soil-capped stone walls and rock ledges, and in soil-filled rock clefts. The soil colonized is probably always acid, at least mildly so.

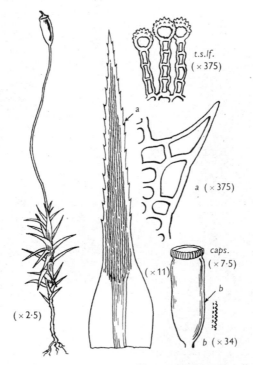

Fig. 8. *Polytrichum urnigerum*: at *t.s.lf.* are seen three lamellae from a transverse section of a leaf; *caps.* capsule without lid.

9. **POLYTRICHUM JUNIPERINUM** Hedw. (Plate I)

This medium-sized species—the leafy shoots seldom grow above 6 cm.—is a common plant of heaths. It often occurs in great abundance on ground that has recently undergone clearance by burning.

The upright stems are bright red, strong, and about 1 mm. thick. The lanceolate leaves are drawn out into sharp reddish tips. The greater part of the leaf margin is quite without teeth, and this feature, clearly seen with a hand-lens, at once distinguishes the species from *P. commune* and *P. formosum*. A strong inrolling of the leaf margins gives the leaves a bristle-like character in the dry state. In this condition the whole plant looks quite different, the leaves being appressed to the stem and somewhat twisted. The reddish leaf tip is a useful field character.

Male plants in spring are conspicuous with their bright orange or red open 'inflorescence' of flower-like aspect. The leaves which surround the archegonia differ from the vegetative leaves in their broad membranous margins and whitish tips. The capsule is borne on a bright red seta 2–5 cm. long. It is 4-angled, and has a deep crimson lid, with a short beak.

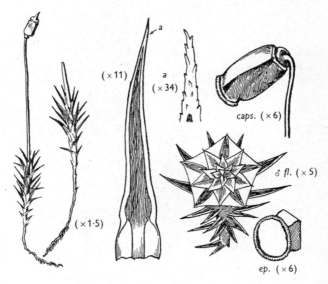

Fig. 9. *Polytrichum juniperinum*: of the two plants drawn, that on the right is in the very early fruiting condition, that on the left bears a mature capsule; ♂ *fl.* male 'inflorescence', *caps.* capsule, *ep.* epiphragm over mouth of capsule after fall of lid.

Ecology.   Its most characteristic habitats are acid grass heath and bare patches on *Calluna* heathland. It also occurs in other habitats on rather dry acid soils, as on some fixed dunes. *Polytrichum juniperinum* follows *Funaria* in the succession on burnt heathland. In north and west Britain it occurs where a thin crust of peaty soil covers stone walls or boulders. Thus it grows in better drained places than those favoured by *Polytrichum commune*.

10. **POLYTRICHUM PILIFERUM** Hedw.

This species is found in very similar situations to *P. juniperinum* and is about equally common. Like that species it is often listed among the early colonists on freshly burnt heathland.

Plate I. *Polytrichum juniperinum* with male 'inflorescences' (natural size)

Plate II. *Polytrichum commune* with capsules (natural size)
Also seen are several pale shoots of a species of *Sphagnum*

It is at once distinguished from *P. juniperinum*, which it otherwise resembles, by the greyish white, not red-brown, toothed points which extend outwards from the leaf tips to a distance often equal to about a third of the length of the leaf. In the dry state, with the leaves closely appressed to the stem, and the hoary points consequently drawn together, the whitish terminal 'brush' thus formed is particularly conspicuous. Further, the plants tend to be shorter than those of *P. juniperinum*, and the leaves are more evidently crowded into a short terminal tuft at the end of the stem. Not uncommonly the stem is quite bare of leaves for 2 or 3 cm., whilst the rest of the stem, often less than 2 cm., bears very numerous close-set leaves. This gives the plants a very distinct character, quite unlike that of *P. juniperinum*.

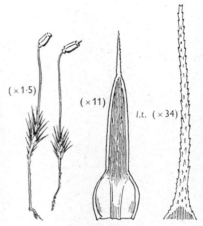

Fig. 10. *Polytrichum piliferum*: *l.t.* top of leaf showing hair point.

P. piliferum closely resembles the last species in the characters of male and female 'inflorescence' and capsule structure. It has, however, a shorter seta (1·5–3 cm.) and smaller capsule.

Ecology. It occurs in similar habitats to the last species. On East Anglian heaths it is a colonist of looser sands, *P. juniperinum* favouring those more stabilized.

11. POLYTRICHUM FORMOSUM Hedw.

This is a common species of woods and hedge banks; it also occurs on heaths. When growing luxuriantly it may be taken for *P. commune*, to which it bears a strong general resemblance.

The stem in *P. formosum* is never as tall as in well-grown plants of *P. commune*. It is usually between 5 and 10 cm. The leaves, when expanded in the moist condition, do not tend to spread quite so widely as in *P. commune*, i.e. they are not so squarrose. In form and in their toothed margins they are very like the leaves of that species, and (as pointed out under *P. commune*) the rounded terminal cells of the leaf lamellae form the most reliable character of *P. formosum*. Thus, in determining doubtful sterile material of these *Polytrichum* species, transverse sections of the leaves should always be cut.

In the 'fruiting' condition *P. formosum* is separated without great difficulty from *P. commune* by the sum of a number of small points: (i) the seta is not above 7–8 cm. in length; (ii) the capsule is 5- or 6-angled usually, rarely 4-angled; (iii) the apophysis is less distinct and

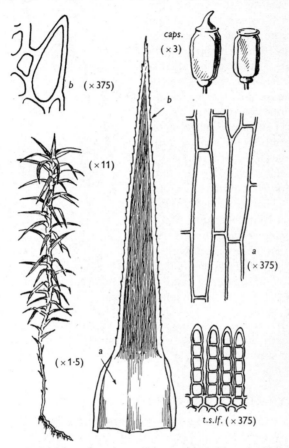

Fig. 11. *Polytrichum formosum*: *caps.* capsules with and without lid, *t.s.lf.* part of transverse section of leaf showing four lamellae.

less disc-shaped than in *P. commune*; and (iv) the lid of the capsule has a longer beak (about 2 mm. against 1 mm. in *P. commune*).

Ecology. Its typical habitat is deciduous woodland on mildly acid soil, e.g. sessile oak woods and beech woods on leached soils. Sometimes it is plentiful on well-drained areas of upland heath. In the Reading district *P. formosum* has been recorded from woodland soils ranging from strongly acid to nearly neutral.

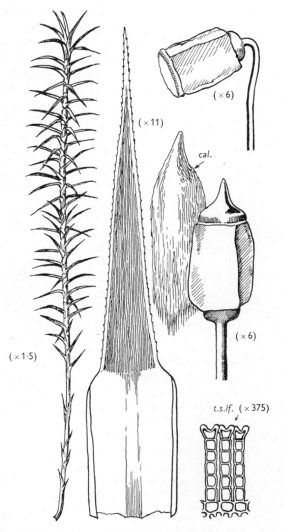

Fig. 12. *Polytrichum commune*: *cal.* calyptra, *t.s.lf.* part of a transverse section of leaf showing four lamellae.

12. **POLYTRICHUM COMMUNE** Hedw. (Plate II)

This is the best known and most generally abundant species of *Polytrichum*. It is a plant of damp moorland, where it often covers considerable areas with its luxuriant, deep green growth. At these times *P. commune* is a very conspicuous plant, attracting the attention of even the casual observer.

The strong stems commonly reach a height of 15–20 cm., and may be found much taller on occasion. The leaves are 8–12 mm. long, with glossy sheathing bases and fine points. The margin is beset with numerous teeth, which are seen readily with a lens. A tall-growing *Polytrichum* with a toothed leaf margin is most likely to be this species or *P. formosum*. The distinction between the two, as regards vegetative characters, depends upon a somewhat minute point—the shape of the terminal cells of the leaf lamellae (as seen in cross-section). In this species these cells are notched, in *P. formosum* they are rounded.

As in other species of *Polytrichum*, the terminal part of the male shoot forms a conspicuous almost flower-like open cup. After the antheridia have been produced there is often a proliferation of the vegetative shoot upwards from the male 'inflorescence'.

The capsule in *P. commune* is borne upon a longer seta than in any other species of the genus—6–12 cm. in length. The capsule itself has the form of a four-sided box when ripe. Its length is little greater than its breadth, so that it becomes in some cases almost cubic. Other 'fruiting' characters are best seen at a slightly younger stage, when the lid with its short beak, and the long, golden brown calyptra are intact. At the base of the capsule, and separated from it by a very narrow constriction, is a clearly marked swelling, the apophysis, which is more strongly developed here than in any other British species of the genus.

Ecology. *P. commune* is most abundant on wet, highly acid moorland or bog. It is frequent on the wetter parts of acid 'white' moor, dominated by matgrass (*Nardus*); it also occurs on the *Rhacomitrium* moor of exposed mountain tops. It is occasional in open woods, its presence indicating strongly acid soil.

ADDITIONAL SPECIES

To recognize the genus *Polytrichum* is not generally difficult; to distinguish the individual species is much less easy, for (as pointed out already) the leaf section has often to be cut. Furthermore, besides those species described here, there are several others which are not rare.

P. nanum Hedw. is a dwarf species which differs from *P. aloides* in its non-papillose and much shorter, wider capsule (1–2 mm. long). It occurs in similar habitats.

P. alpinum Hedw. is a not uncommon moss of mountains. It is normally much taller than *P. urnigerum* and lacks the glaucous green of that species. The leaf agrees with that of *P. urnigerum* in the swollen, papillose end-cells of the lamellae; but the capsule of *P. alpinum*, although rounded in cross-section, is distinct in being wider, asymmetrical, and narrow at the mouth.

P. alpestre Hoppe (*P. strictum* Brid.) is closely related to *P. juniperinum*, but differs in its taller stems (commonly 8–12 cm.) and proportionately shorter leaves. The white tomentum that covers the stems is also characteristic, as is the much wetter habitat of boggy moorland.

P. gracile Sm. is a plant of heaths and acid woodland. It is very like small forms of *P. commune* and *P. formosum*, but the translucent leaf margin is 4–6 cells wide—

wider than in either of these—and the capsule less sharply angled. The shorter, wider cells of the leaf base (30–60μ long) are, however, the best way of distinguishing it from *P. formosum* (some of whose leaf-base cells are 100–120μ long), which it resembles in the almost smooth, rounded end-cells of the lamellae.

BUXBAUMIALES

13. DIPHYSCIUM FOLIOSUM (Hedw.) Mohr

This is our only common British representative of the peculiar group of mosses to which the two rare species of the aberrant genus *Buxbaumia* belong. As in *Buxbaumia*, the leafy shoot of *Diphyscium* is little developed and the capsule disproportionately large.

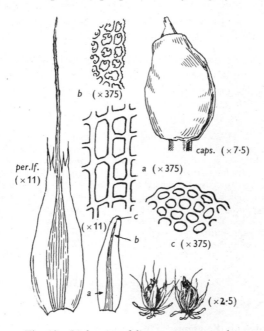

Fig. 13. *Diphyscium foliosum*: *caps.* capsule, *per.lf.* perichaetial leaf.

In the barren state *D. foliosum* can be overlooked very easily, the small plants with narrow strap-shaped leaves having nothing very notable in their structure to distinguish them from some species of *Trichostomum*, *Barbula* or *Tortella*. In fruit, however, this species is unmistakable. The capsule has been aptly likened by Dixon to a grain of wheat, and looks indeed incongruous upon so small a plant. This moss forms dark or brownish green patches or carpets, often intermixed with other bryophytes, on banks in hilly or mountainous regions. Only the female plants attract the eye, the males being distinct and much smaller. Indeed, it is the capsule, surrounded by long, torn perichaetial leaves, that will first be noticed.

Not only in shape, but also in cell structure, the leaves resemble those of such a plant as *Trichostomum tenuirostre*. Thus, the cells of the leaf base are long, rectangular and hyaline, whilst those in the upper part of the leaf are short and rounded, obscure in their outlines owing to an abundance of papillae. The perichaetial leaves are much larger, have a broad, excurrent nerve, and are often torn at the apex into numerous long fine shreds.

The large size of the capsule has already been mentioned. It is also remarkable in other ways, for it is almost sessile, and asymmetrical in shape; moreover, it has a peculiar whitish peristome membrane, which is seen after the fall of the conical lid.

Ecology. It grows on banks and soil-covered boulders on moorlands and in woods, on acid soil. Also, it occurs on soil-capped rocks in the alpine rock-ledge community.

RELATED SPECIES

Buxbaumia aphylla Hedw. is mentioned chiefly because of its interesting structure. Several of the remarkable features of *Diphyscium* are here carried to an extreme degree. The plant consists chiefly of the much-branched protonema, and true leaves, as distinct from perichaetial bracts, are lacking. There is very little green colour in the leafy plant, and the capsule, which is raised on a seta 1 cm. long, is even larger in relation to the rest of the plant than in *Diphyscium*. The male plants are microscopic. *Buxbaumia aphylla* is very rare in Britain and occurs sporadically, chiefly on woodland banks, among other mosses. In its nutrition it seems to be largely dependent on organic matter, i.e. it is to a great extent saprophytic, although there is some green colour in the protonema.

FISSIDENTALES

FISSIDENS

General notes on the genus. The genus *Fissidens* is recognized without difficulty by the strictly two-ranked arrangement of the leaves, the shoots resembling miniature fern fronds. Also, the leaf itself is of a peculiar form not paralleled in any other genus. Each leaf consists essentially of a boat-shaped clasping portion, beyond which there extends a wing of tissue, the nerve being prolonged out into this wing. Thus when a single leaf is removed and viewed under the microscope it will be divisible into three parts: (i) a region of double thickness, the now flattened boat-shaped part of the leaf (the *sheathing lamina*); (ii) an extensive wing of tissue running from base to apex, *below* the boat-shaped part (*inferior lamina*); and (iii) a smaller wing of tissue immediately beyond the apex of the boat-shaped part, and separated from (ii) by the nerve; because it lies *above* the nerve this third region is known as the *superior lamina*. Much use is made of precise measurements of these three parts of the leaf in the identification of critical species of the genus, but for the determination of the common species considered here such details will seldom be necessary.

Other leaf characters that will be found important are (i) presence or absence of a thickened border, (ii) nerve ceasing in the leaf apex or extending beyond it, (iii) cell size.

This genus, which comes near to *Dicranella* in the structure of capsule and peristome, is remarkable in that the fruit arises terminally in some species, in others laterally.

14. **FISSIDENS BRYOIDES** Hedw.

F. bryoides is by far the commonest of several small species of *Fissidens* that occur in Britain. It forms dense colonies on shaded banks or patches of exposed soil, the stems commonly measuring 1–1·5 cm. in height.

The leaf arrangement and the form of the individual leaves are those of the genus; but a feature of this species (in common with some rarer species) is the clearly defined border at the edge of the leaf. This border is composed of long, narrow, thick-walled cells of fibre type, the cells throughout the rest of the leaf laminae being rounded hexagonal and measuring 6–10μ across. The fibrous leaf border will distinguish *F. bryoides* from all the larger species of the genus except *F. crassipes*, which grows in quite different habitats. *F. bryoides*, however, is itself variable in size, small forms occasionally approaching *F. exilis* in minuteness, whilst taller plants, with 8–10 pairs of leaves on the stem, come nearer to *F. taxifolius*; but both these species are without the fibrous border to the leaf.

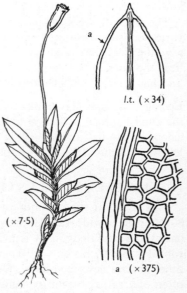

l.t. (× 34)

(× 7·5)

a (× 375)

Fig. 14. *Fissidens bryoides*: *l.t.* tip of leaf.

F. bryoides is autoecious, and the bud-like male inflorescences are found in the axils of the leaves. The capsule is minute (1 mm.), erect and symmetrical. It is borne on an orange-red seta, 5–10 mm. long. The seta normally arises terminally in *F. bryoides*, an important distinguishing mark from *F. taxifolius*, in which it arises laterally.

Ecology. It occurs on soil or decaying wood, in deciduous woodland and on bare waste ground about buildings. It favours neutral loam soils but will tolerate mild acidity.

15. FISSIDENS TAXIFOLIUS Hedw.

This is a rather small species, often less than 1·5 cm. high, sometimes slightly taller, and found in a wide variety of situations. It is a common lowland plant, particularly of clay banks, but is also found on mountain slopes.

A much shorter plant than *F. adianthoides*, and characteristically broader in the leaf than *F. bryoides*, this species can often be identified in the field; but the identification should be confirmed by microscopic examination. It will then be seen that the leaf lacks both the thickened border of *F. bryoides* and the clear marginal rows of cells seen in *F. adianthoides*. Further, the nerve is shortly excurrent, a unique feature among British species that lack the thickened border.

The male flowers are borne on short basal branches. The capsule varies considerably in form, being erect or inclined, symmetrical or asymmetrical, but is always borne on a longish red seta which arises near the base of the stem (contrast the terminal capsule of *F. bryoides*), and has a large, bright red peristome.

Ecology. Besides clay banks (mentioned above and sometimes given as the chief habitat) it occurs in oak, ash and beech woods on neutral to calcareous soils, and on chalk grassland. Its presence on mountain ledges at about 2000 ft. probably indicates local calcareous patches.

16. FISSIDENS ADIANTHOIDES Hedw. (Plate III)

This is by far the commonest of our larger species of *Fissidens*, and is found in a wide variety of habitats, including boggy ground, river banks and rock ledges. It is most plentiful in mountain districts. The robust habit, with stems 3–10 cm. in height, prevents confusion with the other species described here. Dense and conspicuous tufts are formed, often of a rather bright or golden green.

Under the microscope the leaf shows a toothed apex, with the nerve ceasing there and not excurrent as in *F. taxifolius*. The leaf is without the thickened border of *F. bryoides*, but 2–4 rows of cells along each margin are more translucent than the rest, and so give the effect of a pale marginal band. (In the less common *F. cristatus* these cells form an even more distinct light marginal band.) The leaf cells are 10–18μ across.

Small axillary male flowers are produced, and on the same plants the archegonia and capsules normally occur, for this species (like the

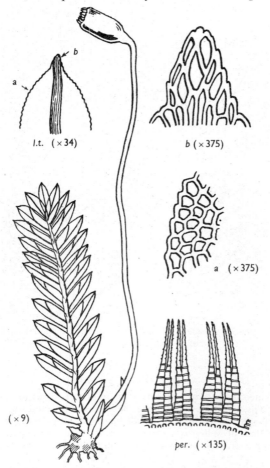

l.t. (×34)

b (×375)

a (×375)

per. (×135)

(×9)

Fig. 15. *Fissidens taxifolius*: *per.* part of peristome (inside view), *l.t.* tip of leaf.

two preceding, but unlike some other members of the genus) is autoecious. The capsule is of somewhat variable form, and is borne laterally on a long, stout, red seta. Often the capsules occur in abundance, when the very numerous scarlet setae and blackish red 'fruits' render the moss conspicuous.

Ecology. *F. adianthoides* is remarkable for the diversity of its habitats, which include boggy ground on moorland, wet soil-capped ledges on mountains, wet hollows in sand dunes, loose soil of shaded river banks, and open chalk grassland.

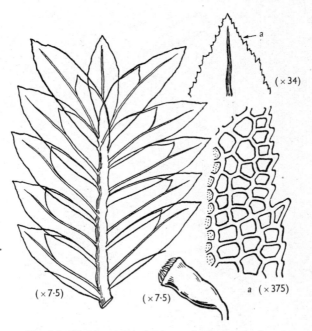

Fig. 16. *Fissidens adianthoides*: only the terminal part of the leafy shoot is shown.

ADDITIONAL SPECIES

Although seventeen species of *Fissidens* occur in Britain, many are rare.

F. pusillus Wils. ex Milde is a minute plant of shaded sandstone and limestone rocks. The leafy stems (a few millimetres tall) have only 3–6 pairs of narrowly bordered leaves—fewer than in well-grown plants of *F. bryoides*—the uppermost pair being very long; moreover, it is often dioecious, and the male flowers are basal, not axillary.

F. incurvus Starke ex Web. & Mohr is another very small species, with bordered leaves. It grows on shaded banks and is distinct in its curved, almost horizontal, capsule.

F. crassipes Wils. has stems that are longer (1·5–5 cm.), and the plant is a characteristic dark green, with large, opaque leaf cells (12–18 μ against 8–10 μ in *F. bryoides*). It occurs on stones, in or by calcareous streams.

F. exilis Hedw. differs from all our other small species in lacking a leaf border. Its minute stem bears only 2–4 pairs of leaves and would be readily overlooked but for the erect capsules that are sometimes produced in abundance on orange-red setae 5–6 mm. long. It grows on shaded clay banks.

Allied to *F. adianthoides* are two rather less common species:

F. osmundoides Hedw. is sometimes a taller plant (5–15 cm.) with shorter leaves that give the shoot a narrower outline. The only species combining tall habit with a terminal fruit, it occurs on moist rock ledges in hilly districts.

F. cristatus Wils. (*F. decipiens* De Not.) is usually shorter than *F. adianthoides*, and is known by the more distinct marginal band of clear cells in the leaf and the smaller cell size (6–10μ against 10–18μ). Like *F. adianthoides* it is found in chalk or limestone grassland, and on rock ledges.

DICRANALES

17. **PLEURIDIUM ACUMINATUM** Lindb. (*P. subulatum* of Dixon, *Handbook*, not of Rabenh.)

This is one of our small and rather inconspicuous ground mosses, which may be found on patches of exposed soil on heaths, on banks, or in open woodland. Despite its small size it is a species which, once known, is readily recognized. In vegetative characters and habit it

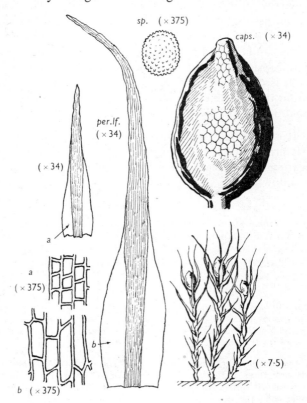

Fig. 17. *Pleuridium acuminatum*: *per.lf.* perichaetial leaf,
caps. capsule, *sp.* spore.

bears some resemblance to other small members of the Dicranaceae; also, more distantly, to *Leptobryum pyriforme*, which has similar fine, hair-like (setaceous) leaves. When in fruit, however, it is unlike any other common moss, with its neat, orange-brown, almost spherical capsules, each nearly sessile in the centre of a cluster of long, fine, perichaetial leaves. Other cleistocarpous mosses, such as members of the genus *Phascum*, show a very different leaf shape, and cell structure.

The stems, about 1 cm. tall, are generally unbranched, and the leaves are narrowly lanceolate. A conspicuous and useful diagnostic character is the great length of the perichaetial bracts compared with the normal vegetative leaves. The leaves are not curved to one side (falcato-secund) as in *Dicranella heteromalla*.

A microscopic character to note is the rather broad nerve, which, however, is neither so broad nor so well defined as in *Campylopus*. The cells are very shortly rectangular in the small vegetative leaves, but are long and narrow in the elongated perichaetial leaves. The long leaf tip is entire (contrast the denticulate leaf tip of *Dicranella heteromalla* and *Campylopus pyriformis*) and is almost wholly occupied by the nerve.

Antheridia can be found by dissection, in the axils of the perichaetial bracts. The capsule is ovoid to globose, with a short blunt point at the apex. It is cleistocarpous, and is borne on a stalk so short that it ripens within the rosette of perichaetial leaves. The spores are notably large (about 30μ across).

Ecology. It is essentially a colonist of bare patches, on soils deficient in lime. It grows, for example, with *Leucobryum glaucum* on leached soil in clearings of 'plateau' beech woods in southern England; it also occurs in fallow fields.

ADDITIONAL SPECIES

Pleuridium subulatum (Hedw.) Lindb. (*P. alternifolium* (Dicks.) Brid.) grows in wet fallow fields and at the margins of pools. It may be distinguished from *P. acuminatum* by the bud-like, axillary male flowers and by the perichaetial leaves which are abruptly narrowed just above the base to form long fine points.

RELATED SPECIES

Archidium alternifolium (Hedw.) Schp. lacks the rigid habit of *Pleuridium acuminatum*, but, like that species, is a small cleistocarpous moss of bare ground. It is quite distinct in its perfectly spherical capsule, which contains only about 16 remarkably large polyhedral spores. It is not a common plant.

Pseudephemerum nitidum (Hedw.) C. Jens. (*Pleuridium axillare* (Dicks.) Lindb.) is a minute, sometimes glistening, pale green plant of moist bare ground and the margins of pools. It differs from *Pleuridium* spp. in the perichaetial bracts, which are no longer than the upper leaves. It is cleistocarpous, and the oval, shortly pointed capsule is borne on a seta so short that it appears almost sessile.

18. **DITRICHUM HETEROMALLUM** (Hedw.) E. G. Britton
(*D. homomallum* (Hedw.) Hampe)

This is a small, inconspicuous moss of bright yellowish green colour and silky texture. It resembles *Dicranella heteromalla* (with which it is often associated), and, as in that plant, the leaves tend to be curved in one direction, although less markedly so. *Ditrichum heteromallum* is a low-growing plant, not forming the dense tall cushions of *Distichium*

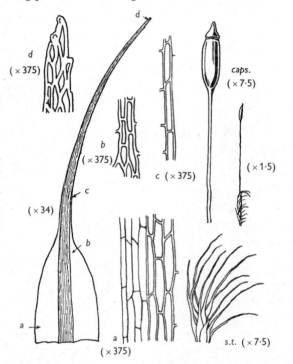

Fig. 18. *Ditrichum heteromallum*: *s.t.* tip of shoot, *caps.* capsule.

capillaceum and much less robust than *Ditrichum flexicaule*. The bare patches of stony or sandy ground where it grows may indeed require careful scrutiny if the tiny scattered plants, scarcely 1 cm. tall, are not to be overlooked.

Microscopic examination shows a leaf with a broad base, which tapers gradually above to the finely drawn-out tip. The nerve is rather broad, but not very well defined. The leaf apex is smooth or faintly denticulate. Perhaps the best character lies in the structure of the cells, which are narrow and elongate both at the base and higher up the leaf.

The antheridia are formed on separate, slender plants. The symmetrical, cylindrical fruit is held upright on a reddish or purple seta 1–1·5 cm. tall. In form and peristome it closely resembles the capsule of *Distichium capillaceum*. In their long narrow form and erect position the capsules of these two genera contrast sharply with those of *Dicranella* and other allied plants.

Ecology. It occurs almost always with *Dicranella heteromalla*, *Polytrichum aloides*, etc. (and the liverwort *Diplophyllum albicans*), on acid sandy or gravelly soil, e.g. in old quarries, or by the roadside in hilly districts. It is sometimes found on rock ledges on mountains. It avoids lime.

19. DITRICHUM FLEXICAULE (Schleich.) Hampe

D. flexicaule has the tall habit and exceedingly long slender leaves of *Distichium capillaceum*, but is darker and duller in colour and lacks the 2-ranked leaf arrangement of that plant. It is found, moreover, in somewhat different habitats and is even more definitely calcicole. It occurs on limestone rocks and soil, and on sand dunes where the lime content is high. The dense tufts are commonly 4–7 cm. tall, and this makes the moss conspicuous. Sometimes on less favourable sites, however, one finds stunted specimens of dull brownish tint, and then the long narrow flexuose leaves will aid in determination.

A short distance above the broad base the leaf narrows suddenly. Above this it tapers more gradually to the very long fine point. A few minute teeth occur at the apex. Apart from a few elongate cells close to the nerve and along the extreme margin of the expanded base,

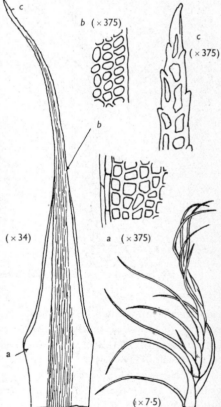

Fig. 19. *Ditrichum flexicaule*: only the terminal part of the leafy shoot is shown.

the areolation is shortly rectangular or oval throughout the leaf. These short rounded cells stand in strong contrast with the narrowly elongate cells of *Ditrichum heteromallum*. The absence of distinct cells at the basal angles of the leaf separates the present species from the rather rare *Dicranodontium denudatum*.

Separate slender male plants occur, but are rare. *Ditrichum flexicaule*, as known in Britain, is almost invariably without fruit. The capsule is not unlike that of *D. heteromallum*.

Ecology. As mentioned above, it always indicates calcareous conditions (calcicole). On limestone hills in northern England it sometimes grows in association with *Rhytidium rugosum*. On calcareous sand dunes, on chalk downs and the margins of beech woods, a commonly associated species is *Camptothecium lutescens*.

ADDITIONAL SPECIES

Ditrichum cylindricum (Hedw.) Grout (*D. tenuifolium* Lindb.) is the only one of the other five species of *Ditrichum* that needs to be mentioned. It has the long narrow leaf and smooth cylindrical capsule of the genus, but is distinct in the way in which the leaves stand out at a wide angle with the stem (squarrose). *D. cylindricum* has been regarded as rare, but it is certainly more widespread along woodland rides than has hitherto been suspected.

20. DISTICHIUM CAPILLACEUM (Hedw.) B. & S. (*Swartzia montana* (Lam.) Lindb.)

This is one of the most beautiful and distinctive species occurring on mountain ledges and rock clefts. It does not normally appear below about 1800 ft., but, locally in the west and north-west, it descends to sea-level.

The long fine points of the pale green leaves and the erect cylindrical capsule on a reddish seta will help to identify it; but the surest character is provided by the 2-ranked (distichous) arrangement of the leaves on the stem (cf. *Ditrichum flexicaule*). The plants may grow tall, with stems 12–15 cm. in length, but are commonly much shorter and more compact. The lower parts are much interwoven with brownish radicles, the plants forming dense, soft tufts.

The leaf is 3–4 mm. long. Rather less than half this length consists of a sheathing base (shining when dry), in which the lowest cells are long and narrow but those farther up are shortly rhomboidal. Above this the leaf narrows abruptly and consists mainly of the nerve. The leaf tip is rough with strongly papillose cells. These microscopic details serve but to confirm a determination which can be made in the field, without much difficulty, owing to the very characteristic flattened appearance of the shoot as a whole.

Antheridia occur in the axils of the uppermost leaves. This moss is most commonly found with abundant fruit. The reddish seta is 1·5–2 cm. long; the capsule is erect or slightly inclined, usually narrowly cylindrical, and light brown, with a reddish peristome of 16 teeth.

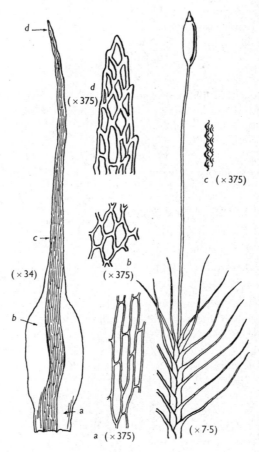

Fig. 20. *Distichium capillaceum.*

Ecology. It indicates the presence of at least some lime. It colonizes cracks, soil caps and wet rock faces in the alpine zone; it also occurs on mountain-top detritus. I have found it near sea-level in Wester Ross.

21. CERATODON PURPUREUS (Hedw.) Brid.

C. purpureus and *Hypnum cupressiforme* are probably the two commonest mosses in this country, and the collector is likely to find them very frequently; but owing to their variability both these mosses may continue to confuse the beginner until they have been collected and examined many times. *Ceratodon purpureus* is a conspicuous plant in its typical state in spring, when patches of bare ground or burnt heathland are often purple with the countless setae of fruiting *Ceratodon*. At other times it is less noticeable, although it is usually possible in

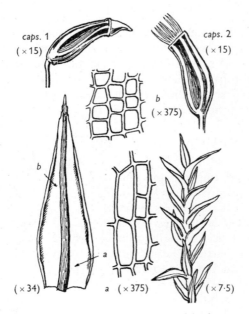

Fig. 21. *Ceratodon purpureus*: *caps.* 1 and 2, ripe capsules before and after removal of the lid.

winter to see the young capsule, borne on a delicate wine-coloured seta which has not yet lengthened to its full extent. *Ceratodon* is extremely catholic as regards habitat. The erect, slightly branched, leafy shoots are normally 1–3 cm. tall, and very variable in colour. The leaves are lightly curled when dry.

With a little experience *C. purpureus* may be identified in the field by an appearance which is characteristic but difficult to describe. Under the microscope identification may then be confirmed by the following combination of leaf characters: (i) broad base tapering to rather fine point, which is often curved to one side a little; (ii) apex with a few distinct teeth; (iii) nerve strong, running in a channel the whole length of leaf; (iv) margin recurved, from base to within a short distance of apex; (v) cells very distinct, never obscure or papillose, broadly elongate below, short and nearly square (10–15μ across) above; (vi) leaf length about 1–1·5 mm.

This species is dioecious, with rather slender male plants. The 2 cm. long purple-red seta it shares with certain species of *Barbula* and others, but the form of the capsule is a good guide—at first nearly upright and almost symmetrically ovoid to ellipsoidal with a curved conical lid, but when mature nearly horizontal, dark chestnut-brown and deeply furrowed. There are 16 peristome teeth, cleft nearly to the base.

Ecology. Typical states are found most abundantly on bare acid soil on grass heath, lowland *Calluna* heath and upland heather moors. But it tolerates a wide range of soil reaction, and occurs in many habitats, including wall-tops, rotten wood, fallow fields, duneland, shingle beaches and soil-capped rock ledges at high altitudes. It is often common in large towns, even in smoky situations.

22. SELIGERIA CALCAREA
(Hedw.) B. & S.

As its name implies, this plant grows upon calcareous rocks. It is indeed very restricted in its habitat, occurring almost exclusively on the exposed vertical faces of chalk or limestone cliffs and quarries. The individual plants are minute, but they form wide, dark green, velvety patches which, unless examined carefully, might be overlooked or mistaken for an alga. The shoots are unbranched, and in height do not exceed a few millimetres. In its own specialized habitat, particularly in the south and east of England, *S. calcarea* is locally common.

Closer examination shows a leaf (up to 1 mm. long) of characteristic shape, with a rather broadly oblong, somewhat concave base, which narrows suddenly above to form a tip that is wholly occupied by the rather stout, but not very sharply defined, nerve.

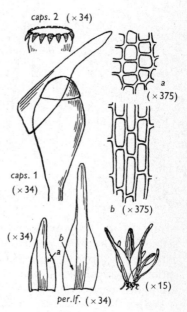

caps. 2 (×34)

a (×375)

caps. 1 (×34)

b (×375)

(×34)

(×15)

per.lf. (×34)

Fig. 22. *Seligeria calcarea*: *caps.* 1, capsule with calyptra; *caps.* 2, mouth of old capsule showing peristome; *per.lf.* perichaetial leaf.

The tip is somewhat obtuse, and without teeth. The cells are short and rather ill defined above, but towards the base of the leaf become clear and shortly rectangular.

S. calcarea, like all the British species of *Seligeria*, is autoecious. The capsule, at the end of a short (up to 2 mm.), straight seta, is minute, erect and symmetrical; it is somewhat pear-shaped when ripe, and has 16 distinct peristome teeth.

A patch of a minute, dark green moss on otherwise bare chalk will be sufficient to make one suspect *Seligeria*, and close examination of the leaves under the microscope should enable one to say whether the specimens are of this genus or an immature state of some other moss.

Ecology. It is a species characteristic of chalk pits in southern England, though extending to other related types of habitat in the north. It is a strict calcicole. *S. calcarea* is essentially a pioneer colonist of the exposed chalk surface.

ADDITIONAL SPECIES

In addition to *S. calcarea* there are six other species of the genus found in Britain, all of very small size. The majority of these are rare.

S. pusilla (Hedw.) B. & S. differs from *S. calcarea* in the rather longer (up to 2 mm.), denticulate leaves, with more acute tips.

S. recurvata (Hedw.) B. & S. is a plant of shaded sandstone rocks; the upper leaves are again as much as 2 mm. long, but when in fruit the species is easily recognized by the seta which becomes strongly curved in moist conditions.

23. **BLINDIA ACUTA** (Hedw.) B. & S.

In the mountainous regions of the north and west *B. acuta* is a not uncommon moss of wet rock ledges and rocks by streams. Occasionally it will grow to 6–8 cm. in height, but is normally much smaller, with rather small leaves (2–3 mm.) which taper to long fine points in the manner of so many of the Dicranaceae. In colour the tufts of *Blindia* are normally a dull or olive green, sometimes with a tinge of reddish brown.

From the above it will be seen that *Blindia* bears some resemblance to a dull-hued, short-leaved *Dicranella heteromalla* but with leaves less bent to one side. Beneath the microscope, however, the leaf is at once distinct because it has well-marked orange-brown patches on either side of the midrib at the leaf base—the inflated, coloured alar cells. These are a good diagnostic character, and even with a lens they can be detected as deep orange-coloured spots. The nerve, which is narrow (contrast the broad nerve of *Campylopus*), occupies the whole of the upper part of the fine leaf tip. The extreme apex is not very acute, and is marked by a few faint teeth. The cells, especially near the leaf base, are very long and narrow, often with greatly thickened walls, giving a rather characteristic areolation.

Although dioecious, *Blindia acuta* is commonly seen in fruit. The capsule is borne on a seta of variable length (3–10 mm.). It is shortly pear-shaped, with a beaked lid, and a peristome of 16 broad red teeth.

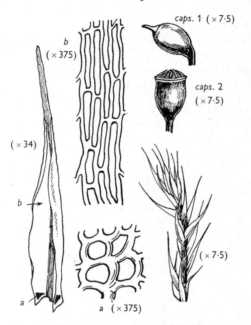

Fig. 23. *Blindia acuta*: *caps.* 1 and 2, capsules with and without a lid.

Ecology. As a mountain plant it grows on very wet rock and peat ledges; also on slabs of rock subject to a constant drip of water, and in tiny streams with a rocky or gravelly bed. Old Red Sandstone rocks and cliffs constitute another, quite different habitat.

24. DICRANELLA SQUARROSA (Starke) Schp.

Deep loose tufts or carpets of moss beside some mountain waterfall or rivulet, conspicuous for their brilliant golden green colour, and drenched with moisture, are likely to be this species, for it is quite common in the hill districts of the north and west. *D. squarrosa* is quite unlike any other species of its genus as regards general habit and leaf characters.

In its tall growth (2–10 cm.) and broad, blunt-tipped, spreading (squarrose) leaves it comes closer to robust forms of *Dichodontium pellucidum*, but its bright or (more rarely) dull yellow-green colour will in most instances identify it at sight.

With so well marked a moss as this the study of microscopic detail is necessary only to confirm an identification already suspected. Thus, the nerve, extending almost to the extreme tip of the leaf, will prevent

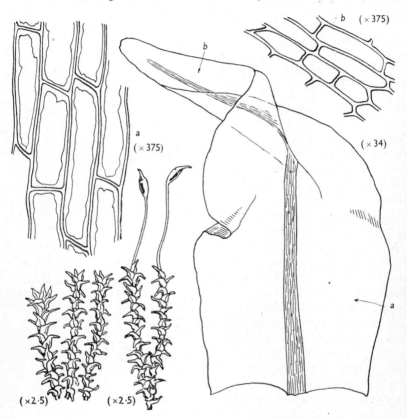

Fig. 24. *Dicranella squarrosa*: the extreme bases of the plants on the left are not shown.

confusion with certain pleurocarpous mosses such as *Rhytidiadelphus squarrosus*, which it resembles in the shape and spreading limb of its leaf. The non-papillose cells, wide and elongate in the broad leaf base, will separate it from the more dully coloured *Dichodontium pellucidum*.

The fruit is rarely seen, but when found it is usually on plants of rather dwarf habit. The fruiting characters show the affinity of *Dicranella squarrosa* with other species of *Dicranella*. The capsule wall is smooth, the seta bright scarlet, stout and 1–1·5 cm. long. Thus the fruiting characters resemble those of *D. varia*, but with all structures three times the size.

Ecology. Highly characteristic of mountain situations where there is a constant trickle of running water, it occurs typically with *Philonotis fontana*, *Bryum pseudotriquetrum* and others. A rocky or sandy substratum is favoured.

25. DICRANELLA VARIA (Hedw.) Schp.

This species is our commonest lowland *Dicranella* with a red seta, but it does not approach *D. heteromalla* in abundance or form such extensive patches. Indeed, when we come upon it on some clayey or chalky bank, it is not a conspicuous moss, and growing commonly to a height of less than 1 cm. it may easily be overlooked. The shoots are much less curved to one side, and the leaves (1–2 mm.) are shorter than those of *D. heteromalla*, but resemble them in their bright green colour and silky texture.

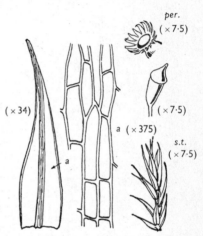

Fig. 25. *Dicranella varia*: *per.* mouth of ripe capsule showing peristome, *s.t.* terminal part of leafy shoot.

The more minute leaf characters require careful attention if this plant is to be distinguished in the barren state from allied species. The leaves, which are lanceolate, curved and tapering, and up to 3 mm. long, lack the sharply saw-edged tip of *D. heteromalla*. Also the nerve occupies about ⅙ instead of nearly ⅓ of the breadth of the leaf base, and the margins of the leaf are narrowly revolute for much of their length. The cells, even in the upper part of the leaf, are long and narrowly rectangular in outline.

Like most species of *Dicranella*, *D. varia* is dioecious, but the delicate male plants are far less likely to attract the eye than are the orange-red setae and capsules of fruiting material. Then, viewed with a lens, this small moss becomes a striking object. Having a red seta it may be confused only with certain rarer species of the genus, and the smooth (not furrowed) surface of the curved capsule separates it from all of these except *D. rufescens*, a species usually well marked by the reddish hue of the leafy shoots themselves.

Ecology. It occurs chiefly on clayey and chalky soils. Thus it will colonize bare clay patches, or exposed slabs of chalk in and about chalk pits. Dr F. Rose notes that in Kent *D. varia* is consistently associated with lime-rich conditions, occurring on formations other than the chalk only where lime has been dumped.

26. DICRANELLA HETEROMALLA (Hedw.) Schp.

The bright green, soft tufts or patches of this very common moss are likely to be met with in a variety of situations; on heathland, in woods or on mountain slopes—but almost always on acid soil. It resembles *Dicranum* spp. in its tapering, finely pointed leaves—all curved to one side; but the small size of *Dicranella heteromalla* will distinguish it at once from most of these. It is more likely to be confused with *Campylopus pyriformis*, or occasionally perhaps with *Blindia acuta*, but when covered with orange capsules, each on a pale yellow seta, the

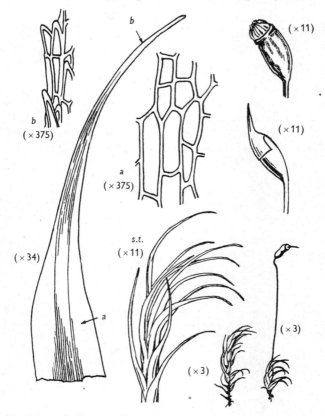

Fig. 26. *Dicranella heteromalla*: *s.t.* shoot tip. The lower capsule shows the calyptra; the upper shows the peristome after removal of the lid.

present plant is not usually hard to identify. It is by far the most abundant of our species of the genus.

The individual leaf is 3–4 mm. long, and usually curved like a sickle. Microscopic characters which help greatly to separate it from other small allied plants are the shape of the leaf base, with rather broad nerve (see fig. 26), the toothed leaf tip and the narrowly rectangular cells throughout. The absence of distinct swollen cells at the basal angles will separate it from small species of *Campylopus*, as also from *Blindia acuta*.

The male organs occur on separate and different-looking plants, which, with their leaves spreading and not falcate-curved, and with the antheridia in a tight spherical mass at the summit, may be taken by the unwary for young fruiting material of a species of *Pleuridium*.

The pale greenish yellow seta is usually 1–1·5 cm. long, and the curved, pear-shaped capsule is almost horizontal when ripe. The old capsule is bright or deep orange in colour, strongly furrowed, with tapering neck and oblique mouth, and red peristome teeth. The absence of a definite swelling on the neck of the capsule serves to distinguish this plant from the only other British species of *Dicranella* with a yellow seta, *D. cerviculata*.

Ecology. *D. heteromalla* favours acid conditions, growing on soils which range from near neutral to markedly acid. It occurs on sandy and gravelly soils, sandstone rock, peat banks and ledges; also on raw humus in most types of woodland.

ADDITIONAL SPECIES

D. cerviculata (Hedw.) Schp., which is a somewhat smaller plant than *D. heteromalla*, is the only other British species with a yellow seta. It is not uncommon on peaty banks and is quite distinct in its capsule, which has a small round pimple-like swelling on its neck.

D. rufescens (Sm.) Schp., the smallest species of the genus, is recognized by its reddish tint, which is accentuated on drying; moreover, the leaf margins are plane (not recurved as in *D. varia*). It is frequent on clay banks.

RELATED SPECIES

Rhabdoweisia denticulata (Brid.) B. & S. is our most frequent species of a genus which, in Britain, comprises three mountain-ledge mosses. It is like a small species of *Weissia*, but has obtuse, toothed leaves and a furrowed capsule with well-developed peristome.

Cynodontium bruntonii (Sm.) B. & S. is the most widespread representative of a genus which contains several rare British mountain mosses. It occurs in rock clefts and is distinguished by its toothed leaves with recurved margins, and capsules not unlike those of *Dicranoweisia cirrata*, but shorter.

27. **DICHODONTIUM PELLUCIDUM** (Hedw.) Schp.

This plant, which occurs most commonly on loose soil beside upland streams, varies so much in habit that it is not always easy to recognize. Tall forms occur which approach *Dicranella squarrosa* in size and appearance, whilst at the other extreme are minute tufts, 1 cm. tall, resembling some small species of *Barbula* or forms of *Ceratodon purpureus*. A useful constant character, however, is provided by the

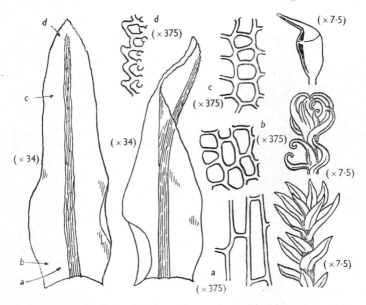

Fig. 27. *Dichodontium pellucidum*: the upper of the two drawings of the tip of the shoot shows its appearance in the dry state.

rather widely spaced leaves on the stem. When dry the leaves are curled, but not tightly so; when wet they are widely spreading and their broad rounded tips are evident. The colour varies, but is usually a dull green. The var. *flavescens* (Turn.) C. Jens., which has often been regarded as a distinct species,* is a plant of taller habit, with longer, narrower leaves.

The leaf under the microscope is at once distinctive in its closely toothed margins, towards the tip. Individual cells, both at the margin and at the back of the nerve, are shaped like the teeth of a saw and project in a characteristic way. Further points to note are the nerve, strong below yet disappearing in the leaf tip, and the rather wavy leaf margin, often infolded or recurved in places. The areolation is shortly

* *Dichodontium flavescens* Lindb. of Dixon, *Handbook*.

rectangular below, quadrate to round above, less clearly defined than in *Ceratodon*, but not so obscure as in most species of *Barbula*.

The fruit of this dioecious moss is not very common. The capsule is shortly ovoid, and curved like that of a *Dicranella*. The lid has a prominent beak and the seta is pale yellow, about 1 cm. long.

Ecology. It grows principally on soft rock, gravel or sand in or by streams. Thus it is normally a colonist of relatively unstable bank habitats, chiefly in mountain districts but by no means confined to them.

28. DICRANOWEISIA CIRRATA (Hedw.) Lindb.

The neat round cushions or extensive patches of this moss are among the first bryophytes to attract attention on the trunks or larger branches of trees in lowland woodlands. Of a deep, slightly yellowish green colour, with leaves much crisped when dry, and with an abundance of pale brown capsules on yellow-green setae, it is not usually a difficult species to identify. The species of *Ulota*, so common on trees in the north and west, are distinct in their hairy calyptra, whereas *Weissia* and *Barbula* spp., which may approach the present plant in habit and general appearance, do not commonly grow on trees. *Dicranoweisia cirrata*, however, is not strictly confined to trees (see below).

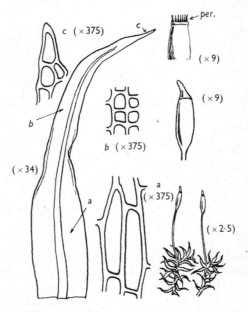

Fig. 28. *Dicranoweisia cirrata*: *per.* peristome; the lower capsule shows the lid intact.

Individual plants are about 1 cm. tall, with wavy, channelled, lanceo-late leaves 1·5–2 mm. in length. Microscopic leaf characters to note are the recurved margins (especially in the middle of the leaf), the well-defined, non-papillose areolation, elongate in the leaf base, short towards the tip, and the absence of all distinctive features such as leaf teeth, unusual breadth of nerve or prominently enlarged alar cells.

D. cirrata is autoecious, and capsules are normally plentiful. In late winter the capsule is seen at its best, bright green ripening to pale brown, symmetrical and ovoid-cylindrical. It is borne erect on a pale yellow-green seta 0·5–1 cm. long, and has the lid drawn out to a longish oblique point. A red rim distinguishes the narrowed mouth, on which the 16 undivided peristome teeth are set.

Ecology. It is most typically an epiphyte on the trunks or branches of trees, where in many lowland districts it is often the principal cushion-forming species associated with *Hypnum cupressiforme* var. *filiforme*. It also occurs on oak palings and gate posts. In the south-east *Dicranoweisia cirrata* is relatively rare on rocks and walls, but in some districts (e.g. in Cornwall) they are its normal habitat. Dr F. Rose has noted it on thatch in Kent.

ADDITIONAL SPECIES

D. crispula (Hedw.) Lindb., a moss of mountain rocks, is the only other British species. The leaves do not have recurved margins and the alar cells are inflated. It is usually black below, and superficially may suggest a species of *Grimmia*.

29. DICRANUM FUSCESCENS Turn.

Although a variable moss in some respects, especially in the micro-scopic details of the leaf and the shape of the capsule, *D. fuscescens* is usually fairly distinct in its dull yellowish green colour and narrow, much-curled leaves. Thus, as one comes upon wide patches of it on some exposed moorland or mountain top it will bear a general re-semblance to *D. scoparium*, but will strike one immediately as a darker plant, with more black about the lower parts and with the upper leaves (if dry) twisted and curled to an extent never seen in the more familiar species. Away from moors and mountains it will not be found. In height (2–8 cm.) and in leaf length (5–7 mm.) this species comes close to *D. scoparium*, but the outline of the leaf is narrower and the apex much finer.

Under the microscope the leaf may be identified by a combination of the following characters: (i) a long, fine, curved tip beset with spiny teeth all round, much as in *D. majus*; (ii) cells in the upper part of the leaf always short and without pores, variable in outline (quadrate, shortly rectangular or with sharp angles), and always to some extent papillose; (iii) a fairly broad nerve (about $\frac{1}{5}$ the width of the leaf base); but as all these characters are subject to variation they require to be considered as a whole; no single point may safely be used as an absolute criterion.

The capsule, which is not rare, is, however, too variable in this species to help much in identification. It is like that of *D. scoparium*, but is usually stouter, darker in colour, and distinctly furrowed, especially when old.

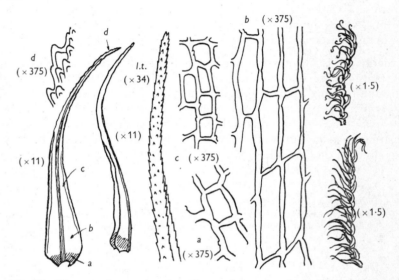

Fig. 29. *Dicranum fuscescens*: alar cells are shown at *a*; *l.t.* leaf tip; the upper of the two shoot drawings shows its appearance when dry.

Ecology. It occurs in various habitats in hilly or mountain districts, on substrata ranging from mildly to strongly acid. It grows typically on peat or rock surfaces, less commonly on rotten wood. *D. fuscescens* ascends to the summits of the highest British mountains where it is often a member of a summit plateau flora of tundra type; but it will also grow at sea-level, at least on the west cost of Scotland, where I have found it associated with *Grimmia maritima* on the rocks.

30. **DICRANUM MAJUS** Turn.

D. majus, which is typically a plant of mountain woods, differs from *D. scoparium* in its looser and taller habit, with the leaves much longer (10–14 mm.) and more strongly and uniformly curved in one direction.

These characters combine to give it a distinct appearance, so that only poorly grown specimens will present any difficulty.

Where some doubt arises in the separation of this species from *D. scoparium*, the leaves may be examined microscopically. The porose areolation and enlarged alar cells are much as in that species; but the

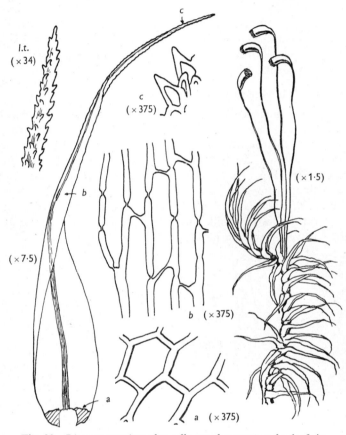

Fig. 30. *Dicranum majus*: alar cells are shown at *a*; *l.t.* leaf tip.

very long, fine leaf tip of *D. majus* is a good character, both in its length and in being sharply toothed almost all round. Thus the teeth are smaller, sharper and more numerous than in *D. scoparium*, so that the leaf tip has a totally different appearance and is, indeed, more like that of *D. fuscescens*.

At most times this is a fine moss, but especially so in the fruiting state, which is by no means rare. The unusual and striking feature here is that any number from 1 to 6 capsules, each on a straw-coloured seta several centimetres long, arise normally from a single 'inflorescence'. Each capsule is curved as in *D. scoparium* but is rather shorter and broader in shape, dark greenish brown, and lightly furrowed. No related British moss has capsules arising in a bunch in this way.

Ecology. It is commonest in the west and north, where it is especially characteristic of *Quercus petraea* woods, but occurs in many types of woodland and mountain-ledge habitat. It is rather uncommon in south-east England, but is found, for example, in the Chiltern beech woods, in Epping Forest, and in the plateau heaths and woods on the Greensand in Kent.

31. DICRANUM BONJEANI De Not.

This moss, though common in many districts, is not generally so abundant as the closely related *D. scoparium*, which it resembles in robust habit and large silky leaves of prevailing yellowish green colour. In habitat it shows certain differences (see below), though the two species overlap. For field recognition of *D. bonjeani* the best character is the distinct transverse undulation of the leaf surface, especially towards the tip. This is particularly evident when moist, and, combined with the rather looser mode of growth, softer texture and usually straighter leaves, generally identifies this species quite readily.*

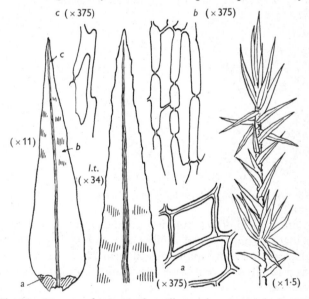

Fig. 31. *Dicranum bonjeani*: alar cells are shown at *a*; *l.t.* leaf tip; on the right is shown the terminal part of a typical leafy shoot.

* But cf. straight-leaved forms of *D. scoparium*.

Under the microscope the leaf resembles that of *D. scoparium* in its 'porose' areolation and enlarged alar cells. But the toothed leaf tip is usually of a rather different shape, remaining wide until quite near the extremity, when it tapers off into a delicate point; and the back of the nerve lacks the rows of sharp teeth found on that of *D. scoparium*. The narrow nerve and porose cell walls will prevent confusion with tall states of *Campylopus flexuosus*, which this species slightly resembles in habit and in its straight leaves.

The capsule is rarely seen. In form it is very like that of *Dicranum scoparium*, but smaller and faintly furrowed longitudinally. The seta is paler and lacks the reddish hue.

Ecology. Found principally, though possibly not exclusively, on soils ranging from basic to mildly acid, it favours open situations; marshland is a frequent habitat, and it also occurs in the turf on chalk grassland.

32. DICRANUM SCOPARIUM Hedw.

D. scoparium is among those mosses which, by their large size and general abundance, will be first to attract the notice of the beginner. In its robust, erect habit it approaches *Polytrichum* spp., but lacks the stiff leaves of that genus. In texture it is soft and glossy, in colour typically a bright yellowish green, though variable. It varies too in the extent to which the shoot tips are curved to one side. Normally this is a feature of *Dicranum scoparium* (though less pronounced than in *D. majus*), but forms occur in which the leaves are almost straight. It forms dense tufts, 2–10 cm. tall, and often covers wide areas in woodland clearings and on heaths.

The stems are sometimes covered with whitish or orange tomentum. The leaves are lanceolate, 4–8 mm. long, and have conspicuous orange-brown patches at their basal angles. The narrow nerve at once separates it from *Campylopus* spp., and the leaf surface is not normally transversely undulate as it is in *Dicranum bonjeani*.

Under the microscope the leaf cells are seen to be elongate-rectangular below, shorter and variable in shape near the apex; the alar cells are large and nearly square in outline. Throughout most of the leaf the cell walls are thick, with thin places at frequent intervals. This 'porose areolation' is not common among British acrocarpous mosses but is found in a few allied species of *Dicranum*. When they are well developed, as they usually are in those forms of *D. scoparium* with undulate leaves, the three or four toothed ridges at the back of the nerve are a good character for separating this species from *D. bonjeani*. The leaf tip (which is variably toothed) is intermediate between the wide flat leaf tip of *D. bonjeani* and the narrow, much longer one of *D. majus*.

D. scoparium is dioecious, and will very commonly be found barren; but capsules are not rare. In fruit this moss is striking, with its reddish yellow seta, 3–4 cm. long, bearing a curved, cylindrical capsule with a long, obliquely beaked lid.

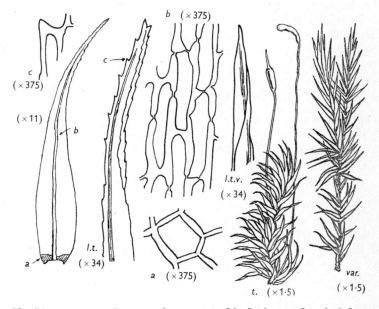

Fig. 32. *Dicranum scoparium*: *t*. and *var*., part of leafy shoot, of typical form and of a straight-leaved variety, respectively; *l.t.* leaf tip in side view, showing one of the rows of teeth at the back of the nerve of typical *D. scoparium*; *l.t.v.* leaf tip of straight-leaved variety lacking teeth.

Ecology. It appears to be tolerant of a wide range of conditions, from strongly acid to around neutral. Thus it occurs on acid raw humus in woods, on the bases of trees and, most abundantly perhaps, on heathland. But forms also occur in marshes, in the chalk hill flora, and on mountain ledges. It is quite likely that distinct ecotypes are involved, a point which would merit investigation.

ADDITIONAL SPECIES

Of nearly twenty species of *Dicranum* found in Britain many are rare. Besides those described, there are two that are most likely to be met with.

D. falcatum Hedw. is a small species (1·5–4 cm. tall) marked by sickle-shaped leaves that are very regularly turned to one side; it is found on detritus and in rock crevices, mainly on the higher mountains of Scotland.

D. scottianum Turn. occurs on rocks in woods, chiefly at low altitudes and mainly in western districts; it resembles *D. fuscescens* but the leaf apex is almost without teeth.

33. CAMPYLOPUS PYRIFORMIS (Schultz) Brid.

Although structurally very closely allied to *C. flexuosus*, this species is in size comparable with *Dicranella heteromalla* and *Leptobryum pyriforme*. It is a small, fine-leaved moss of peaty ground, growing commonly to a height of 1–2 cm. It is often found in association with *Dicranella heteromalla*, but differs in its straight or wavy, not strongly curved leaves, and in forming flat carpets rather than tufts. It also

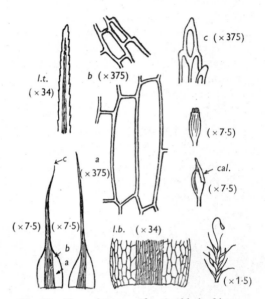

Fig. 33. *Campylopus pyriformis*: *l.b.* leaf base, *l.t.* leaf tip, *cal.* calyptra.

shows a very characteristic tendency to break off shoots and leaf tips; such light green fragments are often plentiful on the surface of the dark green carpets of the barren moss, and serve as a means of propagation.

The leaves, which taper from broad bases to long fine points, are those of the genus, and they are marked by the same broad nerve that is found in *Campylopus flexuosus*. Indeed, small forms of *C. flexuosus* are with difficulty separated from it. The leaf of *C. pyriformis*, however, tends to be more suddenly narrowed above the broad base, to form the long, fine, channelled upper part; and the cells in the base tend to be wider, thinner-walled and more hyaline than in *C. flexuosus*. The best confirmatory character will be found in the extreme leaf base, which lacks the very pronounced swollen cells of *C. flexuosus*; instead, the slightly altered cells of the extreme base pass over rather gradually into those just above them, and obvious auricles are not formed. Moreover, in *C. pyriformis* the radicles which form a loose tomentum are more or less confined to the base of the plant, and hence are not seen when leaves are detached from farther up the stem.

The capsule is smaller than that of *C. flexuosus*, ovoid to pear-shaped and pale-coloured; and the seta is strongly curved like a swan's neck (cygneous), at least at a fairly young stage in development.

Ecology. The habitats favoured agree rather closely with those of *C. flexuosus*, but probably show less latitude. By far the most characteristic are peat banks and ledges, or, indeed, wherever exposed patches of peaty soil occur, even in the comparatively deep shade of pine plantations. It is equally prevalent on lowland heaths and upland peat cuttings. *C. pyriformis* also occurs on raw humus in woods, e.g. on old tree stumps in a late stage of decay.

34. CAMPYLOPUS FLEXUOSUS (Hedw.) Brid.

The specific epithet 'flexuosus' derives from the wavy outlines of the leaves of this moss in the dry state. When moist the shoot tips usually have something of the straight, sharply pointed character seen so well in larger species of *Campylopus*, such as *C. atrovirens*. The present

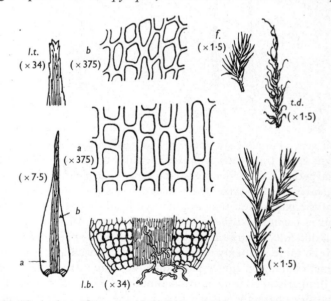

Fig. 34. *Campylopus flexuosus*: *l.b.* leaf base showing alar cells and radicles; *l.t.* leaf tip; *f.*, *t.* and *t.d.* shoots of dwarf form, typical form (moist) and typical form (dry) respectively.

species is, however, only moderately robust, 2–4 cm. being a common height, although not infrequently tufts may grow twice as tall or in other states be only half this size. Commonly bright mid-green above, and warm brown below, it is unfortunately variable in colour too. In general proportions, e.g. leaf length (4–5 mm.) and breadth, it is intermediate between the robust *Dicranum scoparium* and the delicate *Dicranella heteromalla*. It occurs in a wide range of habitats, including wall-tops and boulders, open moors, marshes and peaty woods.

Under the microscope the broad nerve ($\frac{1}{3}$ to $\frac{1}{2}$ of the total width of the leaf base) confirms the genus. The presence of rather abundant tomentum on the stems and the greatly enlarged, hyaline or reddish brown cells at the basal angles of the leaves indicate *Campylopus flexuosus*. These characters are of greater diagnostic value than the areolation—elongated below, shorter and somewhat oval above—or the leaf tip, which is generally marked by a few obscure teeth. In typical *C. flexuosus* the transition from the normal green cells to the swollen basal cells is very sharp, and since these inflated cells and the associated tomentum are commonly coloured orange-brown, this feature may be apparent even from the naked-eye examination.

Like all the other British species *C. flexuosus* is dioecious, but capsules are commoner here than in any other species of the genus except *C. pyriformis*; the capsule is ellipsoidal and borne on a curved seta.

Ecology. As indicated above, the precise habitat is very varied, but *C. flexuosus* probably always grows on substrata which are acid in reaction, commonly strongly so. When growing on rock habitats it requires a covering of humus or peaty material; thus soil-capped wall-tops in mountain districts are favoured; it is less usual than *C. pyriformis* as a colonist of bare peat, where associated species are *Pohlia nutans* and *Polytrichum* species.

35. CAMPYLOPUS ATROVIRENS De Not.

The darker forms of this moss appear as patches or wide mats of 'black velvet' on the wetter parts of many upland habitats. Sometimes a wide expanse of moor or mountain plateau will be splashed with these dark patches, and in such conditions *C. atrovirens* is a species which, once known, is not readily forgotten. Of a characteristic blackish hue and with its very long, fine leaves ending in silvery white hair-points, it can scarcely be mistaken for any other species. At times, however, the blackness is replaced by olive or yellowish green, and then a more thorough examination may be necessary to separate it from some other less common species of the same genus. It is usually 5–10 cm. tall, but may grow much taller. Frequently the white hair-points of the leaves will appear at first sight to be wanting. This is because they tend to fall off readily and are also rather undeveloped on the youngest leaves. In shape lanceolate, straight, with broad bases and long fine points, the leaves are typical of the genus. In length they are commonly 6–9 mm.

Under the microscope the very broad nerve is seen to occupy fully half the width of the leaf base. Inflated alar cells form well-marked auricles, and the hair-point at the leaf tip is strongly and regularly

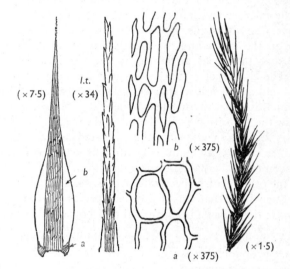

Fig. 35. *Campylopus atrovirens*: alar cells are shown at *a*; *l.t.* leaf tip.

toothed. The areolation is long and narrow. These elongated cells, even in the upper part of the leaf, taken together with the hair-point and the very large inflated auricles, will distinguish *C. atrovirens* from any other species of *Campylopus*.

The fruiting state of this dioecious moss is unknown in Britain.

Ecology. A plant of oceanic type of distribution, it is plentiful in many parts of north and west Britain, but rare elsewhere. It is probably confined to acid substrata, though these may be peat or rock. Moisture seems to be essential; thus dripping rock faces are favoured, and very wet ground on moors, where it is often closely associated with *Sphagnum* spp.

ADDITIONAL SPECIES

Several additional species are not uncommon.

Campylopus brevipilus B. & S. is distinguished from *C. atrovirens* by its narrower nerve and shorter hyaline leaf tip.

Plate III. *Fissidens adianthoides* (natural size)

Plate IV. *Leucobryum glaucum* (natural size)

C. fragilis (Turn.) B. & S. is a variable plant, allied to *C. pyriformis*, but with a still broader nerve and a narrower leaf base.

C. schwarzii Schp. is a tall golden-green plant of mountain rocks, distinguished from pale forms of *C. atrovirens* by short cells and absence of hyaline leaf tip.

RELATED SPECIES

Dicranodontium denudatum (Brid.) E. G. Britton (*D. longirostre* (Starke) B. & S.) is a plant of rotting wood and the humus-rich forest floor in mountain districts. It resembles some forms of *Campylopus flexuosus*, with which the leaf agrees too in its broad nerve and inflated auricles. In *Dicranodontium denudatum*, however, the leaf is commonly curved (falcate) and narrows abruptly just above the base to form a very long bristle-like point.

36. LEUCOBRYUM GLAUCUM (Hedw.) Schp. (Plate IV)

L. glaucum is a moss that can be recognized, not merely with the naked eye, but at a distance; it is so completely unlike any other British species. It combines the habit of a robust 'Dicranoid' moss with a porose, absorbent leaf structure like that of a *Sphagnum*. The dense cushions which it forms vary from a few centimetres to nearly 1 m. across, whilst the height of the plants varies from less than 3 cm. to over 10 cm. Commonly there is a massive semi-decayed whitish lower part to the shoot, which ends in a short glaucous green tip. Even when moist these shoot tips are a comparatively pale, almost greyish, green; when dry they become distinctly white-looking, and it is this whiteness that makes for such ready recognition. The wide cushions or mats of this moss will often be found covering quite extensive areas of moorland or forest

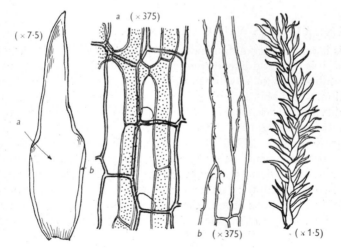

Fig. 36. *Leucobryum glaucum*: hyaline cells (clear) and green cells (stippled) can be seen at *a*.

floor, and, on examination, it will be noticed that the moss tufts are only very loosely attached to the humus or peat on which they are growing. Whole cushions readily become detached, and may continue to live lying free on the surface of the ground. If turned over, they will, in time, form almost perfect 'moss balls'.

The lanceolate leaves vary in length, the longest reaching 8–10 mm.; they are broad at the base and taper to a not very fine point. Under the microscope the exceedingly wide nerve is evident. It occupies the greater part of the breadth of the leaf base, and is flanked by only some 6–8 rows of narrow, elongated hyaline cells. The structure of this very broad, thickened nerve is complex, a layer of green photosynthetic cells being sandwiched between two layers of wide hyaline cells with porose walls.

Leucobryum glaucum is dioecious, and the rather small, curved capsule is rare. However, a form of vegetative reproductive unit occurs. This consists of a dense cluster of rhizoids, appearing dull purplish in hue, and occupying the centre of a terminal rosette of leaves. From such fragments new plants arise.

Ecology. It has two principal habitats: (i) on the woodland floor, under beech, oak, or conifers, where dense readily detachable cushions are formed on relatively raw leaf mould; (ii) wet moorland situations, on peaty slopes and in bogs, intermixed with other vegetation. Both habitats are alike in being strongly acid in reaction. In wet districts it is not uncommon on trees.

POTTIALES

37. ENCALYPTA VULGARIS Hedw.

This is our only common species of *Encalypta* which regularly fruits. Indeed, it is the hood-like, extinguisher-shaped calyptra of the capsule— quite unlike that of any other British genus—which is apt to attract attention. In habit (forming loose tufts or cushions) and in their partiality for calcareous walls *E. vulgaris* and *E. streptocarpa* are alike. When dry the leaves have the same curled appearance, but in *E. vulgaris* they are somewhat shorter (about 3 mm. long) and the whole plant is less robust, the stems being generally under 1·5 cm. tall. Both leaf shape and areolation are very much those of *E. streptocarpa*, but with these two distinctions: (i) the nerve here is commonly, though not invariably continuous beyond the leaf apex, running out to form a short protruding point, and (ii) the tooth-like papillae, which make the back of the nerve rough in *E. streptocarpa*, are here very few or absent altogether. In the wide, hyaline basal cells and in the short upper cells with 'hobnail' papillae the two species agree very closely.

An autoecious moss, *E. vulgaris* is often found abundantly fertile. Each capsule is borne erect on a red seta usually about 5 mm. in length, and is quite enveloped by the symmetrical hood-like calyptra, which is itself about as long as the seta. The lid of the capsule ends in a beak more than half as long as the body of the capsule itself. At maturity the capsule wall remains smooth, never becoming strongly furrowed as in some of the rarer species of the genus. The peristome is ill developed, often lacking altogether.

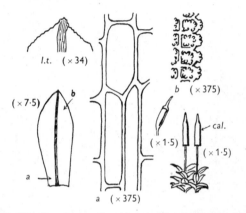

Fig. 37. *Encalypta vulgaris*: *cal.* calyptra enveloping capsule, *l.t.* leaf tip.

Ecology. It occurs chiefly on soil-capped ledges and crevices in limestone walls; also on chalky banks. It is probably absent from all non-calcareous habitats.

38. ENCALYPTA STREPTOCARPA Hedw.

This moss forms loose tufts or cushions, vivid green when moist, on calcareous walls and rocks, where it gives the impression of a robust species of *Tortula* without a hair-point to the leaf. The stems are commonly less than 3 cm. tall, but the obtuse, tongue-shaped leaves are large (3–6 mm. long) and tend to form open rosettes at the ends of the crowded shoots. Thus, when moist, it becomes quite a conspicuous moss. In the dry state the leaves are twisted and curled (cf. species of *Tortella* and *Trichostomum*), and then the nerve forms a prominent pale rib along the back of the leaf.

Encalypta streptocarpa is not usually difficult to distinguish in the field; in any case it is readily identified by the microscopic characters of the leaf. The areolation is highly characteristic. At the base are wide, oblong, hyaline cells, which pass over abruptly to the short chlorophyllous cells that occupy the greater part of the leaf. If some of these cells be examined at the leaf margin they will be seen to be studded with peculiar hobnail-like papillae, whilst the back of the nerve is rough with prominent papillae of a more tooth-like character. These features, taken in conjunction with the robust habit and the nerve which vanishes just below the leaf tip, should make identification simple.

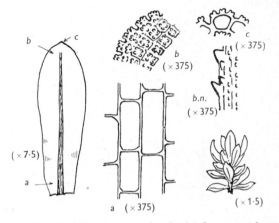

Fig. 38. *Encalypta streptocarpa*: *b.n.* back of nerve; at bottom right is shown the terminal part of a shoot.

Only among other species of *Encalypta* and some species of *Tortula* will the peculiar hobnail-like papillae be found, and none of these has the large obtuse leaf and vanishing nerve of *Encalypta streptocarpa*.

This moss is dioecious, the capsule being exceedingly rare. It is, however, the spiral striations on the capsule wall that give the species its name.

Ecology. It is characteristic of mortar-filled chinks in walls; indeed, it often becomes the dominant moss in this calcareous habitat—even in a district otherwise non-calcareous. It also occurs on shaded limestone rocks and on sheltered grassy slopes on the chalk, and is an important colonist of limestone screes.

ADDITIONAL SPECIES

E. ciliata Hedw. is a rather rare plant of base-rich alpine rock ledges. It is quite distinct from *E. vulgaris* in the fringed (ciliate) base of the calyptra, and paler seta.

39. TORTULA RURALIS (Hedw.) Crome

Generally distributed though far less abundant than *T. muralis*, this species is distinguished in the field by its more robust habit (stems 1·5–8 cm. tall) and by its colour, which is a rather bright golden-green above and reddish brown below. It forms irregular swollen cushions on old walls, overgrown roof-tops, and among boulders. Thus the

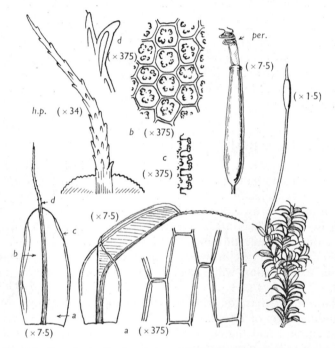

Fig. 39. *Tortula ruralis*: *h.p.* hair-point of leaf; the enlarged capsule has the lid removed showing the peristome (*per.*).

habitat is distinct from that of *T. laevipila*, whilst the tall growth and long hair-point will prevent confusion with *T. subulata*. *T. ruralis* is the most robust of our common species of *Tortula*, and the broad, obtuse, recurved leaves are densely arranged on the stems.

Under the microscope the best diagnostic feature of the leaf lies in the hair-point, which in this species is covered with small spiny out-growths. This feature immediately separates *T. ruralis* from all except the two following plants and one much rarer species. The strong red-brown nerve is a further point of difference from *T. laevipila*. The areolation is of the usual *Tortula* type, but the papillose character of the cells is especially well marked on the back of this leaf, towards the apex, where the papillae approach the 'hobnail' appearance of those found in *Encalypta* species. The margin of the leaf is recurved from just above the base to near the apex.

Although *Tortula ruralis* is dioecious, and most commonly barren, the fruit is not rare. The strong reddish seta may reach 2·5 cm. in length, the slightly curved, narrowly cylindrical capsule 6 mm. In capsule shape, however, as also in its long twisted peristome, tubular at the base, *T. ruralis* comes very close to related species of *Tortula*.

Ecology. It is notably a plant of old roofs, especially those that are thatched. It also occurs on stony ground and wall-tops, chiefly where these are calcareous; thus in many limestone districts it becomes a locally dominant moss. More rarely *T. ruralis* grows with *T. ruraliformis* on sand dunes.

40. **TORTULA RURALIFORMIS** (Besch.) Dix.

This plant is one of the most characteristic mosses of sand dunes, where it forms, in places, very extensive patches of loose growth. These patches are conspicuous by their colour, dull brown when dry, golden-green or orange-brown when moistened by rain. The leaves are strongly squarrose-recurved, as in *T. ruralis*, and crowded at the shoot tips, which often alone protrude through the loose sand. When dry, with leaves curled and twisted, the brownish plants may not attract attention, but a closer examination will reveal the long silvery hair-points of the leaves.

T. ruraliformis has sometimes been regarded as scarcely more than a sand-dune form of *T. ruralis*. Indeed, in structure it comes very close to that species, and the colour difference sometimes quoted is not in itself reliable. However, not only does *T. ruralis* have a different habitat range and a rather denser cushion-forming habit, but there is one microscopic character that distinguishes the leaf of *T. ruraliformis* with certainty. The leaf tip, instead of being blunt and rounded, tapers gradually into the hair-point, which thus has a broad insertion quite different from that of the hair-point in *T. ruralis* and all other British species of *Tortula*. One is reminded rather of the broad hyaline leaf apex that is found in some species of *Rhacomitrium*.

The fruit, which is not common, closely resembles that of *Tortula ruralis*.

Ecology. Its typical habitat is unstable ('yellow') coastal sand dunes, where it often plays the pioneer role in places where fresh sand has accumulated. It may become the dominant bryophyte in such places, or at least locally abundant. It continues to play a part in more stable parts of dune systems, but disappears as the sand becomes acid and the cover of vegetation more continuous. *T. ruraliformis* is found all round our coasts where sand dunes occur, and is also known from sea shingle and from some inland habitats, such as the calcareous sand of the East Anglian Breckland.

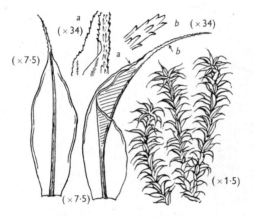

Fig. 40. *Tortula ruraliformis.*

41. TORTULA INTERMEDIA (Brid.) Berk.

Intermediate in many of its characters between *T. laevipila* and *T. ruralis*, this species is a not uncommon plant of wall-tops and similar habitats. It is usually marked by its dull green or golden-brown colour. Normally it is less robust than *T. ruralis* (1·5–4 cm.), and the leaves are almost straight and scarcely concave, not strongly squarrose-recurved and deeply concave as in that species. This gives the plant quite a different appearance, by which (combined with its duller, browner colour) it may usually be distinguished fairly readily.

The microscopic features of the leaf are in general agreement with those of *T. ruralis*, but the leaf margin is less strongly recurved (and only in the lower half of the leaf), and the hair-point is less spiny; it is, indeed, intermediate between the smooth hair of *T. laevipila* and the very long, coarsely spiny hair of *T. ruralis*. The small papillose cells of the upper part of the leaf are smaller and less distinct than in *T. ruralis* (9–10μ against 12–16μ), and the papillae are less prominent than in that species. The marginal band of clear cells found in *T. laevipila* is lacking here.

Like *T. ruralis*, this species is dioecious. The capsule is about 4 mm. long, the seta 1–1·5 cm.—both somewhat shorter than in that plant.

Ecology. Favouring limestone walls and rocks, it overlaps in its range the closely related *T. ruralis*. It grows more rarely on soil. It is probably not exclusively calcicole, though certainly most common in calcareous districts. *T. intermedia* is comparatively rare in many parts of the north and west.

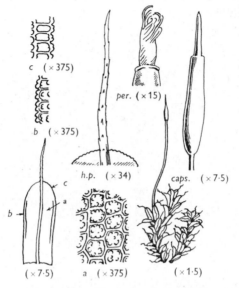

c (×375)

b (×375)

per. (×15)

h.p. (×34)

caps. (×7·5)

c

a

b

(×7·5)

a (×375)

(×1·5)

Fig. 41. *Tortula intermedia*: *h.p.* hair-point of leaf, *per.* peristome, *caps.* capsule with lid still intact.

42. TORTULA LAEVIPILA (Brid.) Schwaegr.

This species is akin to *T. subulata*, but is at once distinguishable from it by the long smooth hair-point to the leaf. Moreover, it will be found most frequently on the trunks or branches of trees. It differs from that species, too, in its rather taller stems, tufted habit, and leaves which become more strongly curled and twisted on drying. It is also closely related to *T. ruralis* and *T. intermedia*, but neither of these grows on trees.

The leaf of *T. laevipila* under the microscope is seen to conform to the usual *Tortula* shape, whilst the areolation, elongate below, short and papillose above, does not differ markedly from that in related species. In the upper part of the leaf, however, a few rows of clear but not elongate cells form a fairly definite border. The leaf margin is only very slightly recurved, about mid-leaf, a useful diagnostic character separating it both from the smaller *T. muralis* and from the more robust species such as *T. ruralis*. In the smooth or only faintly denticulate hair-point it differs strongly from *T. intermedia* and *T. ruralis*.

This species is autoecious and capsules typical of the genus (long and narrowly cylindrical) are common. The seta is not very long (8–15 mm.) and is yellowish in colour. The lid of the capsule is long-beaked; the peristome teeth are united at their base into a tube, which is, however, not so long as in *T. subulata*.

Ecology. It is an epiphyte on the trunks and larger branches of trees, typically in rather dry sites. Professor Richards considers it to be most frequent on ash, poplar and elm; but it is regularly present in the exceptionally rich epiphytic flora of old elders, and is recorded from many other trees.

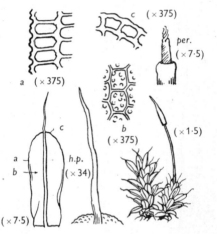

Fig. 42. *Tortula laevipila*: *h.p.* hair-point, *per.* mouth of capsule showing peristome.

43. **TORTULA SUBULATA** Hedw.

A not uncommon plant of sandy banks and rock ledges, *T. subulata* is remarkable in combining an extremely short, erect, leafy shoot (about 5 mm. tall) with a long, strong seta and very long capsule. It is thus usually the capsules which attract attention, and only a closer inspection reveals the rosettes of bluntly oblong leaves. In the dry state the leaves become somewhat curled and twisted and the strong yellowish nerve is prominent. The leaves, 3–6 mm. in length, are tongue-shaped like those of *T. muralis*, being usually broadest above the middle, but the nerve projects only as a very short point beyond the blunt leaf apex.

A number of long, narrow cells form an irregular border to the leaf. This border is several cells wide near the leaf base, but disappears altogether towards the apex. Otherwise, no special diagnostic value attaches to the areolation, elongate and well defined in the leaf base, short, obscure and somewhat papillose above. There are a few obscure teeth near the leaf tip.

This plant is autoecious and the capsules are common. The fully grown, dark red seta may reach 2–3 cm. in length, and supports a narrowly cylindrical, slightly curved capsule, itself up to 9 mm. long. The lid has a distinct beak. The peristome teeth are united at the base into a spirally striated tube. This feature is characteristic of one section of the genus *Tortula*.

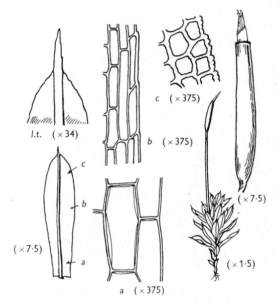

Fig. 43. *Tortula subulata*: *l.t.* tip of leaf; the capsule on the right has the lid removed and shows the peristome.

Ecology. It occurs chiefly on steep banks, wall-tops and rocky ledges, but grows always on soil rather than on bare rock. It is possible that a light or loose soil texture is especially favoured; thus many observers have noted its fondness for sandy soils, but it will also grow on well-drained loam and chalky soil.

44. TORTULA MURALIS Hedw.

One of the commonest mosses of walls in almost all parts of the country (even in large towns), *T. muralis* is usually quickly recognized by the combination of long, upright, cylindrical capsule and hair-pointed leaves. Many species of *Barbula* have not dissimilar capsules but lack the hyaline points to the leaves. *Grimmia* species, most of which have hair-pointed leaves, have shortly ovoid capsules. *Tortula muralis* is likely to be confused only with certain other species of *Tortula*, none of which approaches it in abundance. The plants typically form neat cushions (less than 1 cm. tall) which when dry appear grey with the hair-points of the curled and twisted leaves. When moist the leaves tend to form neat rosettes; but were it not almost always fertile *T. muralis* would not be a conspicuous plant.

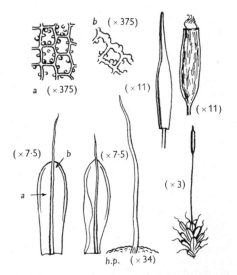

Fig. 44. *Tortula muralis*: *h.p.* hair-point; two enlarged capsules are shown, left, unripe with calyptra intact, right, ripe showing peristome.

The broadly tongue-shaped leaf, with hair-point projecting far beyond the rounded tip, is found in many species of *Tortula*. Elsewhere, only *Bryum capillare* approaches this leaf character, and that is a far more robust plant, with a totally different cell structure. The best microscopic characters for distinguishing *Tortula muralis* from other species of *Tortula* are the recurved leaf margins and the absence of sharp teeth or spines on the hair-point. In cell character, elongate near the leaf base, short, obscure and papillose above, it resembles many related mosses.

This moss is autoecious, and the capsules are normally produced in abundance. Narrowly cylindrical in shape, the capsule is borne on a seta 1–2 cm. long; yellowish at first, the seta turns purplish red with age. There are 32 twisted peristome teeth, which are united at the base in a very short tubular membrane.

Ecology. Its principal habitats are brick and stone walls. It is one of the few mosses that are common in towns. It is less common in 'natural' habitats such as limestone and sandstone rocks. On wall-tops *T. muralis* is very commonly associated with *Grimmia pulvinata*, often also with *G. apocarpa* and *Bryum capillare*.

ADDITIONAL SPECIES

Tortula marginata (B. & S.) Spruce occurs on sandstone rocks and walls, and is distinguished from *T. muralis* by its much shorter, not hyaline, leaf point and by the presence in the leaf of a well-defined marginal band consisting of several rows of elongated cells. The longest leaves reach 2·5 mm.

T. latifolia (Bruch) Hartm. (*T. mutica* (Schultz) Lindb.) is known at once by its bluntly rounded leaf apex, with no trace of hair-point. It grows on tree roots liable to flooding by rivers.

T. papillosa Wils. ex Spruce is a species occurring not uncommonly on tree trunks. It comes near to *T. laevipila* but may be known by the long and prominent papillae which cover the back of the nerve, also by the numerous rounded multicellular gemmae which are usually present on the younger leaves.

RELATED SPECIES

The genus *Aloina* includes several species formerly included in *Tortula*, but differing in the inrolled margins of the fleshy-looking leaves, and broad nerve rendered thick and indistinct by upgrowths of tissue which remind one of the lamellae of the genus *Polytrichum*.

Aloina rigida (Hedw.) Kindb. (*Tortula rigida* (Hedw.) Schrad.), which occurs on mud-capped calcareous walls, has the leaves very short (usually 1·5–2 mm.) and rounded, and the capsule ovoid, with a beaked lid.

Aloina ambigua (B. & S.) Limpr. (*Tortula ambigua* B. & S.), a rather commoner plant of similar habitats, has the leaves distinctly longer (up to 3 mm.) and less rounded at the apex, and the capsule is long, straight and cylindrical.

Aloina aloides (Schultz) Kindb. (*Tortula aloides* (Schultz) De Not.), which may be found in similar places, has the leaves still more acute, and the long narrow capsule held at a slight angle.

Desmatodon convolutus (Brid.) Grout (*Tortula atrovirens* (Sm.) Lindb.) is found chiefly in bare places by the sea. It approaches *Pottia lanceolata* in its oblong leaves with nerve shortly excurrent, and in the shape of its short capsule; but in the leaf the strongly revolute margin and thick nerve are characteristic.

Pterygoneurum ovatum (Hedw.) Dix. (*Tortula pusilla* (Hedw.) Mitt.) is another small moss which resembles *Pottia lanceolata* in the form of its capsule, but the stem is very short and the leaves concave. The leaf is remarkable in bearing lamellae on the upper part of the nerve, which itself protrudes beyond the leaf tip in a longish point. From the lamellae arise groups of chlorophyllous cells which make the leaf opaque.

45. **POTTIA LANCEOLATA** (Hedw.) C.M.

This species forms to some extent a link between the small species of *Pottia* and the related genus *Tortula*. It grows in small, rather dense tufts, with stems 2–5 mm. high; compared with *Pottia truncata* it is rather more a plant of wall crevices, although also occurring on bare ground. The leaf, which is 2–3 mm. long, is of the same general shape as in most other species of *Pottia*, broad, tapering slightly at the apex, with the nerve running out into a point of variable length.

The leaves have recurved margins, and the upper cells are always to some extent papillose (cf. *P. truncata*). These rounded hexagonal or very shortly rectangular upper cells commonly measure 15–18 μ across.

As in other autoecious species of *Pottia*, the capsules are often produced in abundance. The orange-red seta (4–9 mm. long), the ovoid

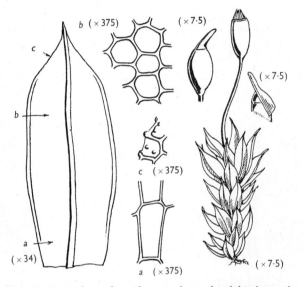

Fig. 45. *Pottia lanceolata*: the capsule on the right shows the calyptra intact, that on the left shows the lid, and that in the centre the peristome.

to ellipsoidal capsule, and the well-developed peristome are not matched in any other common British species of *Pottia*. When young, the long-beaked lid and long, asymmetrical, smooth calyptra remind one of *P. truncata*, though the body of the capsule is much longer than in that species; when older, the long peristome teeth are seen to be erect or spreading, not twisted as in *Tortula* species; the very old, empty capsules finally become almost purplish brown in colour and somewhat contracted at the mouth. The spores are only very slightly rough with minute, scarcely raised papillae, and measure 20–24 μ.

Ecology. It colonizes soil on banks or wall-tops, and is strongly, but not exclusively, calcicole. Thus, it occurs with *Encalypta vulgaris*, *Aloina* spp. and other mosses on old, soil-capped limestone walls. In

Berwickshire Mr J. B. Duncan has recorded it chiefly from calcareous ground on the coast whilst in Kent Dr F. Rose has noted it on ant-hills. *Pottia lanceolata* is absent from many parts of west and north-west Britain.

46. POTTIA TRUNCATA (Hedw.) Fürnr. (*P. truncatula* Lindb.)

This is our commonest species of a rather large genus, most members of which are small plants growing on patches of bare soil, in fallow fields or gardens. They are apt to be overlooked owing to their small size, but *P. truncata* is a plant which, once known, will be found to be neither

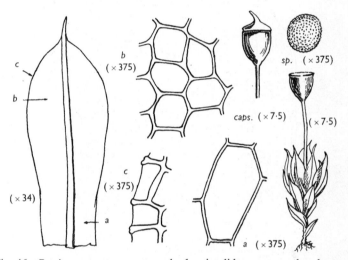

Fig. 46. *Pottia truncata*: *caps.* capsule showing lid, *sp.* spore; the plant on the right shows an old capsule from which the lid has fallen.

uncommon nor difficult to recognize. The upright, clustered or scattered stems are only 3–5 mm. tall, but as *P. truncata* is usually fertile, the numerous short, very wide-mouthed capsules without peristomes will generally serve to identify it.

The broadly oblong leaf is about 2 mm. long and is widest just above the middle; the nerve projects as a short point beyond the leaf apex. In *P. davalliana* the leaves are often only half this length, whilst *P. lanceolata* has a longer capsule, with a peristome.

Under the microscope the leaf margin is seen to be plane or nearly so, not strongly revolute as it is in the other two species described here; it is without teeth, but the cell walls near the apex may project slightly. The cells in the leaf base are elongate-rectangular; above they are hexagonal or shortly rectangular, large (about 20μ across), clearly defined and without papillae (cf. *P. davalliana*).

The capsule is borne erect on a yellowish seta 3–6 mm. long. When young it is very shortly ovoid-cylindrical (up to 1 mm. long), the oblique beak of the lid giving it a characteristic appearance; the asymmetrical calyptra is smooth. When empty and old the capsules are wide-mouthed and almost funnel-shaped. The spores measure 22–30μ across, and are covered with very minute papillae; hence they appear slightly rough, not strongly papillose as in *P. davalliana*.

Ecology. Like *Phascum cuspidatum* it is one of the principal members of a well-defined stubble and garden community of heavy soils, which reaches its peak of development in autumn and winter. Woodland rides are another habitat.

47. POTTIA DAVALLIANA (Sm.) C. Jens. (*P. minutula* (Schleich.) Fürnr.)

Found growing in similar situations to the last, *P. davalliana* is in many respects a diminutive replica of that species, being like it in habit, but only about half the size in all its parts. In addition to its smaller size, however, it has numerous well-defined differences of structure.

The stems are rather less than 3 mm. in height, to which is added a further 3–4 mm. for the seta and capsule; for, like *P. truncata*, this species is commonly very fertile. The leaf is ovate and acute, but similar in general shape to that of *P. truncata*; typically it is only about 1 mm. long.

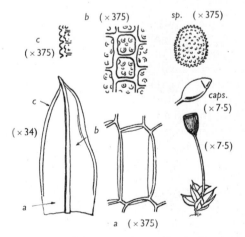

Fig. 47. *Pottia davalliana*: *caps.* capsule in an earlier stage than that shown on the whole plant below it; *sp.* spore.

Under the microscope the nerve of the leaf is seen to protrude into a short point, much as in *P. truncata*; but in *P. davalliana* the margin is recurved for most of its length, and the cells in the upper part of the leaf are markedly papillose. In addition, an average diameter for the upper cells is only 12–14μ, as compared with 20μ in *P. truncata*.

The capsule is short and broad, becoming cup-shaped when empty. It is without peristome teeth. The lid is broadly and bluntly conical, not long-beaked as in the last species, and the calyptra is rough. The spores are about the same size (25–30μ across), but differ in being covered with sharp spine-like papillae.

Ecology. *P. davalliana* has usually been recorded as a colonist of bare soil, chiefly clay, with the implication that it belongs to the same fallow field and garden community as *P. truncata* and *Phascum cuspidatum*. Mr E. A. Schelpe informs me, however, that—unlike these two species— it is, in his experience, definitely calcicole. It sometimes occurs on the soil of pots in greenhouses.

ADDITIONAL SPECIES

Pottia recta (Sm.) Mitt. is a minute cleistocarpous moss of bare ground, somewhat resembling *Phascum curvicollum*, but known by its straight, not curved seta.

Pottia heimii (Hedw.) Fürnr., chiefly a maritime species, is more robust than most members of this genus (up to 8 mm. tall); the leaf is distinguished by its toothed apex, and the stout orange-red seta is 8–14 mm. long.

P. intermedia (Turn.) Fürnr. is closely allied to *P. truncata*, but differs in its taller habit, leaf margins revolute in the middle, and longer capsule (1·5 mm.).

48. PHASCUM CUSPIDATUM Hedw.

This is probably the commonest British cleistocarpous moss. It grows in small patches (2–3 mm. high) on bare soil in gardens and fields, each stem bearing a dense tuft of leaves. The capsules, which are almost spherical, except for a small apical prominence, lie immersed among the crowded leaves of the bud-like shoot tips. Although very dwarf in habit, this moss is altogether more robust than the species of *Ephemerum*, whilst the broad leaves prevent confusion with *Pleuridium* and related cleistocarpous mosses. The light twisting of the leaves when dry affords a good character for distinguishing this species from *Acaulon muticum* (which may further be known by its toothed perichaetial leaves).

A well-developed leaf is about 2 mm. long, very wide in the middle and tapering towards the tip, where the nerve runs out into a point of variable length. The lower leaves are smaller. The cells are relatively large (20μ across) and clearly defined, elongate in the leaf base, shorter above. The margin is recurved in the mid-region of the leaf. Thus in shape, as in microscopic details, the leaf comes near to that of several species of *Pottia*. The upper cells bear prominent papillae, which are well seen when the back of a folded leaf is viewed under high power. *Physcomitrella patens* differs in its toothed leaves and in cell structure.

The antheridia occur on axillary branches. The almost sessile capsule distinguishes *Phascum cuspidatum* from all members of the genus *Pottia*, to which *Phascum* is nearly related. The nearly spherical, bluntly

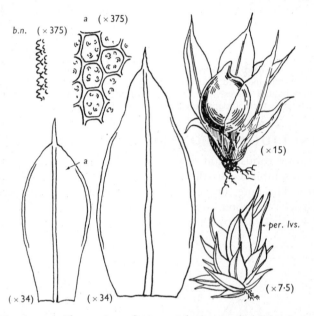

Fig. 48. *Phascum cuspidatum*: two plants are shown on the right, the lower with perichaetial leaves (*per. lvs.*), the upper with a fully developed capsule; the larger leaf (centre) is a perichaetial leaf; *b.n.* back of nerve showing papillae.

apiculate capsule, immersed among the leaves, is like that of other cleistocarpous mosses, such as *Pleuridium* and *Pseudephemerum* spp. Indeed, the combination of such a capsule with the *Pottia* type of leaf will usually serve to identify the present species. The capsule is not completely concealed by the perichaetial leaves, as it tends to be in *Acaulon muticum*.

Ecology. It occurs very regularly as a member of the bryophyte community which develops on fallow soil in field and garden. This community is perhaps most characteristic of a fairly heavy, retentive soil, and is seen at its best during autumn and winter. Besides *Phascum cuspidatum*, it includes species of *Pottia* and other mosses, and liverworts of the genus *Riccia*. *Phascum cuspidatum* is also common in woodland rides and as a colonist of bare soil generally.

ADDITIONAL SPECIES

P. floerkeanum Web. & Mohr differs from *P. cuspidatum* in being still more minute, with shorter leaves (all 1·5 mm. or less) and leaf cells only 15 μ across; the whole plant has a reddish brown colour.

P. curvicollum Hedw. is also somewhat smaller than *P. cuspidatum*, and brownish in colour. It is known at once by the fact that the capsule is borne (almost pendulous) on the end of a curved seta several millimetres long.

RELATED SPECIES

Acaulon muticum (Brid.) C.M. is another fairly common cleistocarpous moss of bare ground. It is bud-like and also very minute; it may be recognized by its jaggedly toothed perichaetial leaves, with nerve protruding beyond the rounded apex, and by the perfectly spherical capsule without an apical prominence.

49. CINCLIDOTUS FONTINALOIDES (Hedw.) P. Beauv.

This rather robust plant of streams and lakes takes its specific name from its resemblance to the still larger aquatic mosses of the genus *Fontinalis*. The long, branched stems, which grow on frequently submerged rocks or wood, are 5–18 cm. in length, and bear rather long (about 4 mm.), narrow, tongue-shaped leaves. When dry the plant

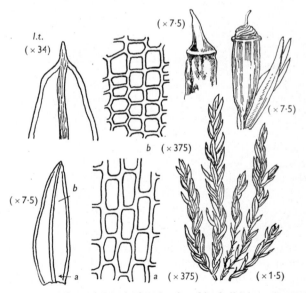

Fig. 49. *Cinclidotus fontinaloides*: *l.t.* tip of leaf; the capsule on the right has the lid removed, showing the peristome.

resembles *Grimmia alpicola* var. *rivularis* and *Orthotrichum rivulare*, the leaves then appearing shrivelled and somewhat twisted; but in the wet state the size and shape of the leaves distinguish it, and under a lens the thickened leaf margin provides a further good character. In colour the plant varies from dull olive to deep blackish green.

Under the microscope the strongly thickened leaf margin, composed of several layers of short cells, and the rather blunt apex beyond which the nerve usually extends as a short point, are easily seen and highly diagnostic. The leaf cells as a whole are distinctive, being almost uniformly short with rather thick walls and dense chloroplast content; nearly square in outline (8–10μ), they become slightly elongated only in the extreme base of the leaf.

Although *Cinclidotus fontinaloides* is dioecious, the capsule is not uncommon. It arises on a lateral branch, on a very short seta, so that it is partly concealed among the leaves. It is yellow-brown, with a reddish rim. The lid is long-beaked; the peristome of 16 slender red teeth is spirally wound when dry.

Ecology. It is common on rock and wood in streams and rivers, both in lowland and hilly districts. It is also often found on boulders around the margins of lakes. It flourishes best perhaps in places where it is submerged periodically rather than continuously.

ADDITIONAL SPECIES

C. mucronatus (Brid.) Moenk. & Loeske (*C. Brebissonii* (Fior. Mazz.) Husnot) is the only other species likely to be met with. It is not uncommon on rocks and tree stumps by rivers and slow streams. It forms short, erect tufts and does not closely resemble *C. fontinaloides* to the naked eye; microscopically it differs in its highly papillose leaf cells. The seta is 6–10 mm. long.

50. **BARBULA CONVOLUTA** Hedw.

This plant—one of our commonest blunt-leaved species of *Barbula*—occurs on bare ground and wall-tops, where at times it will grow in great profusion, the massed effect of the pale yellow setae of fruiting material rendering the moss quite conspicuous. When barren (the commoner condition) it is rather lacking in distinctive features, but its low habit (stems about 1 cm. tall), small leaves (usually less than 1·5 mm. long), and bright yellowish green colour are all fairly characteristic. It is, in fact, the smallest and shortest in the leaf of our common British species of *Barbula*. When dry the leaves are much curled, and may approach the corkscrew spirals of the next species, *B. unguiculata*, but on a smaller scale.

Under the microscope the blunt leaf tip, with short protruding point (apiculus), is a good character. The margin may be slightly recurved near the base; the nerve may extend into the apiculus but normally ceases just below this. In its areolation, obscure and papillose above, clear and elongated towards the base, it agrees with many other species of *Barbula*.

The specific epithet refers to the folded, sheathing perichaetial bracts, which are much larger than the other leaves. These are absolutely diagnostic and may be seen, even with the naked eye, as a shining sheath around the base of the seta. Although this species is dioecious, capsules are not uncommon; the erect capsule has the long twisted peristome found in many species of *Barbula*; but the pale yellow seta (red in *B. unguiculata*) is a good specific character.

Ecology. Soil-filled chinks in walls or similar situations on paved paths or rockeries are common habitats. It will also grow on bare soil (including sand dunes) and on rock ledges. I have found this species and *Ceratodon purpureus* together practically covering a disused hard tennis court of the red 'ash' type. It is probably absent from all extremely acid habitats, but the extent of the calcicole tendency in this and some other species of *Barbula* would be worth investigating. Dr E. W. Jones tells me that in his experience *B. convoluta* is very abundant on garden paths, cinder paths and waste ground in general, provided that the substratum is not poor in mineral nutrients.

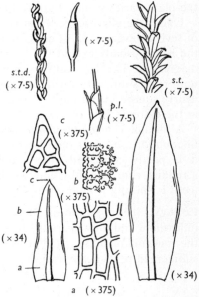

Fig. 50. *Barbula convoluta*: *p.l.* perichaetial leaves surrounding base of seta; *s.t.* and *s.t.d.*, terminal part of shoot in moist and dry conditions respectively. Two leaves are shown to give some idea of the range in size.

51. **BARBULA UNGUICULATA** Hedw.

B. unguiculata is in many respects a considerably more robust replica of *B. convoluta*. Like the latter it grows on wall-tops and waste ground, where it is a very common plant. When dry it can often be recognized at sight by the strong, corkscrew spirals into which the leaves become contorted, the effect being accentuated by the thick and shining yellowish nerve. The upright stems (usually 1–3 cm. tall) are loosely held together to form untidy tufts or patches, somewhat dull green above, brownish below. The leaves are 1·5–2 mm. long.

Under the microscope the oblong blunt-tipped leaf is very different as a rule from that of all other species of *Barbula* except *B. convoluta*. The nerve is extended into a short point beyond the leaf apex. In this feature, as in leaf shape, it comes nearer to some species of *Trichostomum*, such as *T. brachydontium*, than to most others of its own genus. In *Barbula unguiculata*, however, the leaf margins are markedly recurved

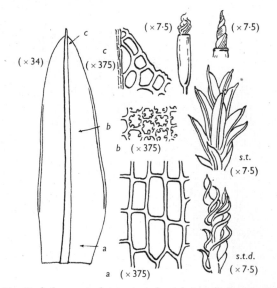

Fig. 51. *Barbula unguiculata*: *s.t.* and *s.t.d.*, terminal part of shoot in moist and dry conditions respectively. At the top right are seen two views of the peristome.

(in *Trichostomum brachydontium* they are plane), and the elongated cells of the leaf base pass over rather gradually into the short, obscure, papillose upper areolation.

In the characters of the fertile plant this dioecious species is very different from *Barbula convoluta*. Thus, the perichaetial bracts are not folded together, and the seta is purplish red, not yellow. The peristome teeth are very long and spirally twisted.

Ecology. It is often a soil-patch colonist, like *B. fallax*, occurring on banks, in fallow fields, in chinks of walls, and on bare stony ground. The range of habitat is quite as extensive as that of *B. convoluta*, possibly more so; indeed, it would be interesting to know whether, as seems likely, this species is absent from strongly acid soils.

52. BARBULA FALLAX Hedw.

This common moss of bare ground and soil-capped walls is somewhat lacking in obvious distinguishing characters. It may indeed be confused at first sight with the still commoner *Ceratodon purpureus*, which it resembles in size (erect stems 1–4 cm.) and in its habit of growing in loose tufts or patches. In colour it varies from yellowish green through brownish green to dark olive-brown, and thus differs in colour range from *Ceratodon*, which is usually mid-green tinged in varying degree

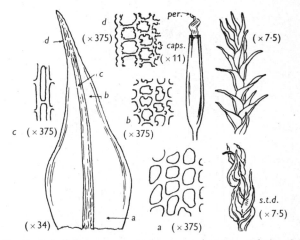

Fig. 52. *Barbula fallax*: *caps.* capsule, *per.* peristome, *s.t.d.* tip of shoot in dry state, that shown above it being fresh (moist).

with wine-red. The leaves tend to be less crowded than in *Ceratodon* and, becoming incurved and lightly twisted when dry, often give the shoots a characteristic chain-like form.

The leaf, 1·25–2 mm. long, tapers from a broad base to an acute tip. The leaf margin, as in *Ceratodon*, is revolute for much of its length, but the leaf apex is never toothed, and the areolation is quite different both from that plant and from most other species of *Barbula*. Thus, the cell walls in the leaf of *B. fallax* tend to be much thickened at the corners so that the cell cavities are rounded (not 'square' as in *Ceratodon*), and the cells are papillose to a variable extent. If, however, the cells immediately overlying the nerve be examined, they will be found to be elongated, even well up towards the leaf tip, contrasting with the short cells of the adjoining lamina tissue. This last character is a certain means of separating *Barbula fallax* from *B. vinealis*, *B. cylindrica* and *B. trifaria*, in none of which such elongated cells occur.

Although *B. fallax* is dioecious the capsule is not rare. It is borne on a red seta about 1 cm. long. The 16 long, twisted peristome teeth, extending from the mouth of the capsule as a delicately curled, pale orange-red brush, prevent confusion with any totally unrelated genus, although they may lead the beginner to identify the plant as a species of *Tortula*. In the younger capsule, the long beak to the lid affords a useful character.

Ecology. It is essentially a colonist of bare patches of soil, especially perhaps on chalk and wet calcareous clay. In this role it often appears in soil-filled crevices or on soil-capped rocks and walls, sometimes on crumbling mortar, but never on the bare rock. Dr E. W. Jones considers that it demands a fairly high concentration of nutrient salts in general; but, as with other species of *Barbula*, it would be interesting to know its precise range of tolerance in regard to soil reaction. *B. fallax* is widespread generally, but is least common in some parts of north-west Britain.

53. BARBULA RECURVIROSTRA (Hedw.) Dix. (*B. rubella* (Hoffm.) Lindb.)

The rusty-red colour of the lower parts of plants of this species will usually serve for identification. In other respects it is not markedly different from a number of other mosses (cf. *Ceratodon*), with which it agrees in forming loose or moderately dense cushions, in the upright stems 1–4 cm. tall, and in the narrowly lanceolate tapering leaves; but in no other common British species of similar size and habit is the brick red colour characteristically produced. *Barbula recurvirostra* is a very common moss on walls, rock ledges and bare soil, perhaps most plentiful in hilly calcareous districts. When, as occasionally happens, the brick red is replaced by dingy brown, microscopic characters will be necessary for the determination.

Under the microscope the tapering leaf (about 3 mm. long) is seen to have the margins narrowly recurved, from just above the base almost to the extreme apex. The whole leaf is somewhat channelled in form, the strong reddish nerve forming the base of the channel. The nerve disappears among the short, obscure, highly papillose cells of the acute leaf tip, which is marked by a few indistinct teeth. Although a variable character, these teeth will, when present, separate *B. recurvirostra* from all other British species of *Barbula*. The cells in the rather wide, sheathing leaf base are long and narrow, but pass over gradually into the shorter cells above, not abruptly as in species of *Tortella*.

The narrowly cylindrical capsule is borne erect on a red seta 7–10 mm. in length. The lid has a short oblique beak. The 16 peristome teeth are short.

Ecology. It grows on rock and soil, probably much more commonly on soil-capped ledges or in soil-filled crevices than on bare rock. It is a frequent member of the wall-top community and is common on the mortar of walls. Calcareous habitats are favoured, but *Barbula recurvirostra* is probably not a strict calcicole. It is perhaps most common in limestone districts but is widespread generally.

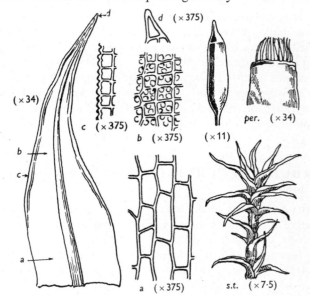

Fig. 53. *Barbula recurvirostra*: *per.* part of capsule showing peristome, *s.t.* tip of leafy shoot.

ADDITIONAL SPECIES

Several other species in this rather difficult genus should be mentioned as by no means rare; indeed, on calcareous walls in certain districts some of them are very common.

B. revoluta Brid. is a very distinct species, resembling *B. convoluta* in dwarf habit and in leaf shape, but at once recognized by the very broad nerve and the leaf margins so widely revolute that they almost reach the nerve.

B. rigidula (Hedw.) Mitt. comes near *B. fallax*, but the leaves tend to be narrower, tapering to rather thick, blunt tips; and the cells overlying the nerve are not elongated as they are in *B. fallax*. Also the habitat is almost always different, for *B. rigidula* forms dense olive-green cushions on stony substrata, not on soil. Nearly spherical, multicellular gemmae are quite often present and afford a useful character.

B. trifaria (Hedw.) Mitt. (*B. lurida* (Hornsch.) Lindb.) differs from *B. fallax* in its broader, more concave leaves, which are, moreover, held erect and appressed when dry, not incurved as they are in *B. fallax*. *B. trifaria* differs too in the lack of elongated cells overlying the nerve of the leaf and in its habitat, for it grows almost always on stone, rarely on wood and never on earth.

B. tophacea (Brid.) Mitt. has leaves relatively narrow in outline and tapering so little that the apex is notably broad and rounded. The nerve is lost well below the apex and the areolation is very well defined throughout the leaf and scarcely at all papillose; indeed, the outlines of the cells in the leaf apex are clearer than in any other British species of *Barbula*. *B. tophacea* forms olive-brown tufts on damp calcareous walls and in lime-rich springs.

B. cylindrica (Tayl.) Schp. is the longest in the leaf (upper leaves 2–4·5 mm.) of the commoner British species of *Barbula*. The cells in the leaf base differ from those of *B. fallax* in their 'square' rather than rounded cavities, whilst the cells in the very acute leaf apex are highly papillose, much more so than in *B. fallax*; moreover, there are no elongated cells over the nerve. The leaves are much twisted and curled when dry, and even when moist have a characteristic, slight, spiral twist. It grows on earth or rock, always in fairly sheltered places.

B. vinealis Brid. is allied to *B. cylindrica* and has the cell structure of that species. The noticeably curved leaves are about as long as those of *B. fallax*, but the lack of elongated cells overlying the nerve in the upper part of the leaf will distinguish *B. vinealis* with certainty from that species. It is locally very common on walls, forming dense rounded cushions that are bright green at the tips and light reddish brown below. It almost always grows on stone, more rarely on wood and only exceptionally on soil.

54. GYMNOSTOMUM AERUGINOSUM Sm. (*Weissia rupestris* (Schwaegr.) C.M.)

Most plentiful on rock ledges in limestone hill districts, this moss grows in dense, compact, deep green tufts or cushions which at times

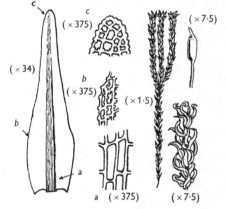

Fig. 54. *Gymnostomum aeruginosum*: the drawing on the extreme right bottom corner shows the appearance of the shoot when dry.

look rather like those of *Amphidium mougeotii*, another rock-ledge species. In height (2–8 cm.) and in mode of growth, the two species are alike,* but *A. mougeotii* has longer leaves which taper to finer points. The leaves of *Gymnostomum aeruginosum* are only 0·75–1·5 mm. in length and, tapering only slightly from base to apex, end in rather blunt tips.

* *G. aeruginosum* is, however, remarkably variable, and at times one sees capsules arising from leafy shoots that are barely 1 cm. tall. Then it looks a very different plant.

Under the microscope the strong nerve (70μ wide at base), usually orange or reddish in colour in older leaves, combined with the rounded leaf apex in which the nerve disappears, make recognition fairly easy. The areolation is rectangular and clear cut at the base, small (7–12μ) and made obscure by abundant papillae towards the leaf tip. Even near the base of the leaf projecting papillae often impart an irregular outline to the leaf margin.

G. aeruginosum is dioecious and is more often found barren than with fruit. The capsule is borne erect on a short (3 mm.) seta and is ovoid-ellipsoidal, with a beaked lid. It lacks a peristome.

Ecology. Its chief habitat consists of wet base-rich rocks where it occupies sites similar to those taken by *Amphidium mougeotii* on siliceous rocks, i.e. dripping rock faces and permanently moist ledges and clefts. Sometimes, in these rather specialized calcareous rock habitats, it becomes one of the dominant bryophytes.

ADDITIONAL SPECIES

Gymnostomum recurvirostrum Hedw. (*Weissia curvirostris* (Ehrh.) C.M.) is a not uncommon species of base-rich mountain rocks. Closely allied to *Gymnostomum aeruginosum*, it differs in its more narrowly tapering, acute-tipped leaves, with margins recurved below, and in its more clearly defined cell structure. Also, the seta is 8–10 mm. long and the spores 15–22μ (against 10–12μ in *G. aeruginosum*).

RELATED SPECIES

Eucladium verticillatum (With.) B. & S. (*Weissia verticillata* With.) is a characteristic tufa-forming species of wet limestone rocks, where it is recognized by its pale green colour, and leaf margin noticeably toothed near the base.

55. ANOECTANGIUM COMPACTUM Schwaegr.

This species is not very common, but where it is fairly plentiful, as in shaded rock clefts on some of our mountains, it can hardly fail to attract attention, owing to the peculiarly light, vivid green colour of the shoot tips. The erect stems are crowded, sparingly branched and matted together with a variable amount of red-brown tomentum. Thus they form dense, deep cushions which are bright green above but vary from red-brown to very dark brown below.

The small size of the lanceolate, keeled leaves (about 1 mm. long) will distinguish *A. compactum* from most of the mosses that commonly form cushions in similar places. Among these only *Gymnostomum aeruginosum* has equally small leaves, and in that species the shoot tips are never of that light but vivid shade of green found in *Anoectangium*; moreover the leaf is proportionately narrower and its apex is more obtuse.

The cells in the leaf base are thick-walled and shortly rectangular in outline, whilst those throughout the rest of the leaf are isodiametric but so densely covered with papillae as to make their outlines obscure. Where the leaf narrows at its acute apex a few clearly defined cells may usually be seen. The leaf margin is plane; the nerve forms an unusually deep channel on the upper surface of the leaf and is correspondingly prominent on its underside.

A. compactum is dioecious, but fertile plants are not very rare. The small ovoid to ellipsoidal capsule is borne on a seta 7–10 mm. long, that arises laterally from a position some distance down the stem. This

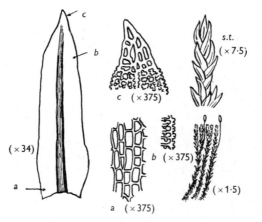

Fig. 55. *Anoectangium compactum*: *s.t.* shoot tip.

is a most unusual feature among mosses of the growth form of *Anoectangium*, and it has caused much speculation as to the true systematic affinities of the plant. There is no peristome.

Ecology. It grows in rock clefts and on soil-covered ledges on mountain cliffs, from about 1500 ft. upwards. Shady situations are favoured, and it is most typically a plant of slightly basic, siliceous rock. A commonly associated species is *Amphidium mougeotii*.

56. TORTELLA TORTUOSA (Hedw.) Limpr. (*Trichostomum tortuosum* (Hedw.) Dixon)

The dense tufts or cushions of *Tortella tortuosa*, rather pale green above and light yellowish brown below, are a common and characteristic feature of walls and rock ledges in limestone districts. It also ascends into the alpine zone on mountains, wherever there is an outcrop of basic rock. From other related mosses, such as various species of

Barbula, *Weissia* or *Trichostomum*, it is usually distinguished quite readily by its robust habit (2–10 cm. tall) and crowded, very long, narrow leaves (up to 6–7 mm.), which become greatly curled and twisted when dry. Almost linear in shape, tapering to long fine points and with the margins noticeably wavy, these leaves are not matched in any other common British moss. Once known, *Tortella tortuosa* is generally easy to recognize in the field. *Ptychomitrium polyphyllum* has much con-

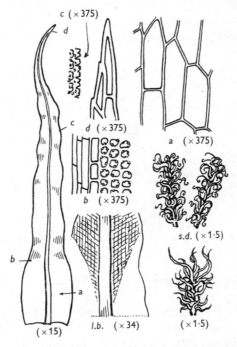

Fig. 56. *Tortella tortuosa*: *l.b.* part of leaf base to show well-defined line between hyaline cells (clear) and green cells (cross-hatched); *s.d.* shoots in a dry state (in these and in the moist shoot below it, the basal part of the stem is omitted).

torted leaves when dry, but the plants are black rather than light brown below, and it is usually abundantly fertile, whereas the present species is rare in fruit.

Under the microscope the prominent yellowish nerve is seen to continue to the apex of the leaf, and the undulations of the margin are very evident; but the best diagnostic character lies in the areolation. Elongated, hyaline cells occupy the leaf base, then pass over abruptly to short, obscure, chlorophyllous cells, and the boundary line where these two types of cell join runs obliquely up from nerve to leaf margin. The result is that the clear, hyaline cells occur far up the leaf at the margins, but near the nerve are confined to the leaf base. The effect under the microscope is most characteristic. The short upper cells, their outlines obscured by the numerous papillae, are not notably different from those occurring in a number of related mosses.

This moss is dioecious and the cylindrical capsule, with long-beaked lid, and long, twisted peristome, is rare.

Ecology. It will grow on rock or soil; thus, while most typically a plant of calcareous rock ledges in hill districts, it will also appear in the grass of the hillside itself. It is probably always associated with calcareous or near-neutral conditions. In north-west Scotland I have found it, apparently on moorland, but in fact only where the acid substratum was overlaid by highly calcareous shell sand blown up from adjoining dunes.

ADDITIONAL SPECIES

Tortella flavovirens (Bruch) Broth. (*Trichostomum flavovirens* Bruch) somewhat resembles *Trichostomum brachydontium*, but the leaf is distinct in its inrolled margins and in the fact that the hyaline cells of the base extend obliquely up the edges of the leaf. It occurs chiefly on rocks and sand near the sea.

57. **TRICHOSTOMUM BRACHYDONTIUM** Bruch (*T. mutabile* Bruch ex De Not.)

A common plant of wall-tops and stony ground, especially near the sea, *T. brachydontium* is not altogether an easy moss to define. It lacks any conspicuous diagnostic feature and is almost always barren. Usually between 1 and 4 cm. tall, it looks a little like *Barbula unguiculata*, which it resembles, when dry, in the much curled leaves with prominent yellowish nerves. It is, however, a much more robust plant than

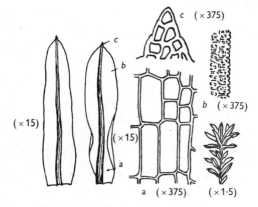

Fig. 57. *Trichostomum brachydontium.*

B. unguiculata, with the leaves 2–4 mm. in length, nearly twice as long as in that plant.

From related species with fairly long leaves *Trichostomum brachydontium* differs in leaf shape. The tongue-shaped leaves taper little and end in blunt tips; but the nerve runs out to form a short protruding point. The leaf margin may be somewhat wavy, especially near the base, but is quite plane, not recurved as in *Barbula unguiculata*. The cells are large and hyaline $(8–12 \times 20–60\mu)$ in the leaf base, short and much obscured by papillae in the upper part of the leaf. The area of hyaline cells is not as sharply defined as in the much larger *Tortella tortuosa*, nor does it extend obliquely up the edges of the leaf. Care is needed to distinguish *Trichostomum brachydontium* from the allied *T. crispulum*, which has, however, a characteristic hood-like leaf apex.

The ovoid-ellipsoidal capsule, borne on a yellowish seta, is rarely found, for *T. brachydontium* is dioecious and the fruit is seldom produced.

Ecology. It occupies a variety of habitats, rock (including both chalk and granite), soil-capped ledges and wall-tops, and sometimes bare soil. It is most plentiful near the sea, on the west coast.

ADDITIONAL SPECIES

Trichostomum crispulum Bruch is closely related to *T. brachydontium* and not always easily distinguishable from it. Rather less common than *T. brachydontium*, it may generally be recognized by the hood-like (cucullate) leaf tips.

T. tenuirostre (Hook. & Tayl.) Lindb. is a long-leaved species resembling *Tortella tortuosa*. It lacks, however, the obliquely ascending character of the hyaline cells in the leaf base, and the leaf margin is notched near the apex. It occurs on wetter rocks than *T. tortuosa*, and is not a calcicole.

Trichostomum sinuosum (Wils.) Lindb. (*Barbula sinuosa* Wils.) occurs on shaded walls and stones in calcareous districts, chiefly in the midlands and south. The stems are only 0·5–2·5 cm. tall and do not form dense tufts as in the last species. It thus differs from *Trichostomum tenuirostre* in habitat and mode of growth, but resembles that species in its long (up to 2·5 mm.), narrow leaves, distinctly notched near their tips. The leaves are notably fragile (some usually have the tips broken) and the longest cells in the leaf base only reach 40–50 against $70–80\mu$ in most other species of the genus.

58. WEISSIA CONTROVERSA Hedw. (*Weissia viridula* Hedw.)

If a lowland bank, with bare patches of earth, be searched for mosses, this species is very likely to be found. Being very small (about 1 cm. tall), barren plants are not conspicuous, but they tend to form loose tufts or wide patches, and when these patches are liberally dotted with capsules they will catch the eye. In colouring the tufts are of a vivid mid- or yellowish green; in mode of growth they suggest a *Barbula*, but the leaves are longer than in most of the low-growing species of that genus. They are lanceolate, up to 3 mm. long, and much curled when dry.

Under the microscope the leaf is seen to have a rather broad base, then to taper to a fine point, the nerve running out into a sharp cusp at the leaf tip. The areolation, somewhat elongate, clear and hyaline below, becomes suddenly short and very obscure in the upper part of the leaf. Perhaps its best distinguishing feature, however, lies in the leaf margin which is very strongly and obviously inrolled throughout almost the whole length of the leaf, except the sheathing base. This makes the upper part of the leaf concave and hood-like.

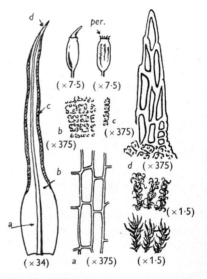

Weissia controversa is autoecious, and capsules are often plentiful. Each is ovoid-ellipsoidal, and is held erect on a delicate yellowish seta about 4 mm. in length. The lid has a long oblique beak. The peristome is developed to a very variable extent; thus while an incomplete series of short teeth can usually be distinguished, they are at times almost wanting. Old, empty capsules become light brown, and slightly furrowed. They are wide-mouthed, unlike those of the closely allied

Fig. 58. *Weissia controversa*: *per.* peristome; the upper group of shoots shows the appearance when dry.

W. microstoma. The spores are usually 15–18 μ across, against 18–20 μ in *W. microstoma*.

Ecology. Its principal habitat appears to be on the bare soil of loamy banks. It also occurs in rock crevices, but probably always where some soil or humus has collected.

ADDITIONAL SPECIES

W. microstoma (Hedw.) C.M. is distinguished from *W. controversa* by the very narrow mouth of the capsule and complete absence of peristome. It is locally common, especially in calcareous places.

W. tortilis (Schwaegr.) C.M. is a somewhat larger plant, with more broadly pointed leaves and a stout reddish nerve (50–80 μ wide). It is a rather rare plant of calcareous places.

W. crispa (Hedw.) Mitt. is a minute cleistocarpous moss of calcareous ground. The very long perichaetial leaves are the most conspicuous feature of the plant, and, with their incurved margins and projecting nerve, will distinguish it from other British cleistocarpous mosses.

Leptodontium flexifolium (Sm.) Hampe is the most frequent of the three British species of the genus. It occurs on bare gravelly soil or peat, and in leaf structure resembles *Dichodontium pellucidum*. The present species is, however, more slender in habit and lacks the expanded leaf base of *D. pellucidum*. In general appearance it is like *Barbula convoluta*.

GRIMMIALES

59. GRIMMIA MARITIMA Turn.

This plant is aptly named, for it is almost the only truly maritime moss, being confined to rocks by the sea. It is recognized by its neat cushions (1–4 cm. tall), of rigid texture and dark olive green colour, blackish in the lower parts. When dry, the leaves become much in-curved, but only slightly twisted. It differs from most species of *Grimmia* in the lack of hair-points to the leaves. Few other mosses grow so near to the breaking waves, and there is none that could readily be mistaken for it. Though somewhat local, it is abundant on many parts of the west coast.

The lanceolate leaf, about 2 mm. long, is more suddenly contracted into a long, narrowly acute tip than is the leaf of *G. apocarpa*. Also, the whole of the upper part of the leaf is of a peculiarly solid, opaque texture, being composed of several layers of cells. The margin is lightly recurved near the base, not extensively so, nor thickened as in *G. apocarpa*.

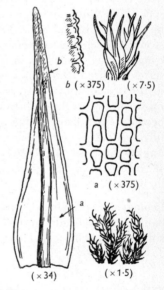

Fig. 59. *Grimmia maritima*: plants at bottom right in a dry state, shoot tip at top right moist.

The thick, reddish brown nerve disappears in the dense, slightly papillose cells of the leaf apex, which is thick and opaque. Lower, where the outline of the rather thick-walled, but not greatly elongated cells can be clearly seen, these cells will be found to lack the wavy walls of *G. apocarpa*.

Although *G. maritima* is autoecious, capsules are not in my experience very common. As in *G. apocarpa*, they are immersed in the perichaetial leaves. In form the capsule is somewhat more rounded, approaching globose. It lacks the bright red of *G. apocarpa*, and the teeth of the peristome are broader than in the allied species.

Ecology. This is the most markedly maritime of all common British mosses; it is never found inland. Its habitat consists of crevices and ledges of rock close to high-water mark; shore rocks and cliffs are equally favoured, but it rarely or never grows on calcareous rock. Its cushions are often found on rock that is bare except for some lichens and scattered plants of thrift (*Armeria maritima*) and sea-campion (*Silene maritima*).

60. GRIMMIA APOCARPA Hedw.

A very common plant of wall-tops and boulders, *G. apocarpa* is notably variable in size, colour and the length of the hair-points on the leaves. Thus one common form is strongly tinged with reddish brown, and each leafy shoot ends in a conspicuous white brush (formed by the hair-points of the terminal tuft of leaves); but other forms occur in which the colour is a deep, dull green and the hair-points are very short or wanting. The stems (1·5–3 cm. tall) are usually massed together to form tufts or cushions. When dry the leaves are lightly twisted together so that the shoot tips become characteristically pointed.

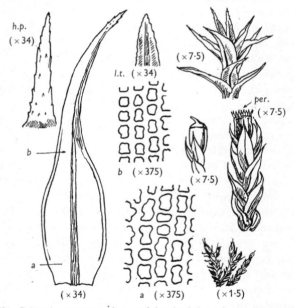

Fig. 60. *Grimmia apocarpa*: *h.p.* and *l.t.*, leaf tips of two extreme forms, the former with evident hair-point; at bottom right are plants in a dry state; *per.* peristome.

The leaf is 1·5–2 mm. long, ovate-lanceolate, and usually channelled. A useful microscopic character lies in the thickened, narrowly recurved leaf margins; in addition much of the tissue in the terminal part of the leaf is usually two cells thick. The variable hair-point is lightly toothed. The leaf cells, shortly rectangular in the leaf base and rounded-quadrate above, have thickened, slightly wavy walls.

An autoecious species, *G. apocarpa* fruits abundantly, and when fertile is not readily mistaken for any other moss. The red lid of the young capsule, and later the red peristome, will make the plants conspicuous, although the seta is so short that the capsules are not raised clear of the perichaetial leaves. The lid has an evident beak; the peristome consists of 16 rather short teeth which are widely spreading when dry.

Ecology. Its chief habitats are boulders and wall-tops, but it is fairly catholic in its choice of these. Thus, while perhaps most abundant in limestone districts, it is still common in many non-calcareous places. It is often associated with *G. pulvinata* and *Tortula muralis* on wall-tops.

61. GRIMMIA DONIANA Sm.

Several British species of *Grimmia* are practically confined to mountains, and of these *G. doniana* is perhaps the commonest. It forms tiny rounded cushions or low tufts, often only 0·5–1 cm. high and about 1 cm. across, on siliceous boulders, where it is in general fairly readily identified by its blackish colour and pale ovoid capsules. The blackish colour, however, is sometimes less marked, and then the long hair-points of the leaves may suggest *G. pulvinata*; but in *G. doniana* the cushions are not dense and mound-like, and the seta is not curved as in *G. pulvinata*. Also, even when the young shoots appear olive green, the plants are still very dark, almost black, below.

The leaf is lanceolate, tapering to a narrow apex, beyond which extends the hair-point of variable length. In some of the upper leaves it may be as long as the leaf itself; in many of the lower leaves it commonly remains short, or may be absent altogether.

Under the microscope the hair-point of the leaf is seen to be lightly toothed, and the cells of the leaf base are found to be much elongated (some reaching $70 \times 10\mu$). The areolation, however, as in most species of *Grimmia*, is subject to a good deal of variation. The short cells towards the leaf apex are in two layers. This imparts a characteristic appearance to the leaf as a whole, making it hyaline and translucent below, opaque above.

An autoecious species, *G. doniana* is often abundantly fertile. The ovoid capsule is borne fairly erect on a very short (2 mm.) seta. It has a bluntly conical lid, which at a certain stage is bright orange in contrast with the greenish body of the capsule. The older, pale brown capsules do not become furrowed as in *G. pulvinata* and *G. trichophylla*.

Ecology. It appears to be especially characteristic of hard siliceous rocks at moderate to high altitudes. Thus, in north Wales, it may grow with *Rhacomitrium heterostichum* and *Grimmia trichophylla* on wall-tops at

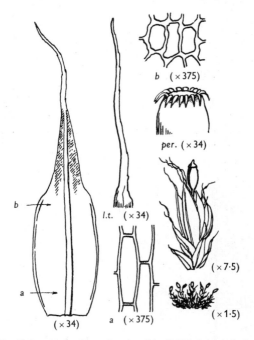

Fig. 61. *Grimmia doniana*: *l.t.* tip of leaf with longer hair-point than that of leaf on left; *per.* peristome.

600 ft., but occurs equally on loose boulders of block scree at over 2000 ft. Professor P. W. Richards has suggested to me that its preference for hard rock may be correlated with a slow growth rate and consequent inability to maintain itself on more rapidly weathering surfaces.

62. **GRIMMIA PULVINATA** (Hedw.) Sm. (Plate V)

The compact, often almost hemispherical cushions of *G. pulvinata*, hoary grey with the long hair-points of the leaves, are one of the most characteristic and familiar adornments of wall-tops in most parts of Britain. Apart from a close similarity to a few much rarer species of

Grimmia, there is no moss with which the present species is likely to be confused, at least when in fruit. Even when barren, the cushions are usually deeper, denser, and more mound-like in form than those of *Tortula muralis*, which *Grimmia pulvinata* resembles in its hair-pointed leaves, and which shares its wall-top habitat. The more loosely growing cushions of *Rhacomitrium heterostichum* are more characteristic of

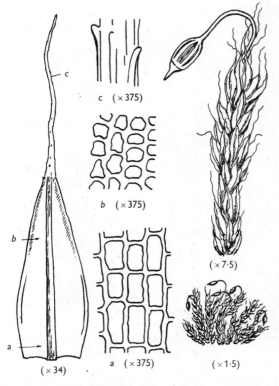

Fig. 62. *Grimmia pulvinata.*

mountain regions, whereas *Grimmia pulvinata* is principally a lowland plant.

The stems are 1–1·5 cm. tall, the leaves oblong-lanceolate in shape, broad at the base and moderately so at the tip, beyond which the silvery hair-point extends for a distance that may equal the length of the leaf.

The most useful microscopic characters are provided by the leaf margin, which is thickened in the upper part of the leaf, and the cell shape, which is remarkably uniform and only slightly elongated throughout the breadth of the leaf base. The short upper cells have

Plate V. *Grimmia pulvinata* with capsules (× 2)

Plate VI. *Pleurozium leuwigeaceum* (natural size)

rather thick, somewhat wavy walls. The hair-point is lightly but distinctly toothed.

The long, much-curved seta distinguishes this plant when in fruit from other species with hair-pointed leaves. Thus at an early stage the seta has the effect of partially burying the capsule amongst the leaves, but it straightens out with age. The capsule is ovoid, with a beaked lid and symmetrical calyptra. It has 8 longitudinal ridges or striations when dry, and a spreading peristome of 16 reddish teeth.

Ecology. This is a typical member of the lowland wall-top community, with *G. apocarpa*, *Tortula muralis*, *Bryum capillare*, and such pleurocarpous mosses as *Camptothecium sericeum*. It will tolerate roof-top slates and brick walls. It perhaps has some preference for limestone (where it is often very abundant); *Grimmia pulvinata* becomes markedly scarcer in some of the upland districts of north and west Britain.

63. **GRIMMIA TRICHOPHYLLA** Grev.

A not uncommon moss of wall-tops, *G. trichophylla* differs from *G. pulvinata* in its looser growth, and more yellowish green colour. Instead of forming dense cushions, grey with the massed hair-points of the leaves, it grows usually in wide irregular patches, in which the underlying green is not so completely masked by the silvery white of the hair-points. Thus, although in size and in many structural details it closely resembles the last species, the general appearance of the plant is quite different.

In length the leaf resembles that of *G. pulvinata*, but it is narrower in outline and, at the apex, tapers more gradually into a hair-point. This is extremely variable in length; in many of the leaves it forms only a short white tip, and, though it often attains a greater length, it is rarely as long as in the leaf of *G. pulvinata*.

Under the microscope the hair-point is seen to be smooth or lightly toothed. The basal cells of the leaf are distinctly longer on the average (usually several reach $60 \times 12\mu$) than the corresponding cells of *G. pulvinata*. The leaf margins tend to be somewhat recurved.

This species is dioecious and is not very commonly met with in fruit. In curved seta and ovoid capsule it comes near to the last species, but the young capsules of *G. trichophylla* have a characteristic pale yellowish colour.

Ecology. It occurs on rocks and walls, preferring siliceous rock in relatively dry situations. Its range overlaps that of *G. pulvinata*, but unlike that species, it is mainly restricted to hilly districts and is never found on limestone.

ADDITIONAL SPECIES

Some thirty species of *Grimmia* occur in Britain, but many of these are rare.

G. alpicola Hedw. var. *rivularis* (Brid.) Broth. (*G. apocarpa* Hedw. var. *rivularis* Web. & Mohr) is found on stones in mountain streams. With stems 4–10 cm. long it has much the habit of *Cinclidotus fontinaloides*, but the leaves (2 mm. long) are shorter than in that plant.

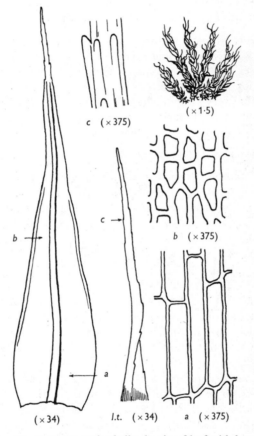

Fig. 63. *Grimmia trichophylla*: *l.t.* tip of leaf with longer hair-point than that of leaf on left.

Grimmia funalis (Schwaegr.) Schp. is an uncommon but rather distinctive moss of alpine rock ledges. With stems 1–5 cm. tall it forms deep round cushions, very black below, dark green or silvery with hair-points at the shoot tips. It may be known by its colour, the tendency of the cushions to fall apart readily when gathered, and the curious twisted, rope-like character of the shoots when dry.

G. torquata Hornsch. ex Grev. is another plant of mountain rock ledges, closely related to *G. funalis*. It is usually brown rather than black below, and olive-green at the shoot tips. When dry it has the twisted rope-like character of *G. funalis*, but the leaves tend to form a looser spiral than in that species. *G. torquata* somewhat resembles *Anoectangium compactum* but is known from it by the short hair-points of the upper leaves, and usually duller colour.

Grimmia hartmanii Schp. is a rather rare plant of siliceous rocks, marked by slender elongated stems and semi-prostrate habit, with leaves slightly curved to one side and often bearing gemmae.

G. decipiens (Schultz) Lindb. is a robust plant of rocks at low altitudes; it resembles *G. trichophylla*, but differs in the strongly toothed hair-points of the leaves. Its loose tufts (stems 2–4 cm.) may suggest *Rhacomitrium heterostichum*, but it lacks the characteristic *Rhacomitrium* type of cells in the leaf base.

Grimmia patens (Hedw.) B. & S., a mountain species, superficially resembles *Rhacomitrium aquaticum*, but is known at once by the two wings of tissue which are developed on the back of the nerve in the upper part of the leaf.

64. RHACOMITRIUM ACICULARE (Hedw.) Brid.

Robust in habit, and dark green in colour, with long forked stems (2–8 cm.), this moss is quite commonly a conspicuous member of the boulder flora in streams in mountain districts. The straggling character of the plant is accentuated by the fact that the leaves often become worn away from the lower parts of the stems. Above they are crowded, appressed to the stems when dry, spreading widely when wet. This plant is the most aquatic in habitat of all our species of *Rhacomitrium*, and is unlike all others in its very obtuse leaves, with their broad, rounded tips (cf., however, *R. aquaticum*).

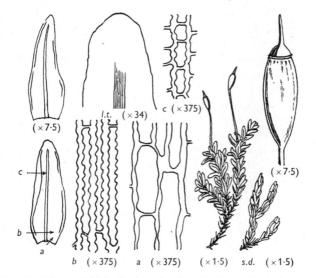

Fig. 64. *Rhacomitrium aciculare*: *l.t.* tip of leaf, *s.d.* shoot in a dry state; two leaves are drawn to show slight variation in shape, especially of leaf tip.

Under the microscope the leaves of all species of *Rhacomitrium* are characterized by the long, narrow basal cells, with thickened, wavy longitudinal, and thin, straight, transverse walls. Such cells serve to separate *R. aciculare* from robust forms of *Grimmia alpicola* var. *rivularis* or any other unrelated moss of boulders in streams. Other microscopic features to be noted in the leaf are the tendency for the margin to be recurved to some extent, the nerve vanishing just below the apex, and the presence, usually, of a few obscure teeth at the tip of the leaf.

Although dioecious, *Rhacomitrium aciculare* is commonly fertile. The capsule is rather large (2·5–4 mm. long), ellipsoidal to cylindrical in shape, and borne erect on a dark brownish seta 5–15 mm. long. The lid has a slender straight beak 1 mm. long. Each of the 16 peristome teeth is divided into 2–3 fine branches.

Ecology. It always grows near water and is often submerged; its principal habitat is on rocks and stones in and about the margins of swift-flowing, non-calcareous streams. It is chiefly a plant of the hill districts of north and west Britain.

65. RHACOMITRIUM AQUATICUM (Brid.) Brid. (*R. protensum* A.Br. ex Hüben.)

A moss of moist rocks in hilly districts, *R. aquaticum* forms luxuriant tufts or patches of a prevailing yellowish or olive-green colour. It differs from *R. fasciculare* in the blunter leaves and much larger branches; this gives it a very different habit from *R. faseiculare*, for in that species the stems bear numerous short, crowded side-branches. The stems of *R. aquaticum* vary from 4 to 12 cm. in length. Since the leaves never have hair-points, it cannot be confused with most of the other robust species of *Rhacomitrium*; but occasionally dark green forms may resemble *R. aciculare* rather closely.

The leaf is broad at the base, tapering above to an apex that is broader than that of *R. fasciculare* but is not as broadly rounded as that of *R. aciculare*. It is about 2 mm. long.

The nerve is very strong at the base, becoming ill-defined only in the leaf tip—a further useful distinction from *R. fasciculare*. The leaf margins are recurved. The upper cells are rather short ($20–30 \times 8–12\mu$) and papillose, with thick, slightly wavy walls; whereas at the leaf base there are cells of the usual *Rhacomitrium* type—very long and narrow—with extremely thick, wavy longitudinal walls.

The ellipsoidal to cylindrical capsule is borne on a seta about 7 mm. long. It agrees fairly closely with that of *R. aciculare* in shape, smooth calyptra and divided peristome teeth.

Ecology. Acidophile, like most other species of *Rhacomitrium*, its favourite habitat is that of wet rock ledges on mountains in the north and west. It is often found near swiftly flowing water, especially on dripping rocks by waterfalls.

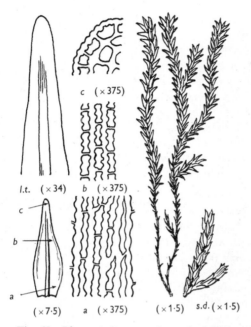

c (×375)

l.t. (×34)   *b* (×375)

(×7·5)   *a* (×375)   (×1·5)   *s.d.* (×1·5)

Fig. 65. *Rhacomitrium aquaticum*: *l.t.* leaf tip,
s.d. portion of shoot in a dry state.

66. RHACOMITRIUM FASCICULARE (Hedw.) Brid.

This species forms extensive patches on rocks in mountainous districts, where, together with *R. lanuginosum* and *R. heterostichum*, it is among the commonest mosses. It may usually be recognized at a glance as a *Rhacomitrium* by its semi-prostrate habit and very numerous lateral branches; it differs from most other species of the genus in its yellowish to olive-green colour and the absence of hyaline hair-points to the leaves. The specific name alludes to the bunched or 'fascicled' effect produced by the numerous short branches.

The leaf is broad at the base, and tapers above to a blunt or fine point. The leaf tip is usually more acute than in *R. aquaticum*, the only other British species of *Rhacomitrium* that is like *R. fasciculare* in colour. *R. aquaticum*, moreover, lacks the numerous short lateral branches of this species and grows in wetter habitats.

The weak, narrow nerve of the leaf is a good microscopic character, for in *R. aquaticum* the nerve is relatively thick and well defined. The long narrow cells with thick, wavy longitudinal walls, characteristic of *Rhacomitrium*, are well seen in the leaf base, and even the upper cells in *R. fasciculare* are slightly elongated.

The erect, ovoid to ellipsoidal capsule is borne on a seta about 1 cm. long. The young seta is pale yellow, but later becomes dark brown, the colour of the ripe capsule. The lid of the capsule has a straight beak, and the calyptra is rough with papillae; the peristome teeth are cleft to their bases.

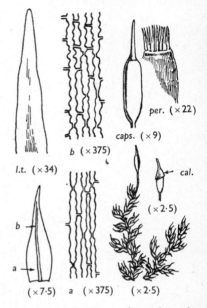

Ecology. *R. fasciculare* grows typically on the surfaces of siliceous boulders and walls in the upland districts of north and west Britain. It is quite absent from limestone, and is very rare in the south-eastern counties. Colonizing relatively bare rock and forming close, flat patches, it is said not to furnish a suitable 'nidus' on which other plants may become established.

Fig. 66. *Rhacomitrium fasciculare*: *l.t.* leaf tip, *per.* mouth of capsule showing peristome, *cal.* calyptra, *caps.* capsule with lid intact.

67. RHACOMITRIUM HETEROSTICHUM (Hedw.) Brid.

This common moss of mountain boulders occurs in two very distinct forms, whilst still other variations must be included in the species in the broad sense. The typical, though not always the most common, form grows in low tufts (2–4 cm. high) or spreading patches, which are greyish green in general colour, with tips made white by the hair-points of the leaves. This plant looks not unlike a very robust form of *Grimmia trichophylla*, but with a lens the rather wider leaf and distinctly broader base to the hair-point can be seen. Another form (var. *gracilescens* B. & S., including *Rhacomitrium sudeticum* (Funck) B. & S.) which is often plentiful on boulders at high altitudes, is marked by its very dark green colour and absence of hair-points to the leaves. In its repeatedly forked branching and in general leaf shape, however, it agrees with the typical form.

The hair-point is toothed and has a very broad attachment to the leaf apex. The cells in the leaf base are of the usual *Rhacomitrium* type, and their long, narrow shape, with unevenly thickened walls, will prevent confusion with any species of *Grimmia* which the plant may resemble in habit. The upper cells are very shortly rectangular, with wavy, thickened walls.

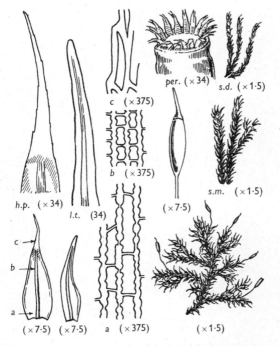

Fig. 67. *Rhacomitrium heterostichum*: the right-hand leaf and leaf tip (*l.t.*) and the shoots, moist (*s.m.*) and dry (*s.d.*), are of the slender variety *gracilescens*; *h.p.* hair-point of leaf of typical form; *per.* mouth of capsule showing peristome. Both capsules shown are drawn from the typical form.

The light brown, ellipsoidal to cylindrical capsule is borne erect on a dark brown seta about 8 mm. long. When fairly young the seta is straw-coloured and the capsule has a long-beaked lid.

Ecology. Its habitat—siliceous rock surfaces in hill districts—is the same as that of *Rhacomitrium fasciculare*, and the two species commonly grow together on boulders. The variety *gracilescens* often becomes abundant at high altitudes, e.g. near the tops of mountains in many parts of north Wales.

68. RHACOMITRIUM CANESCENS (Hedw.) Brid.

R. canescens is more of a lowland plant than most British species of the genus, and its much branched, greyish green tufts, white with the hair-points of the leaves at the shoot tips, may be seen on heaths and roadsides. It is a softer and more slender plant than *R. lanuginosum*, which it resembles in the long hair-points of the leaves, and the branches are commonly shorter and more crowded. Variants occur, however,

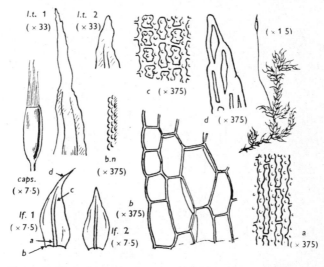

Fig. 68. *Rhacomitrium canescens*: *caps.* capsule showing peristome; *lf.* 1 and 2, leaf of typical form and leaf of variety lacking hair-point, respectively; *l.t.* 1 and 2, leaf tips of same; *b.n.* back of nerve.

with the hyaline points ill developed, and these forms may resemble *R. fasciculare*. Then microscopic examination will alone confirm the identification.

Under the microscope the hair-point is seen to be broad at its base, coarsely toothed and highly papillose. It is the papillose character of the cells of the leaf, however, which forms the surest mark of identification, and which will always separate the present plant from less robust forms of *R. lanuginosum*. A further character is that of the very small but perceptible auricles formed by enlarged thin-walled cells at the basal angles of the leaf.

In most of its capsule characters this species comes near to *R. lanuginosum*, but the seta here is quite smooth, not rough with papillae as in that plant. The peristome teeth are exceptionally long—as long as the body of the capsule.

Ecology. It favours patches of gritty or gravelly soil on roadsides and wall-tops, but also grows among grass on heaths. It is recorded from calcareous sand in the East Anglian Breckland.

69. RHACOMITRIUM LANUGINOSUM (Hedw.) Brid. (Plate VI)

'*Rhacomitrium* heath' is the name given to a plant community of exposed mountain-tops in which this species plays a dominant role. In such places *R. lanuginosum* forms extensive mats, and is not only the most robust and conspicuous bryophyte but also imparts a definite character to the landscape as a whole. It can be recognized at once by

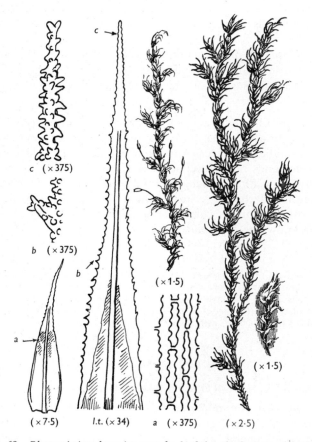

Fig. 69. *Rhacomitrium lanuginosum*: *l.t.* leaf tip; the bottom right-hand sketch shows the effect of the white hair-points (in dry conditions) against a dark ground.

its prevailing greyish green colour, with all the numerous shoots on the much-branched stems terminating in hoary white wisps—the massed hair-points of the leaves. Very stunted forms may resemble *R. canescens* or *R. heterostichum*, but normally, with individual plants attaining 12–25 cm. in length and the cushions or mats of growth 30 cm.–1 m. or more across, *R. lanuginosum* is unmistakable. In mountainous districts it is often extremely abundant, both on open peat moors and on boulders.

Under the microscope the hair-point, which is often as long as the whole of the rest of the leaf, is seen to be very rough with minute teeth that are themselves papillose. The areolation is best distinguished from that of *R. heterostichum* by the elongated upper cells (several times as long as broad), and from that of *R. canescens* by the smooth, not papillose, cell walls. The leaf margin is recurved in the lower part of the leaf.

R. lanuginosum is dioecious and is not very commonly fertile. The ellipsoidal capsule is rather small for the size of the plant, and is raised on a short seta, the surface of which is remarkable in being rough with papillae.

Ecology. It grows equally on rock and on peat surfaces. It is probably always strongly acidophile, its occasional appearance on the tops of chalk downs indicating places where acid conditions have arisen locally. It always demands fairly level surfaces for initial colonization, a fact which, Professor P. W. Richards informs me, is correlated with its poorly developed rhizoid system. On mountain-top detritus it may form almost continuous carpets, thus giving rise to the tundra-like vegetation known as *Rhacomitrium* heath.

ADDITIONAL SPECIES

R. ellipticum (Turn.) B. & S. is found on wet rocks, e.g. by waterfalls, mainly in the west, and is known by its short stems (1–3 cm. tall), rigid habit, thickened leaves—the upper cells are in two layers—and shortly oval capsule of solid texture. The appearance of the plant suggests *Grimmia* rather than *Rhacomitrium*, but the long narrow cells of the leaf base, with thick, wavy, longitudinal walls, are typical of the latter genus.

FUNARIALES

70. FUNARIA HYGROMETRICA Hedw.

F. hygrometrica, commonly taken as a type for the elementary study of the Bryophyta, often grows in extensive carpets on recently burnt land, and is, in general, an easy plant to recognize. The capsules are usually present in profusion, and then the plants may be known by their pale green colour and loosely tufted habit, with the yellowish to orange-brown capsules each raised on a long seta. The green, pear-shaped, immature capsules may be present too, for many different stages will often be found together in the tufts. Immediate post-fertilization stages, however, look very different, with the terminal tuft of leaves (mostly perichaetial leaves) enclosing the very young sporophyte. The shoots are then notable for their bud-like form.

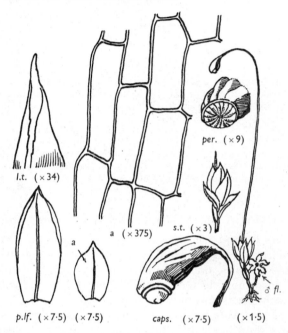

Fig. 70. *Funaria hygrometrica*: *p.lf.* perichaetial leaf, *l.t.* tip of same, *per.* mouth of capsule showing peristome, *caps.* ripe capsule with lid still intact, *s.t.* tip of shoot with very young capsule enveloped in perichaetial leaves, ♂ *fl.* male 'inflorescence'.

The stems of *F. hygrometrica* are short (3–10 mm.), and little branched, the leaves of uneven size, but the upper ones are large, broad and acute at the apex, not unlike the leaves of some species of *Bryum*.

The best microscopic character lies in the very large size of the approximately hexagonal-shaped cells. They measure up to 100–170μ long by 35–50μ broad.

The male 'inflorescences' occur on distinct slender branches and resemble minute green flowers. The capsule is borne on a seta 3–5 cm. long, occasionally even longer. As mentioned above, it is at first green, pear-shaped and curved; later it becomes yellowish to orange-brown, finally dark brown and strongly furrowed, and is held horizontally. The wide mouth is oblique, and bears an outer and an inner peristome, each of 16 teeth. The outer peristome teeth are curved and united by their tips to a minute central disc.

Ecology. It colonizes bare soil in a wide variety of situations, including moorland, open field habitats, woods and gardens. Very often its presence marks the site of a recent fire. Even where no signs of fire are evident, pieces of charcoal can often be found beneath the tufts. On heaths where fires are frequent it is generally very abundant in the first or second season after burning.

ADDITIONAL SPECIES

Only two of the remaining species of *Funaria* are at all common. These are both much smaller plants with erect or nearly erect capsules. They occur chiefly on clayey or peaty banks and the sides of ditches, being commonest in the hill country of the north and west.

F. obtusa (Dicks.) Lindb. (*F. ericetorum* (Bals. & De Not.) Dix.) has a short, wide obovoid capsule and bears a superficial resemblance to *Pottia truncata*. The leaves, however, are strongly bordered and the cells uniformly large (up to $80 \times 30\mu$).

Funaria attenuata (Dicks.) Lindb. (*F. Templetoni* Sm.) has only faintly bordered leaves and a longer, narrowly pear-shaped capsule.

71. PHYSCOMITRIUM PYRIFORME (Hedw.) Brid.

This species is most conspicuous in spring, when the characteristic pear-shaped capsules are ripening. Commonly it forms wide sheets, covered with capsules, on some patch of wet mud by road or ditch. Although conspicuous in the mass when fertile, the individual plants are small, with erect stems only a few millimetres tall.

The leaves are ovate or obovate, always narrowing at the extreme tip to form a short acute apiculus. They are very small near the base of the stem but much larger (3–4 mm.) in the terminal rosette. In leaf shape, as in general habit, *P. pyriforme* comes very near some species of the related genus *Funaria*; from most species of *Pottia* it differs in its much larger size, and in leaf-cell structure.

The most striking features of the leaf under the microscope are the sharply toothed margin and large cells. Those towards the base of the leaf measure about $30 \times 100\mu$ and are rectangular in outline; farther up the leaf they are slightly narrower and appear more nearly rhomboid, but are not greatly altered. The nerve vanishes in the leaf tip.

Physcomitrium pyriforme is autoecious and is often abundantly fertile. The young fruit differs from that of *Funaria* spp. in its symmetrical, lobed calyptra. The ripe capsule is borne on a seta about 1 cm. long. As it ripens it swells to become ovoid or pear-shaped, but unlike *Funaria hygrometrica* it continues to be erect when fully ripe, and the lid has a short beak. Moreover, there is no peristome.

Ecology. It is essentially a colonist of bare mud in lowland situations. Thus it may be found on the bed of a dried-out pond, on the bank of a river, or on wet mud thrown out of a ditch, provided the soil is clayey and moist. It is rather common in many lowland districts, but tends to be scarce in the more mountainous parts of the country.

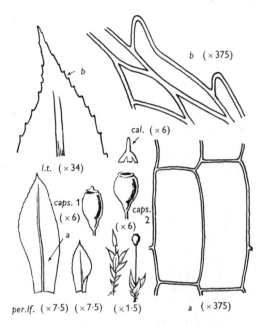

Fig. 71. *Physcomitrium pyriforme*: *per.lf.* perichaetial leaf, *l.t.* tip of same, *cal.* calyptra; *caps.* 1, capsule at time of shedding lid; *caps.* 2, old empty capsule. The left-hand plant at bottom centre shows an earlier stage in the developing capsule covered by calyptra.

RELATED SPECIES

Physcomitrella patens (Hedw.) B. & S. is the only British species of the genus *Physcomitrella*. It sometimes appears in abundance on the dried-up margins of ponds and similar places. It somewhat resembles *Phascum cuspidatum*, but may be distinguished by its leaf margins which are toothed towards the apex, and by the large $(40–80 \times 20\mu)$ uniformly rectangular to hexagonal cells of the leaf.

Ephemerum spp. are extremely small cleistocarpous mosses with a persistent protonema and short-lived leafy shoots. *E. serratum* (Hedw.) Hampe is the commonest species. It may be known by its delicate, coarsely toothed, nerveless leaves. It grows on bare soil in fallow fields or gardens, and by pond margins.

72. SPLACHNUM AMPULLACEUM Hedw.

S. ampullaceum is the commonest British representative of a genus of mosses remarkable for growing almost exclusively, perhaps in part saprophytically, on decaying animal excrement. Dense tufts of this moss may be found, often abundantly fertile, on cattle dung on boggy ground. It is easily identified by the peculiar shape of the erect capsule.

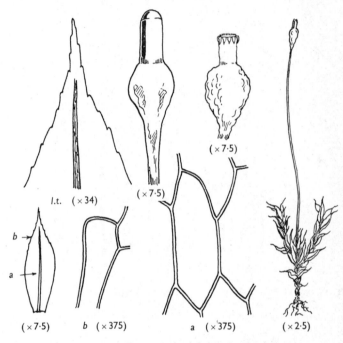

Fig. 72. *Splachnum ampullaceum*: the left-hand enlarged capsule is ripe but with lid intact, the right-hand one is old and shows the reflexed peristome teeth; *l.t.* leaf tip.

There is nothing very striking about the short upright stems and the light green leaves, tapering to acute points and twisted when dry. The leaf, however, is large (3–5 mm.) and is of a somewhat unusual shape, wide in the middle and greatly narrowed at base and apex. The leafy shoots somewhat resemble those of *Funaria hygrometrica*.

Under the microscope the leaf is immediately distinctive in its coarsely and distantly toothed margin, and very large cells. Only *F. hygrometrica*, among common mosses, has equally large cells (about $120 \times 30\mu$), and it lacks the coarsely toothed leaf margin.

Splachnum ampullaceum is autoecious, and capsules are commonly formed in great profusion. Each capsule is borne erect on a red seta 2–5 cm. in length and owes its peculiar form (like a Greek amphora) to the excessive development of the lower part, or apophysis; for in *Splachnum* this is much wider than the spore-containing part of the capsule. In *S. ampullaceum* the apophysis is much swollen and, tapering below, is approximately pear-shaped. The 16 peristome teeth are arranged in pairs and are reflexed over the margin of the ripe capsule in the dry condition. They are pale yellow in colour.

Ecology. It grows principally, though not exclusively, on the dung of cattle; it is certainly always found on an organic substratum of animal origin. It is met with chiefly in the hill districts of north and west Britain, on open moors.

ADDITIONAL SPECIES

S. ovatum Hedw. (*S. sphaericum* (L. fil.) Sw.) is not uncommon at high altitudes. It may be distinguished from *S. ampullaceum* by its scarcely toothed leaves, and its shorter, broader capsule with nearly globose, dark apophysis. The seta is weak and pale (cf. *Tetraplodon mnioides*).

RELATED SPECIES

Tetraplodon mnioides (Hedw.) B. & S. occurs not infrequently on decaying animal matter (chiefly on dead sheep) on mountains, where the massed effect of the stout dark red setae and purplish black capsules, with conspicuous enlarged apophyses, is very striking. The leaf tip tapers to a long fine point which is not toothed as it is in *Splachnum ampullaceum*, and the apophysis of the capsule is not as broad as in that species.

SCHISTOSTEGALES

73. SCHISTOSTEGA PENNATA (Hedw.) Hook. & Tayl. (*S. osmundacea* (Dicks.) Mohr)

In its apparently luminous protonema, and in its nerveless leaves arranged strictly in two ranks, this delicate plant of sandstone caves and rock clefts differs widely from all other British mosses. The slender shoots, with stems 5–12 mm. long, suggest miniature fern fronds (cf. *Fissidens* spp.). They are green towards the tips, but reddish brown below. *Schistostega* forms wide loose carpets rather than tufts. It is very local in distribution, but where it occurs it is sometimes very abundant.

The peculiar light-reflecting power of the protonema, which, as Dixon states, 'gives a beautiful golden-green lustre to the plant and seems to fill with light the crevices and caves where it grows', is due to its remarkable cell structure. Instead of the weft of branching threads that forms the protonema in most mosses, *Schistostega* develops a plate of almost lens-shaped cells. The convex outer walls enable these cells to function as light traps, the green sheen that is produced depending on this fact and on the position of the chloroplasts which are clustered against the 'back' wall of each cell.

The leaf is composed of large, elongated cells with pointed ends; often they are diamond-shaped and measure about 100μ long by 25μ wide. Each leaf is attached to the stem by a broad base with a vertical

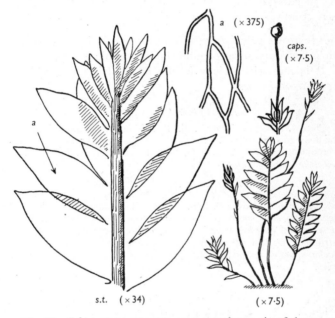

Fig. 73. *Schistostega pennata*: *caps.* capsule, *s.t.* tip of shoot.

insertion which is more or less confluent with the bases of adjacent leaves above and below it. This merging of the bases of the strictly 2-ranked leaves gives the shoot a unique appearance under the microscope.

The fruit of this dioecious moss is borne on a very slender seta, the whole fertile shoot being unusual in that it is almost bare of leaves except at the tip. The capsule is like a tiny pinhead, nearly globose in form and without a peristome.

Ecology. Its most characteristic habitats are sandstone caves, rabbit holes and deep fissures among sandstone rocks—always where the light intensity is very low. It also grows on granite and other siliceous rocks. In west Cornwall it is extremely abundant in old mine shafts.

TETRAPHIDALES

74. TETRAPHIS PELLUCIDA Hedw.

This moss grows in wide patches on peaty banks, and even more commonly on rotten tree stumps and logs. Although the patches may be extensive, the individual stems are very slender and short (0·5–1·5 cm. tall). They grow upright and are only sparingly branched. By far the

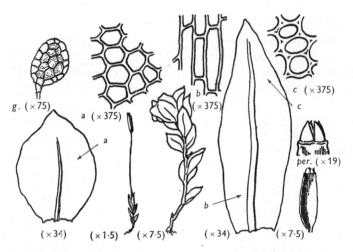

Fig. 74. *Tetraphis pellucida*: *g.* gemma, *per.* peristome; the leaf on the right is from a fertile stem, that on the left from a stem bearing gemmae, such as the drawing in the centre with its flower-like tip.

most conspicuous feature of the species lies in the cup-like endings of most of the sterile shoots. These cups, which contain minute green gemmae, are 1·5–2 mm. across and are formed of much enlarged leaves; when old they are brownish.

The lowest leaves are very small and rather widely spaced. Above they become larger (about 1 mm. long) and more crowded. In outline they are widely ovate, with shortly pointed tips. On fertile stems the leaves are usually longer, narrower in outline and more densely set. Indeed, the two types of shoot are quite distinct in appearance.

The leaf cells are isodiametric, rounded-hexagonal, and the nerve ceases just below the apex. Thus, although the shape of the leaf approaches that of some species of *Bryum*, the cell structure is totally different. The gemmae are rounded, biconvex, many-celled and disc-like.

Tetraphis pellucida is autoecious, but the narrowly cylindrical, upright capsules are not very common, though sometimes present in abundance.

The seta is about 1 cm. long. The capsule of *Tetraphis* is unique among mosses in having only 4 peristome teeth.

Ecology. The chief habitat is rotting tree stumps, but it also grows, more rarely, on peat banks or sandstone rock faces under trees. It is strictly acidophile and is said to require raw humus in the substratum.

EUBRYALES

75. LEPTOBRYUM PYRIFORME (Hedw.) Wils.

This species is most likely to be met with in greenhouses or on cinder heaps, the rather short erect stems and bright green, extremely narrow, finely pointed leaves suggesting a species of *Dicranella*. Indeed *Leptobryum pyriforme* is remarkable in that it combines a *Dicranella*-like leaf

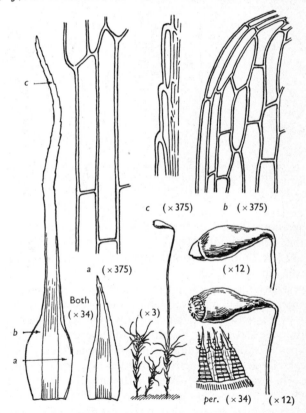

Fig. 75. *Leptobryum pyriforme*: *per.* peristome; the lower of the two enlarged capsules has the lid removed showing the peristome; the leaf on the left is a perichaetial leaf.

with a *Bryum* type of capsule. The upper leaves are much longer than the lower ones, and this might cause barren material to be confused with a species of *Pleuridium*. Capsules, however, are commonly present, and then the plant presents no difficulty.

The annual upright stems are slender and thread-like, 1–2 cm. tall; the leaves forming the terminal tuft may exceed 5 mm. in length and are wavy in outline; each leaf is broad at the base, then abruptly contracted into a very long fine point. The much shorter lower leaves are lanceolate.

The cells throughout the leaf are much elongated (70–140μ by 7–10μ), a character separating this plant from most of the Dicranaceae and their relatives. The nerve is broad (though not as broad as in *Campylopus* spp.) and occupies most of the narrow upper part of the leaf; its limits, too, are ill defined, not clear-cut as they are in *Campylopus*. The leaf margin is lightly toothed, especially near the apex.

Antheridia and archegonia normally occur together in the same 'inflorescence' in *Leptobryum pyriforme*, although at times plants may be found that are male only, or with 'inflorescences' that are predominantly male. The capsule is pear-shaped, horizontal or slightly drooping; in a ripe capsule the long, dark brown neck, the glossy red-brown body and the prominent, pale yellow peristome are characteristic. The orange-red seta is 1–4 cm. long.

Ecology. It is most abundant on cinder heaps, on soil in pots in greenhouses, on burnt peat and other 'man-made' habitats. In natural habitats such as sandstone rocks or peaty banks it is much less common. Its presence may depend to some extent on fires, though it is much less frequent than *Funaria hygrometrica* on burnt ground.

RELATED SPECIES

Orthodontium gracile (Wils.) Schwaegr. resembles *Leptobryum* in its narrow leaves and in size. It is distinct, however, in its rather wide cells and in the capsule which is narrowly cylindrical (2·5 mm. long) with a tapering neck. It is a rare moss of sandstone rocks.

Orthodontium lineare Schwaegr. (*O. gracile* (Wils.) Schwaegr. var. *heterocarpum* W. Watson) has a shorter, broader, somewhat asymmetrical capsule. It might easily be mistaken for a small member of the Dicranales. It was first observed in 1920, near Greenfield, in west Yorkshire. In the last few years it has been found in numerous localities in the southern, midland and northern counties and in East Anglia, and appears to be rapidly extending its range. It grows on rotten wood (usually coniferous) and on peaty banks and is often very abundant where it occurs.

76. **POHLIA NUTANS** (Hedw.) Lindb. (*Webera nutans* Hedw.)

A common plant of heaths and moors, and to some extent of woodlands, *P. nutans* has narrower leaves than most species of *Bryum*, whilst the dull, rather deep green colour is unlike that of most of the less common species of *Pohlia*. Its colour, long reddish seta and drooping capsule make it one of the first mosses of peaty soils that the beginner will learn to recognize. It forms carpets rather than cushions; the rather short (1–2 cm.) erect stems bear lanceolate leaves which are small and widely spaced below, longer and more crowded towards the tips of the shoots. Although variable, the plant usually has a combination of characters by which it may be known without difficulty. Early in the year, with its pendulous, green, unripe capsules,

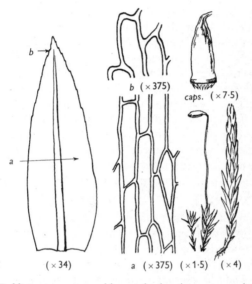

Fig. 76. *Pohlia nutans*: *caps.* old capsule showing outer peristome teeth; the tip of the shoot on the extreme right, with appressed leaves, breaks off readily and serves for propagation.

it might be mistaken in the field for a species of *Bryum*, as indeed it might when the capsules are ripe in summer; late in the season it is often marked by the persistence of the seta after the capsule has ripened and withered.

The nerve ceases in the toothed leaf apex, whilst the leaf margin lacks the border found in many species of *Bryum*. The cells (about 12μ wide) are characteristically narrower than in *Bryum* and elongated-hexagonal in shape. This combination of microscopic leaf characters, indeed, will not be found in any species of that genus.

Pohlia nutans is paroecious, antheridia occurring in the axils of some of the upper leaves of the shoots that bear the archegonia. The capsule (3 mm. long) varies from nearly horizontal to pendulous, and is borne

on a seta 1·5–4 cm. long. The capsule has a distinct neck, though this
is not so long as in most other species of *Pohlia*.

Ecology. A calcifuge, it grows on peaty or sandy banks, or on decaying
wood. It often covers large areas on open moors and heathland, and
is one of the most abundant mosses of such habitats, ascending to high
altitudes. *P. nutans* is an almost constant member of the limited moss
flora of cut peat surfaces in north-west Britain, where it is associated
with *Dicranella heteromalla* and species of *Campylopus* and *Polytrichum*.

77. **POHLIA DELICATULA** (Hedw.) Grout (*Webera carnea* Schp.)

The scattered, or at least never densely tufted, stems of this plant are
usually only about 1 cm. tall and bear delicate, pale green leaves, the
stems and leaf bases often being red. Only in the rather less common
P. albicans and *P. proligera* is a similar colour met with, and these
species are normally much taller. *P. delicatula* is quite common on
lowland clay banks.

The leaf is ovate to lanceolate, those leaves near the base of the stem
being commonly much shorter (under 1 mm.) and more rounded, while
those in the terminal tuft are relatively long (1·5 mm.) and narrow.

P. delicatula is separated from all other species of *Pohlia* except
P. albicans (a taller plant) by its comparatively wide leaf cells (about
20μ); and the nerve ceasing well below the bluntly toothed leaf apex
will distinguish *P. delicatula* from any species of *Bryum* that might
otherwise resemble it.

This species is dioecious, but is abundantly fertile at times. The first
impression given by the plant in fruit, with its pendulous capsules, is
of a small species of *Bryum*, such as *B. bicolor*; but the short seta
(7–12 mm.) has a peculiar, rather fleshy consistency, and the dull greyish

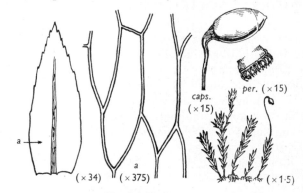

Fig. 77. *Pohlia delicatula*: *caps.* capsule with
lid intact, *per.* peristome.

pink capsule is considerably smaller than in that species; also, a definite neck is present. The old dry capsule is wide-mouthed, with orange-red, spreading peristome teeth, and the part of the seta just below the capsule then becomes strongly hooked or curled in a characteristic way.

Ecology. Wet clay banks, especially by streams, are its principal habitat. It overlaps *Pohlia albicans* in range, but is more exclusively a lowland plant.

ADDITIONAL SPECIES

Besides the two common species described here, several others should be mentioned.

Pohlia elongata Hedw. (*Webera elongata* (Hedw.) Schwaegr.) is often plentiful on banks and ledges in mountainous districts. The capsules, which are excessively long (4–6 mm.) and narrow, with tapering necks, are held horizontally; when young and green they readily catch the eye, for each is borne on a seta 2–3 cm. long.

Pohlia cruda Hedw. (*Webera cruda* (Hedw.) Bruch) is a rather robust species found in rock clefts on mountains, and notable for the opalescent lustre of its pale green leaves.

Pohlia annotina (Hedw.) Loeske (*Webera annotina* (Hedw.) Bruch) and the closely related *Pohlia proligera* Lindb. (*Webera proligera* (Lindb.) Kindb.) are two species found on sandstone rocks and sandy ground, both chiefly remarkable for the fact that they regularly produce gemmae in the axils of the leaves. Those of *Pohlia annotina* are widely ovoid bulbils; those of *P. proligera* have been aptly likened to an empty glove in form.*

P. albicans (Wahl.) Lindb. (*Webera albicans* (Wahl.) Schp.) is a species occurring both on lowland clay banks and on wet mountain ledges, always attracting attention by the strikingly pale, glaucous green of its leafy shoots. It has the wide leaf cells of *Pohlia delicatula*, but the whole plant is taller (2–8 cm.) and the seta much longer (2–2·5 cm.).

RELATED SPECIES

Plagiobryum zierii (Hedw.) Lindb. is a rather striking little moss which occurs on ledges and in crevices of base-rich mountain rocks in the north and west. It bears a strong superficial resemblance to *Bryum argenteum* but has a distinctive pinkish tinge in its lower parts, by which it may usually be known. The cells of the leaf are larger and wider than in *B. argenteum*, and the extremely long narrow capsule, with tapering neck, is unmistakable.

78. ANOMOBRYUM FILIFORME (Dicks.) Husn. (*Bryum filiforme* Dicks.)

This moss resembles the very common *Bryum argenteum* in its silvery appearance and catkin-like (julaceous) branches. The leaves are concave and incurved, and overlap so as to give the whole shoot (especially when dry) a smooth cylindrical outline. This plant, however, has a yellowish tinge that is lacking in *B. argenteum*. *Anomobryum filiforme* is of slender habit, with stems commonly 1·5–4 cm. long. Its habitat is quite different from that of *Bryum argenteum*, for it is a plant of wet gravelly soil in hill districts.

* This is the diagnosis of *P. proligera* as understood by Dixon. More recent work, however, suggests that true *P. proligera* Lindb. is a relatively rare plant distinguished by lustrous leaves and usually having a single point to the bulbil. On this interpretation of *P. proligera*, plants with 'glove-shaped' bulbils are merely forms of *P. annotina*.

The leaf is small (under 1 mm. long), very broad, concave, and obtuse or bluntly pointed at the apex. Under the microscope, the nerve is seen to cease in or below the apex of the leaf. The cells afford a useful confirmatory character, for they are wide (approx. $15 \times 60\mu$) in the leaf base but narrow and finely pointed (5×40–90μ) farther up the leaf. The leaf margin is neither thickened nor recurved and is either entire or very faintly toothed.

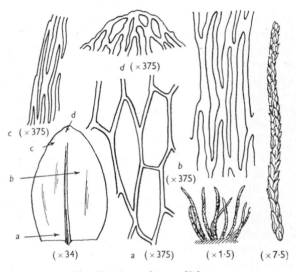

Fig. 78. *Anomobryum filiforme.*

Anomobryum filiforme is dioecious. Antheridia are found on bud-like branches. The capsule is borne on a red seta about 2 cm. long. It is horizontal to pendulous and has a rather long, tapering neck. The small, yellow peristome is perfect.

Ecology. It occurs most often in moist gravelly places in hill country, e.g. stony stream beds partially dried out, quarries, etc. Frequently it grows with *Bryum pallens* and *Dicranella squarrosa* on wet stream-side detritus. It is also fairly common in clefts of wet rocks.

BRYUM

General notes on the genus. The genus *Bryum* may generally be recognized by the combination of ovate or broadly oblong-lanceolate leaf shape and uniform, slightly elongated, hexagonal or rhomboid cells throughout the lamina of the leaf. *Pohlia* spp. have narrower cells as a rule and the leaf margin is usually conspicuously notched near the apex. The capsule in *Bryum* is shortly ovoid to elongate-ellipsoidal in shape, but is always horizontal or pendulous (never erect) when ripe and the lid is never beaked. The peristome is characteristically double, but the inner peristome is imperfect in some species.

The species of this large genus present exceptional difficulties in determination. This arises in great part from the fact that the range of form, in both leafy shoot and capsule, is relatively limited, so that specific distinctions rest on quite slight but constant differences.

Only a few well-defined species (e.g. *B. capillare*, *B. argenteum*, *B. alpinum*) may be safely determined without fruit. Even with fertile plants it should be stressed that it is essential to secure perfectly ripe (but not old and shrivelled) capsules. These should preferably be at the stage when the lid has just been shed, for only at this stage is it possible to be sure of the precise form of the mature capsule. Thus capsules gathered too young will become contracted below the mouth in a way that would not be natural to a perfectly ripened capsule of the same species.

The 'perfect' peristome, found in many species of *Bryum*, consists of outer teeth, inner teeth and, alternating with the inner teeth, a series of slender threads with transverse bars on them; these threads are termed appendiculate cilia. To see these peristome details properly and to ascertain how far any particular peristome is imperfect, it is necessary to make a preparation and examine it under the high power of the microscope (see Introduction, p. 9). Some information, however, may be gained by viewing the mouth of the capsule from above, by reflected light, under a low power. Spore size is sometimes an additional character of value.

79. **BRYUM PALLENS** (Brid.) Röhl.

This moss is extremely variable in its appearance, the most striking feature of *B. pallens* (in its more typical forms) being the prevailing reddish tint. Sometimes it approaches rose red, but more commonly it is pink, and this colouring is often a useful difference from other mosses which grow with it on loose damp soil or detritus, on banks or mountain ledges. The stems may be anything from under 1 to 7 cm. long, and the leaves range in length from 1 to 3 mm. It is wise, therefore, to examine all doubtful material under the microscope.

The leaf has the characteristic large, wide cells of a *Bryum*, the width in mid-leaf being about 20μ. In this *B. pallens* differs from those species of *Pohlia* which it approaches in leaf shape. The nerve reaches, or extends just beyond, the apex, which is bluntly notched as a rule. The leaf margin is both thickened and recurved, and is composed of very long, narrow cells; thus it forms an obvious border which at once distinguishes *Bryum pallens* from any species of *Pohlia* and most other common species of *Bryum*. In *B. inclinatum* and *B. capillare*, which have bordered leaves, the border is not thickened as in typical *B. pallens*; and *B. pseudotriquetrum* is usually more robust than even the largest forms of this species. Abnormal plants occur, however, which have the marginal thickening ill developed, so that identification cannot always be certain without mature fruit.

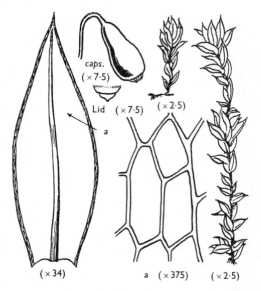

caps. (×7·5)

Lid (×7·5)   (×2·5)

a

(×34)   a (×375)   (×2·5)

Fig. 79. *Bryum pallens*: the shoots drawn represent two extreme forms of the plant; *caps.* capsule with lid removed.

B. pallens is dioecious and fruiting plants are not very common. The capsule is borne on a seta 2–3 cm. long; it is pear-shaped, with a long tapering neck, and varies between horizontal and pendulous. Valuable characters are provided by the pale yellow colour of the outer peristome teeth and the presence of appendages on the cilia of the inner peristome (fig. V, p. 21). In most other species of *Bryum* possessing this so-called 'perfect' peristome the outer teeth are darker (e.g. orange-brown) in colour, at least at the base.

Ecology. Its habitat varies, but it always prefers moist situations. Typically it is found on patches of wet detritus by streams in hill districts (cf. *Anomobryum filiforme*); but it also grows on damp walls, in wet hollows of sand dunes, and in wet places generally. In many lowland districts it shows a preference for clay.

80. BRYUM PSEUDOTRIQUETRUM (Hedw.) Schwaegr.

This is the commonest species of *Bryum* in marshes and bogs, where it may generally be known by its size, colour and habit of growth. The stems are commonly 4–10 cm. tall and are matted together below with brownish tomentum. In colour the shoots vary from light to dark green,

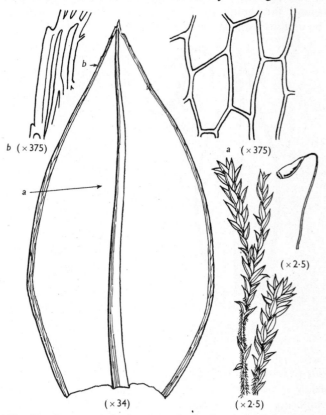

b (×375)   *a* (×375)

(×2·5)

(×34)   (×2·5)

Fig. 80. *Bryum pseudotriquetrum*: in the drawings of the shoot, the extreme bases of the plants are not shown.

but almost always have some tinge of reddish purple. The leaves form a rather wide angle with the stem, except at the tip where they are gathered into fairly conspicuous terminal tufts.

Each leaf is broadly ovate-lanceolate, tapering to a short acute apex; in all superficial characters, indeed, it is an enlarged replica of the leaf of *B. pallens*. Well-developed leaves are 2–3·5 mm. long.

Under the microscope the leaf margin is seen to be slightly recurved, and composed of long, narrow cells. The border thus formed, however, is not thickened as in *B. pallens*. The leaf apex often has a few small, blunt teeth. The strong reddish brown nerve either reaches the apex or ceases just below it. The leaf cells are of the usual wide *Bryum* type, but are rather smaller than in many species of the genus (50–90 × 10–18 μ).

B. pseudotriquetrum is dioecious,* the large terminal male 'flowers' reminding one of those in the common moss, *Mnium hornum*. The capsule, which is not very common, is borne on a long seta (it may exceed 5 cm.) and is more or less pendulous. It has a long tapering neck, and is often asymmetrical, being slightly swollen on the lower surface. The peristome is large and perfect, the outer teeth being yellow at their tips but darkening to orange-brown at the base.

Ecology. It occurs in various types of bog and marsh habitats; thus, it is a frequent species in dune slacks, but is also typical of bogs by springs and by small mountain streams. In these wet flushes on mountains the characteristic species associated with it are *Philonotis fontana*, *Dicranella squarrosa*, and often *Brachythecium rivulare* and the liverwort *Scapania undulata*.

81. BRYUM CAESPITICIUM Hedw.

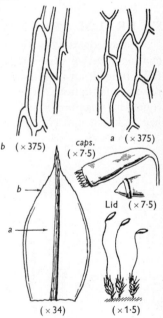

B. caespiticium forms wide patches on bare ground or dense tufts on walls and rocks. It is recognized as a species of *Bryum* by the crowding of the broad leaves into dense tufts at the ends of the close-packed shoots. The species is distinct from most other low-growing members of the genus in its vivid light green colour, rather silky texture and fairly long-pointed leaves. The leaf point, in fact, is longer than in any other common species of *Bryum* except *B. capillare*. The latter is, in general, a more robust plant, and may always be known by its leaves which become corkscrew-curled in the dry state, whereas in the present species the leaves are little altered when dry. The leafy shoot is only about 0·5–1 cm. tall.

* Except in the variety *bimum* (Brid.) Richards & Wallace, which has commonly been regarded as a distinct species—*B. bimum* Brid.

Fig. 81. *Bryum caespiticium: caps.* capsule with lid removed, showing outer peristome teeth.

Under the microscope the nerve is seen to extend beyond the leaf tip as a green, lightly toothed cusp. It is thus not white and hyaline as are the hair-points of the leaves in *Grimmia*, *Rhacomitrium* and *Tortula*. The leaf margins are rather strongly revolute, but the somewhat narrower marginal cells do not form a very well-marked border to the leaf.

This species is dioecious, and is found more commonly barren than fertile. The brownish red seta may reach 3 cm. in length. Useful capsule characters are: (i) the angle at which the capsule is held—more nearly horizontal than in many species of *Bryum*, (ii) its form, pear-shaped, with short neck and a tendency to be slightly swollen asymmetrically, and (iii) the peristome, pale in colour and double, with appendiculate cilia (see Glossary) between the inner teeth (perfect). The spores are small, only 8–14μ.

Ecology. It is, perhaps, particularly characteristic of dry wall-tops, but occurs on bare ground of various types. Thus it grows on rock or soil and appears to show no marked preference for calcareous or acidic habitats. Its requirements, however, would probably repay fuller investigation. Dr E. W. Jones notes that it grows in much more exposed places than *B. capillare*.

82. BRYUM ARGENTEUM Hedw.

At the edges of pavements in towns, where little else grows, the characteristic silvery grey patches of this moss may commonly be found. It is, indeed, an abundant species in a variety of unpromising-looking roadside situations. The short upright stems grow densely packed together, reaching 1–1·5 cm. in height only in robust forms; and the leaves, being concave and crowded, give each shoot a smooth cylindrical

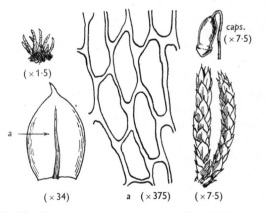

caps.
(×7·5)

(×1·5)

a

(×34)   a (×375)   (×7·5)

Fig. 82. *Bryum argenteum*: *caps.* capsule with lid intact.

outline and narrow catkin-like form (cf. *Anomobryum filiforme*, which differs in yellowish green colour). This unusual form of shoot, coupled with the silvery grey colour, will make *Bryum argenteum* an easy plant to recognize at most times. The only puzzling forms are those that are deep green rather than greyish; these are mostly robust states that occur in moist, shady places.

The microscope shows the leaf to be rounded in outline, but drawn out to a short point at the apex. The nerve, however, ceases well below the apex. The areolation is notably distinct in that the cells are narrow and wavy in outline, with walls much thickened at the angles.

Although *B. argenteum* is dioecious, capsules are quite commonly formed in abundance. As in *B. bicolor*, each capsule is raised on a short seta (about 1 cm. long) and is itself very small (1·5 mm.) and pendulous. In its foreshortened form—passing abruptly into the seta—the capsule closely resembles that of *B. bicolor*, but it ripens more often to a deep reddish brown, less commonly to dark red as in that species. I have, on occasion, found *B. argenteum* with the body of the capsule deep crimson and the lid a bright vermilion, when it has certainly approached *B. bicolor* in intensity of colour.

Ecology. This species is probably nitrophilous, for its chief 'natural' rock habitats are in the neighbourhood of gull colonies, etc.; most typically it grows in 'man-made' habitats, such as cracks in paving stones, roadsides, roofs and wall-tops. It appears to be remarkably tolerant of the atmospheric pollution that occurs in large towns.

83. BRYUM BICOLOR Dicks. (*B. atropurpureum* Web. & Mohr)

This small, low-growing species, with stems often barely 1 cm. high, and leaves gathered in the dense tufts characteristic of the genus, is a common plant of stony or sandy waste ground, where it grows in wide patches, often in association with *B. argenteum*. From the latter it is usually known at once by its green (not silvery grey) colour. The leaf varies from ovate to ovate-lanceolate. Acute-tipped, it is normally longer and narrower than that of *B. argenteum*. It is, in fact, a miniature replica of the leaf of one of the larger species of *Bryum*, such as *B. pallens*.

The microscope will reveal a rather wide, thin-walled type of areolation, a nerve vanishing in the extreme leaf tip or running out just beyond it, and a leaf margin composed of cells which are narrower in form, but not sharply distinct from the rest of the leaf. Gemmae of bulbil form may sometimes be found in the axils of the upper leaves.

A deep purple-red colour normally develops in the ripe capsule of this dioecious moss; and this rather striking colouring sometimes affords a useful diagnostic character. But it is a character that should be taken in conjunction with the shape—short and rounded, with ill-developed neck—a type of capsule differing widely from that found in all the other common species except *B. argenteum.* Not only is the capsule very small (1–2·5 mm.), but the curved seta is short, usually only 8–15 mm. tall. The lid of the capsule is conical, and bright red in colour. The peristome is perfect.

Ecology. It grows typically on bare soil on the ground, occurring in some of the same habitats as *B. caespiticium* and *B. argenteum.* However, it shows a preference for, and reaches its best development on clay. Dr E. W. Jones has noted its preference for wet places and basic soils.

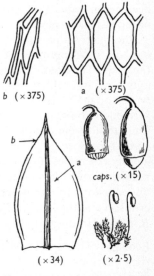

b (×375)　　　*a* (×375)

caps. (×15)

(×34)　　　(×2·5)

Fig. 83. *Bryum bicolor*: the two capsules (*caps.*) show something of the variation in size, that on the left has the lid removed exposing the peristome.

84. BRYUM ALPINUM With.

The bright metallic gloss of the leaves and the deep red or purple-brown colouring (green in var. *viride* Husn.) are the two features which combine to make *B. alpinum* one of the most easily recognized mosses of mountain habitats. It is not confined to high altitudes, but it is only in the wet moorland and mountain country of the north and west that it becomes a common plant. It is one of our taller-growing species of *Bryum*, the densely tufted upright stems at times reaching 7–8 cm., though usually it is considerably shorter. The leaves are rather narrow (oblong-lanceolate) in outline, close-set and not greatly altered by drying. They are somewhat concave in form.

Under the microscope the strong nerve is seen to continue almost to the apex, or even to extend beyond it into a short protruding point. The leaf margin is recurved. The microscopic character, however, which sets *B. alpinum* apart from all other common species with similarly shaped leaves is the extremely narrow areolation. With their thickened walls and narrow rhomboidal cavities the cells here (up to $80 \times 10\mu$) are very different from those in the last few species described. The result is that the still narrower cells of the margin do not form a clear-cut border. The cells, in fact, resemble those of many pleurocarpous mosses of the Hypnaceae.

B. alpinum is dioecious, and the capsule is not common. It is raised on a seta about 2 cm. in length, and is pear-shaped and pendulous (up to

2·5 mm. long), with an evident neck and a short lid; when ripe it becomes deep crimson. The reddish peristome is perfect. In its capsule characters, indeed, the species displays its close affinity with *B. bicolor*.

Ecology. It grows best on moist siliceous rocks and stony ground exposed to full sunlight. Such habitats, in moorland and mountain country, it often shares with *Campylopus atrovirens*. *Bryum alpinum* is not well named, since it is not characteristically an alpine plant.

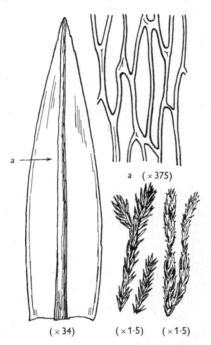

a   (× 375)

(× 34)   (× 1·5)   (× 1·5)

Fig. 84. *Bryum alpinum*: the forked shoot on the right is in the dry state with leaves appressed.

85. BRYUM CAPILLARE Hedw.

This is not only our most widespread species of *Bryum* but is indeed one of the most abundant British mosses. The dense tufts or wide cushions are notable for the corkscrew-like spiralling of the leaves when dry. When moist, fertile material is met with on some wall-top— a favourite habitat—the plants are conspicuous on account of their robust habit, broad leaves and large drooping capsules. The freely branched stems commonly attain 2–5 cm. in height. The leaf, on closer inspection, is seen to be broadened above the middle, and to possess a long, fine, protruding point, which is not, however, hyaline and silvery-looking as in *Grimmia pulvinata* or *Tortula muralis*, two species often found with *Bryum capillare* on wall-tops.

Under the microscope, the leaf cells are seen to be wide, as in most species of *Bryum*, measuring up to $70 \times 35\mu$. The nerve continues into the projecting leaf apex; distinct, narrow, marginal cells form an evident (but not thickened) border; and the leaf margin is to some extent

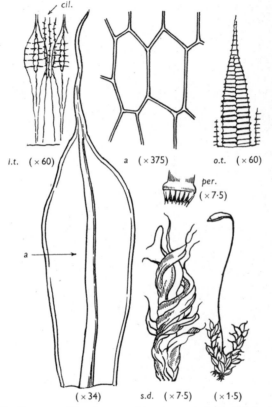

i.t. (× 60) a (× 375) *o.t.* (× 60)

per. (× 7·5)

a

(× 34) s.d. (× 7·5) (× 1·5)

Fig. 85. *Bryum capillare*: *per.* peristome, *o.t.* outer peristome tooth (outside view), *i.t.* two teeth and three appendiculate cilia (*cil.*) from inner peristome, *s.d.* tip of shoot in a dry state.

recurved. But *B. capillare* is a highly variable moss and presents no single microscopic character of absolute diagnostic value. In general, however, the obovate leaf shape and long fine point to the leaf will separate it from all other species of the genus.

B. capillare is typically dioecious but, as indicated above, fertile material is common enough. The capsule is borne on a reddish seta up to 3 cm. in length, and when ripe is inclined or horizontal, and about 4 mm. long. It ripens to a bright or dark brown colour, with blunt orange lid, red mouth and perfect peristome. The outer teeth are yellow or reddish at the base and paler above.

Ecology. It is extremely catholic in its choice of habitat, growing readily on rocks, wall-tops, roof tiles, trees, fences and mountain ledges; in addition, it has been recorded on chalk downs, grassy heaths and coastal beaches, but it attains its most luxuriant development on soil-capped walls. When growing on trees or on mountain ledges it is often very stunted.

ADDITIONAL SPECIES

Several other species of this difficult genus are by no means rare.

B. pendulum (Hornsch.) Schp. and *B. inclinatum* (Brid.) Bland. resemble one another very closely, both being moderately robust plants which occur in heathy and sandy places. The only certain distinction between them lies in the capsule. The cilia of the inner peristome are ill developed or absent in both, but the outer peristome of *B. pendulum* is distinct in having the transverse bars on it connected by an irregular network of vertical lines; and the spores average 30μ in diameter, against 20μ in *B. inclinatum*.

B. intermedium (Ludw.) Brid. is nearly allied to *B. caespiticium*, which it much resembles. It is, however, synoecious, so that it should be possible to find the remains of old antheridia about the base of the seta. Also, the capsule is notably narrow in the mouth and contains large spores, $20–25\mu$ against $8–14\mu$ in *B. caespiticium*.

B. erythrocarpum Schwaegr. and *B. murale* Wils. are the two species which come nearest to *B. bicolor*, and with it compose a natural group notable for the deep red colour of the ripe capsules. In *B. erythrocarpum*, which is a locally common plant of sandy heaths and fallow fields, both capsule (3 mm.) and seta (up to 2 cm.) are longer than in *B. bicolor*, and even when barren it may often be recognized by means of the globular red gemmae which occur at the base of the stems. *B. murale*, which is much less common, occurs sporadically on the mortar of walls, and bears larger capsules even than the last.

86. **RHODOBRYUM ROSEUM** (Hedw.) Limpr. (*Bryum roseum* (Hedw.) Brid.)

This plant may be known at once by the wide and conspicuous rosettes formed by the uppermost leaves. The robust upright stems, which are the upturned ends of long underground runners, often grow to a fair height (4–6 cm.) and sometimes continue above the terminal rosette, so that in time successive rosettes of leaves will be formed. The uppermost leaves are extremely large and proportionately wide, so that well-formed rosettes may be 2–2·5 cm. across and may look almost like flowers with green petals. The leaves are of a rather frail and filmy texture, as in *Mnium*. *Rhodobryum roseum* is not a very common moss, but where it does occur, on some sheltered bank, on sand dunes or in open woodland, it is hardly likely to be overlooked.

Closer inspection shows the leaf to be obovate (broadening above the middle) and to be without the thickened border that is found in all our commonest species of *Mnium*. Superficially, however, the leaf here comes very close to that of *M. stellare*, which it resembles in shape and texture, in the rather distantly toothed apex and in the total lack of a thickened border. But the areolation in *Rhodobryum roseum* is that of a *Bryum*, the cells being elongate-hexagonal in shape, not short as in *Mnium*. This distinctive species, then, may be summed up as one

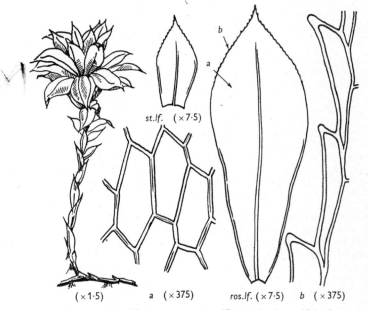

st.lf. (×7·5)

(×1·5)   a (×375)   ros.lf. (×7·5)   b (×375)

Fig. 86. *Rhodobryum roseum*: *st.lf.* stem leaf, *ros.lf.* leaf
from large terminal rosette.

displaying something of the habit and general appearance of a *Mnium*, whilst retaining the cell structure of the genus *Bryum*. The rosettes of *Rhodobryum roseum*, however, are not matched in any species of *Mnium*.

Rhodobryum roseum is dioecious and the fruit is exceedingly rare. When fertile material is found, however, 2 or 3 capsules may be seen arising from a single inflorescence. Each capsule is oblong-cylindrical, and is pendulous on a seta 3–4 cm. long.

Ecology. It seldom forms 'pure' tufts, but occurs as scattered stems amongst grasses and other plants on banks. These may be in woodland, heath or sand-dune communities, and associated bryophyte species will be correspondingly varied. Dixon implies that it prefers shade, but it also occurs, though perhaps less commonly, on exposed sunny banks and in open grassland.

87. **MNIUM HORNUM** Hedw.

The rather dull, dark green tufts of this moss frequently form extensive carpets on the ground in woodland. Indeed, not only is it our most abundant species of the genus, but on many a woodland bank it is the dominant moss. The light green of the young shoot tips in spring

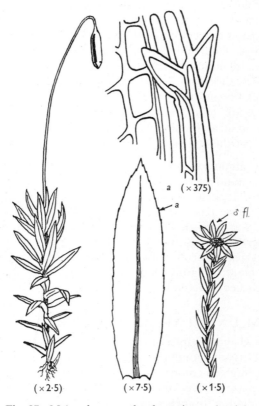

a (×375)

a

♂ fl.

(×2·5)   (×7·5)   (×1·5)

Fig. 87. *Mnium hornum*: the shoot tip on the right
shows a male 'inflorescence' (♂ *fl.*).

presents a striking contrast with the dull green of the older leaves. If a tuft be separated out it will be seen that the individual, upright, sparingly branched stems are matted together below by means of reddish brown radicles. Though rather robust and usually conspicuous—owing to fairly large (4 mm.) crowded leaves—*M. hornum* is seldom more than 4 cm. tall and often much shorter.

A lens will reveal the little teeth which occur all along the margin of the oblong-lanceolate, somewhat filmy, bordered leaves. The leaves become curled inwards (though not tightly crisped or twisted) when dry. A microscope shows that each 'tooth' consists of two separate one-celled projections of the leaf margin, and that the leaf border is composed of long, narrow cells; the remaining leaf cells are fairly large (20–25 μ across) and isodiametric; the nerve ends below the extreme tip of the leaf. Examination of details will seldom be necessary, however, for no other common moss has a leaf of precisely this shape, with strong border and twin teeth, with the exception of *Atrichum undulatum*; and that species is distinct in habit, and in its larger, more translucent, undulate leaves. Confusion can only arise with certain rarer species of *Mnium* (see the notes on additional species, p. 227).

M. hornum is dioecious; but both the flower-like terminal rosettes of the male plants and the fine pendulous capsule, on its reddish seta 2·5–5 cm. long, are quite common. The capsule itself is 5 mm. long, and has a short blunt lid and well-marked tapering neck. It closely resembles those of some of the larger species of *Bryum*.

Ecology. Its habitat range is wide, and it may be that some physical factor of the habitat such as good drainage is common to all the places where it grows. It occurs on humus, peat, rotten wood and rock ledges, in this last position reaching high altitudes on mountains. It seems to be especially characteristic of beech woods, in which it is often locally dominant over wide areas of ground. It is probably always on an acid substratum; thus when on chalk, limestone or chalky boulder clay it will actually be growing on rotten wood, tree bases or on the surface soil locally leached of its lime.

88. **MNIUM UNDULATUM** Hedw. (Plate VII)

A widespread and conspicuous moss of grassy banks, woodlands, and shaded rock clefts on mountains, *M. undulatum* forms extensive patches of a rather light green colour. In well-grown specimens there is usually an erect stem (several centimetres tall), from the summit of which numerous radiating branches may arise, so that the habit becomes 'tree-like' (dendroid). Many of the erect shoots, however, are un-branched and curved; whilst from the base of the plant arise long runner shoots.

The filmy, tongue-shaped leaves vary much in size, but attain 8–12 mm. near the tips of the more robust stems, and are strongly undulate; in this, as in shape and texture, they resemble those of *Atrichum undulatum*, but are less tapering towards the apex. Also, as in that plant, they become much crisped on drying.

The microscope will reveal an absence of the longitudinal plates or lamellae upon the nerve which distinguish the leaf of *Atrichum*. More-

Plate VII. *Mnium undulatum* (natural size)

Plate VIII. *Thamnium alopecurum* (natural size)

over, the nerve here is continuous into the toothed, protruding apex, and the teeth along the bordered leaf margin are single, not double as in *A. undulatum* and *Mnium hornum*. The cells average about 15μ in diameter—smaller than in many species of *Mnium*—and tend to be broader than long.

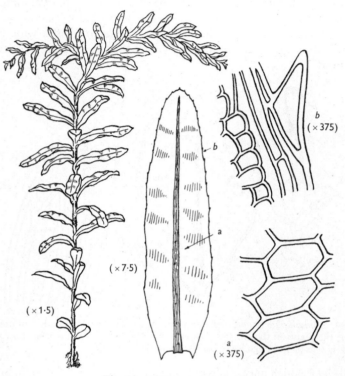

Fig. 88. *Mnium undulatum.*

This species is dioecious, and male plants are striking objects with their wide-open, flower-like tips. The fertile condition is rather rare, but since 2–10 capsules may arise from a single rosette of leaves the plant is very striking when in fruit. Each capsule is ovoid to pear-shaped, pendulous on an orange seta 3 cm. in length.

Ecology. It grows on soil or soil-covered rocks, wherever its shade and moisture requirements are fulfilled. Its luxuriant growth in a wood usually indicates a relatively fertile soil, near neutral in reaction; but it is not strictly confined to such soils. Often it grows in grass on shady banks.

89. MNIUM PUNCTATUM Hedw.

A robust, erect-growing moss of very distinctive appearance, *M. punctatum* occurs commonly in a variety of moist shaded situations, including mountain habitats. The stems are 2–8 cm. tall and matted below with red-brown tomentum. At the base the leaves are small, oval and widely spaced, but they gradually increase in size upwards, so that at the shoot tip the bases of the large, rounded leaves tend to form a wide funnel. The colour is light green in the very young shoots, deep green or reddish when old.

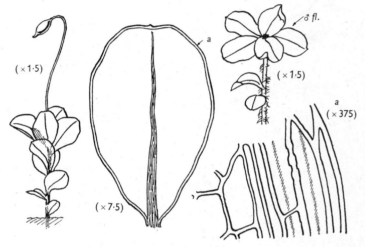

Fig. 89. *Mnium punctatum*: the shoot tip on the right shows a male 'inflorescence' (♂ fl.); the plant on the left has a capsule with the lid intact; the leaf shown in the centre is one of the large leaves of the male 'inflorescence'.

The leaf, which may attain 6–8 mm. in length, and 4–5 mm. in breadth, expands quickly from a very narrow insertion to a wide blunt apex. It is thus broadly obovate in outline, and is further distinguished, with a lens, by the presence of an evident thick border, which is quite without teeth. The rounded leaf apex is usually interrupted by a very short protruding point, or apiculus. *M. pseudopunctatum*, the species which most resembles *M. punctatum*, has the leaf border unthickened.

The microscope will show the border of the leaf to be composed of several layers of long, narrow, thick-walled cells, very different from the roughly hexagonal, isodiametric cells that characterize the leaf as a whole. The single nerve either extends into the apiculus or ceases just below it.

M. punctatum is dioecious, and the male plants, like those of the related species, will attract attention on account of their wide-open terminal rosettes of leaves. The rather narrowly ovoid capsule is borne on an orange-coloured seta 2–3 cm. in length. The lid has a notably long beak.

Ecology. It demands some shade and considerable moisture but will grow on various substrata and is probably indifferent as regards soil reaction. Thus its principal habitats include wet stony ground by streams, deep rock clefts on mountains, and the raw humus of decaying tree stumps in swampy woodland. In Berkshire calcareous bogs are a favourite habitat.

ADDITIONAL SPECIES

Several other species of *Mnium* are by no means uncommon.

M. affine Bland. forms straggling mats in woods, and is recognized by the rather round leaf, prominently excurrent nerve and margin beset with sharp single spines.

M. seligeri (Jur. ex Lindb.) Limpr. (*M. affine* Bland. var. *elatum* B. & S.) is a tall, more erect-growing plant, somewhat resembling *M. punctatum*; it is found in marshes.

M. longirostrum Brid. (*M. rostratum* Schrad.), like *M. affine*, has erect fertile shoots but long, straggling sterile shoots; but the teeth of the leaf margin are short and the nerve is excurrent only in a very short point. The lid of the capsule, moreover, is finely beaked, not conical.

M. cuspidatum Hedw. is smaller than *M. affine* but very like it; it is usually of a characteristic pale green colour and may further be known by its smaller leaf cells (about 20μ, against $25–40\mu$ in *M. affine*).

M. marginatum (With.) Brid. ex P. Beauv. (*M. serratum* Brid.) is a mountain-ledge species, differing from *M. hornum* in its more widely spaced and more decurrent leaves and rounded (rather than angular) cells; also it is more often strongly tinged with dull red than is the commoner plant.

M. stellare Hedw. is a delicate plant of soft texture, dense habit, and characteristic light green colour. It is commonest in limestone districts and may be identified by the complete absence of border to the leaf.

M. pseudopunctatum B. & S. (*M. subglobosum* B. & S.) is a rather rare marsh plant, allied to *M. punctatum*, but with the leaf border unthickened and nerve ceasing well below the leaf apex. The synoecious 'inflorescence' and subglobose capsule are important characters.

RELATED SPECIES

Cinclidium is a genus allied to *Mnium*. *Cinclidium stygium* Sw. is a rather rare plant of deep moorland bogs. It resembles *Mnium punctatum* in appearance but is characteristically of a dark reddish to purple-black colour, and the leaf margin, though bordered, is not thickened as in *M. punctatum*. The peristome is remarkable and differs from that of *Mnium* in that the inner teeth are united at their tips to form an orange cupola.

90. AULACOMNIUM PALUSTRE (Hedw.) Schwaegr.

A rather large and conspicuous yellowish green moss of wet moorland and bog, this common species shares with *Breutelia chrysocoma* the character of having the stems matted with a dense rust-coloured tomentum. This fact, taken in conjunction with the vivid light green shoot tips with leaves held erect or only slightly spreading, usually makes it easily identified in the field. It forms dense tufts, 3–12 cm. tall.

The length of the oblong-lanceolate leaves in the terminal tuft is 4–6 mm., comparable with (though slightly less than) those of a large *Dicranum* such as *D. scoparium*. The lower leaves tend to be shorter. Under the microscope the reddish tomentum is seen to consist of richly branched rhizoids; the nerve of the leaf disappears in the extreme tip; the leaf margin is recurved in the lower part, plane and toothed towards the apex. By far the best microscopic character, however, is the cell

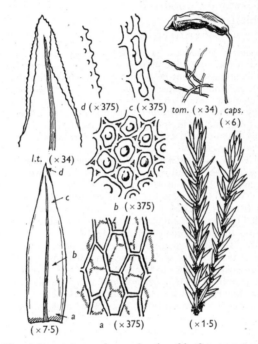

Fig. 90. *Aulacomnium palustre*: *l.t.* tip of leaf, *tom.* tomentum from stem, *caps.* capsule with lid intact.

structure. The cells are rounded, and in the centre of each, as seen in surface view, there is a prominent knob-like papilla; papillae are also conspicuous at the edges of the leaf. There is, in addition, an unusual and characteristic thickening of the corners of the cells. Finally, enlarged, coloured cells at the basal angles of the leaf form orange-brown patches that can be seen even with a lens.

My own experience has been generally to find this dioecious moss in the barren state, but the curved, furrowed capsule, on its 3–5 cm. long seta, is said to be not uncommon. The male plants, with their spreading terminal rosettes of leaves, remind one of the genus *Mnium*.

Ecology. It appears to tolerate a fairly wide range of bog conditions. Often the associated species are *Polytrichum commune* and *Sphagnum* spp. when it grows on wet acid moor; but it occurs also in bogs of nearly neutral reaction, e.g. in the Breckland of East Anglia.

91. AULACOMNIUM ANDROGYNUM (Hedw.) Schwaegr.

Owing to its peculiar structures for vegetative propagation *A. androgynum* could hardly be mistaken for any other British moss. These are gemmae which are borne in dense greenish spherical clusters on the ends of distinct terminal stalks. It is the delicate yellowish green 'drum-

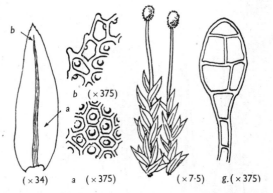

Fig. 91. *Aulacomnium androgynum*: *g.* gemma, with part of its stalk, detached from one of the characteristic 'drum-stick' gemma heads shown in the two shoots drawn.

sticks' so formed that render the moss distinctive and may give the impression, at first glance, that the plants are in fruit. The tufts grow a few centimetres tall, and in their principal habitat of rotting wood stumps are sometimes accompanied by that other delicate gemma-bearing species *Tetraphis pellucida*. *Aulacomnium androgynum* is fairly common in suitable localities.

The leaves are in most respects like those of *A. palustre* but are typically very much smaller (mostly 1–2 mm., but occasionally up to 5 mm.). The nerve that vanishes in the leaf apex, the toothed margin, the rounded papillose cells with thickened corners, all these characters are repeated here on a smaller scale; but the distinct basal bands of orange cells are lacking. The 'head' of the 'drumstick' will be found, under the low power of a microscope, to be composed of hundreds of stalked, 'lemon-shaped' multicellular gemmae, which readily break off for dispersal.

The capsule of this dioecious moss is exceedingly rare. Indeed, its common means of reproduction must be by the gemmae, which are normally present in abundance.

Ecology. Its typical habitat consists of rotten wood; but it appears to be able to extend to humus-rich soil of banks, and occurs occasionally on sandstone rocks. Its habitat range thus follows closely that of *Tetraphis pellucida*, but it probably requires less acid conditions.

92. BARTRAMIA POMIFORMIS Hedw.

The specific epithet *pomiformis* alludes to one of the most notable characters of this moss, the singularly round capsules, looking (when well formed but not yet ripe) like so many miniature green apples.

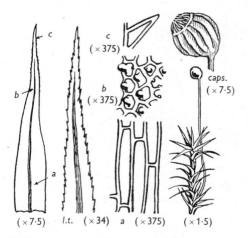

Fig. 92. *Bartramia pomiformis*: *caps.* old capsule showing furrowed wall and teeth of peristome, *l.t.* leaf tip.

The 'apple moss', as it is sometimes called, often grows on rocky ledges on mountains, but it is also widespread on acid lowland banks. The pale glaucous green colour of the tufts is a shade matched by few other British mosses and affords a good mark of recognition.

In habit this species is variable, poorly developed specimens scarcely reaching 2 cm. in height, whereas when luxuriant it will form deep cushions, with the individual shoots 6–7 cm. tall. The lower parts of the stems are always more or less concealed by rust-coloured tomentum. The lanceolate leaves are drawn out to very fine points. They are straight and are held in a star-like spreading manner when moist, but become much twisted and curled when dry. The leaves average 5–6 mm. in length.

Under the microscope a good character is provided by the leaf margin, which is recurved towards the base and sharply toothed towards the tip. Also, the cells are distinctive, being long and narrow at the base, short and obscured by papillae towards the tip. In the less common *B. ithyphylla* all the cells are elongated; in *Plagiopus oederi* the cells lack papillae.

Bartramia pomiformis is either autoecious or synoecious and is commonly very fertile. When young the capsule is almost globose, but when mature it loses this perfect symmetry, and the old brown capsules are deeply and regularly furrowed. The seta is about 2 cm. long.

Ecology. It would seem to grow equally well on sandy banks and on rock ledges; the former often in lowland districts, the latter in mountain country. It will grow successfully on very acid soil, and although I do not know the range of soil reaction which it will tolerate I believe it to be fairly wide, perhaps from extremely acid to near neutral. It does not occur on strongly basic substrata.

ADDITIONAL SPECIES

B. ithyphylla Brid., which grows in clefts of mountain rocks, is known by its very broad, whitish, sheathing leaf bases, and by its uniformly elongated cells.

B. halleriana Hedw. is a rarer plant of shaded mountain rocks. It closely resembles some luxuriant forms of *B. pomiformis*, but is quite distinct in the fruit which, owing to its short curved seta, is hidden amongst the very long, narrow leaves.

RELATED SPECIES

Plagiopus oederi (Brid.) Limpr. (*Bartramia Oederi* Brid.) is a plant of calcareous rocks. Though resembling *B. pomiformis* in a general way, it is more slender and has a smaller capsule. Moreover, it lacks the distinctive pale green colour of *B. pomiformis* and has shorter (3–4·5 mm.), more widely spaced leaves, and cells that are without papillae.

93. PHILONOTIS FONTANA (Hedw.) Brid.

The bright yellowish green or glaucous upright shoots of this moss stand out amongst other species with which it grows in marshy spots on mountains. It is distinctive, too, in the long sparingly branched stems (often up to 8–12 cm.), which are matted together below with red-brown tomentum and bear almost triangular leaves that are notably small for the size of the plant. Once known, it is a species readily recognized, although subject to great variation as regards size. It may be confused indeed only with other less common species of *Philonotis*.

Leaves of the young shoots are barely 0·5 mm. long, but those on the older stems may be more than twice this size; however, under the microscope all the leaves show essentially the same characters. The strong nerve, the deep longitudinal furrows or folds on the leaf surface, the margin toothed and inclined to become recurved in places—all these points should be noted; and, taken in conjunction with the long, narrow cells ($20–50 \times 5–10\,\mu$) and the numerous knob-like papillae that occur at the cell junctions, these microscopic details should readily confirm the identification.

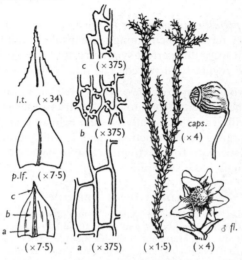

c ($\times 375$)

l.t. ($\times 34$)

b ($\times 375$)

caps. ($\times 4$)

p.lf. ($\times 7.5$)

c

b

a

($\times 7.5$) a ($\times 375$) ($\times 1.5$) ($\times 4$)

♂ fl.

Fig. 93. *Philonotis fontana*: *l.t.* leaf tip, *p.lf.* perigonial leaf, ♂*fl.* male 'inflorescence', with characteristic blunt-tipped perigonial leaves, *caps.* old capsule.
Note. Typical cells are longer than those at *b.*

P. fontana is the central species of a group of dioecious mosses, and the male 'flowers' are not only large and conspicuous, but their leaves (perigonial bracts) provide a good systematic character for distinguishing the species. In *P. fontana* these leaves are obtuse and rounded at the apex, in which the nerve vanishes; in the related *P. calcarea* they are acute and the nerve does not vanish.

The fertile plant is very striking, with its orange setae, 3–5 cm. long, and rather round, deep brown, furrowed capsules.

Ecology. It is a member of that very characteristic moss flora which develops in bog springs on mountains, associated species often including *Bryum pseudotriquetrum* and *Brachythecium rivulare*. It grows chiefly in places where the soil reaction is mildly acid or neutral. Thus in calcareous bogs and springs it is often replaced by *Philonotis calcarea*.

ADDITIONAL SPECIES

P. calcarea (B. & S.) Schp. is the only one of the remaining five British species of *Philonotis* that is at all common. It is usually distinct from forms of *P. fontana* in its longer and more regularly curved leaves. The acutely pointed bracts of the male plant, however, are the diagnostic character. It occurs in calcareous bogs and on wet limestone rock ledges. *P. calcarea* is usually an even more vivid green than most forms of *P. fontana* and the leaf cells are wider (10–15 μ).

94. BREUTELIA CHRYSOCOMA (Dicks.) Lindb. (*B. arcuata* Schp.)

This is among our most distinctive and beautiful mosses of moorland and mountains. In size and habit, as in the widely spreading leaves, it bears some resemblance to a species of *Rhytidiadelphus* such as *R. loreus*, but the presence of a dense felt of rusty brown tomentum on the stems will readily distinguish it. This tomentum is, of course, found in some other genera, such as *Aulacomnium* and *Philonotis*, but they are markedly unlike the present plant in habit and general appearance. The stems are commonly about 5 cm. long, but may reach 12–15 cm. They are but sparingly branched and bright red in colour. The young shoots are like miniature, golden green bottle-brushes.

The leaf is 2·5–4 mm. long, lanceolate, and tapering from a broad base to a long, fine tip. Longitudinal folds (plicae) are prominent on its surface. Under the microscope the long, narrow papillose cells, strong single nerve and toothed margin of the leaf all indicate its affinity with *Philonotis*. The single nerve alone dis-

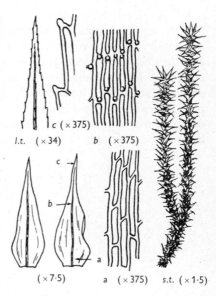

Fig. 94. *Breutelia chrysocoma*: *l.t.* leaf tip, *s.t.* terminal part of leafy shoot.

tinguishes it at once from any species of *Rhytidiadelphus*, and the curious knob-like papillae on the corners of the upper cells will prevent confusion with any other 'hypnoid' plant. The study of minute detail, however, is seldom required in identifying *Breutelia chrysocoma*, for the rusty brown tomentum and the general habit of the plant will usually make it quite easily recognizable in the field.

B. chrysocoma is dioecious, the male plants producing golden brown discoid terminal 'flowers'. The capsule, which is large and strongly furrowed at maturity, is rarely seen. It is like that of *Philonotis fontana* but, being borne on a very short, curved seta, is hidden amongst the leaves and so is easily overlooked.

Ecology. It is a plant of mainly western distribution in Britain. It grows both on moist rock ledges and in bogs, perhaps most typically in the former habitat, where, however, it flourishes equally on acid ledges and on the limestone pavements of western Ireland. In many places its habitat agrees closely with that of *Campylopus atrovirens*, which often grows with it. *Breutelia chrysocoma* avoids deep shade.

ISOBRYALES

95. PTYCHOMITRIUM POLYPHYLLUM (Sw.) Fürnr.

Of this moss Dixon writes: 'the neat cushions with the capsules usually very abundant are conspicuous on nearly every wall in many mountain regions'. Where siliceous rocks prevail in an upland district *P. polyphyllum* is indeed likely to be among the first mosses to attract attention. The cushions are commonly dense and rounded, dull green above, black below, and liberally sprinkled with symmetrically ovoid, pale yellowish brown capsules. The extremely contorted leaves in the dry state may suggest *Tortella tortuosa*, but that moss is never so black below, nor is it often fertile. In colouring and to some extent in habit

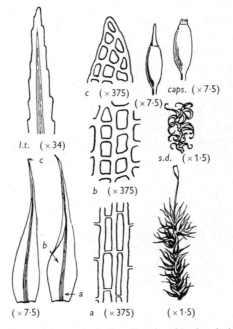

Fig. 95. *Ptychomitrium polyphyllum*: *l.t.* tip of leaf, *s.d.* tip of shoot in a dry state, *caps.* capsules with and without lid.

the present species bears some resemblance to *Dicranoweisia crispula*, but is larger in all its parts than that moss. Although the plants are usually only 2–3 cm. tall, they bear rather long leaves (3–5 mm.).

The lanceolate, tapering leaves possess three well-marked microscopic characters which combine to make identification simple. First, the leaf is traversed from base to near the apex by several well-marked longitudinal folds (deeply plicate); secondly, the apical region is marked by a series of blunt notches, rather distantly spaced, which give the apex a character all its own; and finally, the cells are distinctive, especially those towards the base of the leaf, with their long narrow shape and their thick longitudinal, and thin transverse, walls. Above, they are short and thick-walled, with very small rounded cavities.

This species is autoecious. The seta is pale yellow, 7–15 mm. long, and the capsule is ovoid to cylindrical, symmetrical and upright. At first greenish with a red rim and long-beaked lid, it later becomes pale yellowish brown, but is never deeply furrowed. The fine red peristome teeth are divided to the base. Quite commonly several setae arise from the same 'inflorescence', a most unusual feature among mosses as a whole (cf. *Dicranum majus*, *Mnium undulatum*).

Ecology. It is a characteristic cushion-forming species on hard siliceous rocks. It is often very abundant on wall-tops. It is only in north and west Britain that it becomes an important moss.

96. AMPHIDIUM MOUGEOTII (B. & S.) Schp. (*Zygodon Mougeotii* B. & S.)

When deep, dense cushions of dark green moss occupy moist rock ledges or crevices of the wet rock face in mountain country they are frequently found to be composed of the tightly packed, upright, forked shoots of this species. The colour, dark or olive green above and light yellowish brown to blackish brown below, is fairly characteristic, but the general habit—compact, rounded cushions 3–7 cm. high—is more so. Yet there is nothing very notable about the individual plants, as one separates them from the dense matted tuft, to distinguish them from certain species belonging to such genera as *Gymnostomum*, *Barbula*, *Trichostomum* and *Tortella*. Experience, as in so many cases, is alone a reliable guide for identification in the field. Spreading when moist, the narrow leaves become considerably twisted when dry.

Under the microscope careful attention to the minute details of the leaf will separate *Amphidium mougeotii* from the commoner mosses with which it might be confused. The leaf length, 2–3 mm., is considerably greater than that of *Gymnostomum aeruginosum* (which this plant resembles in habit). Also, the leaves are markedly narrow and acute-tipped (linear-lanceolate), with slightly wavy, partly recurved margins. However, the feature which at once separates it from species of *Barbula* and *Trichostomum* is the areolation, which is clear and rather thick-walled throughout the leaf. Though the upper cells are slightly papillose, the papillae do not obscure the pattern of the cell walls, as they do in *Barbula cylindrica, Gymnostomum aeruginosum, Anoectangium compactum* and species of *Trichostomum* and *Tortella*.

The fruit of this dioecious moss has been found only a few times in Britain. It is borne on a short seta, is ovoid-cylindrical and lacks a peristome.

Ecology. Its favourite habitat seems to be nearly vertical, wet rock faces in sheltered places in hilly districts; here the continuous drip or trickle of water keeps the plant constantly moist. It also occurs in soil-filled crevices of rock in drier situations. *Amphidium mougeotii* is often the principal cushion-forming moss on ledges of acid siliceous rock, but is probably not confined to acid conditions.

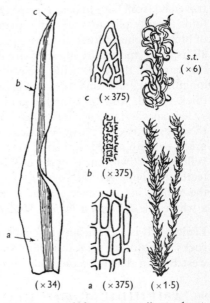

Fig. 96. *Amphidium mougeotii*: *s.t.* shoot tip showing curling of leaves in dry state.

A. lapponicum (Hedw.) Schp. (*Zygodon lapponicus* (Hedw.) B. & S.) is not uncommon in rock clefts on mountains. It forms much smaller tufts than those of *Amphidium mougeotii* (1·5–3 cm. high), and the leaves are more curled and twisted when dry than in that species. Moreover, the plants are usually very fertile, with longitudinally furrowed capsules raised just clear of the leaves.

97. ZYGODON VIRIDISSIMUS (Dicks.) R.Br.

In the moist state the curved, spreading leaves and notably bright (sometimes yellowish) green colour of this moss tend to make its low patches of growth a conspicuous feature on the trunks or branches of trees, where it occurs not uncommonly, together with other epiphytes such as *Orthotrichum* spp. and *Tortula laevipila*. In the dry state the leaves do not become strongly curled, although by becoming appressed to the stem and slightly twisted they have the effect of greatly altering the appearance of the moss. In this state it may easily be overlooked, but on being moistened the leaves spread out remarkably suddenly.

The leaves, rather less than 2 mm. in length, differ from those of the smaller species of *Orthotrichum* in having plane margins. They are of variable shape, more or less ovate-lanceolate, broadest near the middle, and narrowing abruptly at the tip to a short, acute point. Under the microscope it is seen that the nerve is lost at the apex, and does not run

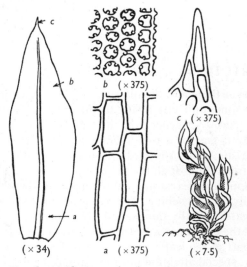

Fig. 97. *Zygodon viridissimus*: the shoot shows the characteristic light twisting of the leaves in the dry state.

out into this sharp point, which is composed of one or a few thick-walled elongated cells. The rest of the leaf is composed mostly of rounded, papillose, but nevertheless clearly defined cells (8–11 μ wide), which give place abruptly to elongated hyaline cells in the leaf base. The leaf margin under the high power shows the prominent papillae. Small club-shaped gemmae often occur on the leaves.

Zygodon viridissimus is dioecious, and the small capsule, raised on a pale seta 3–7 mm. long, is seldom seen. It is oval or pear-shaped, marked with 8 longitudinal furrows, and is contracted at the mouth, which lacks a peristome.

In many of its features, vanishing nerve, areolation, etc., this moss comes very close to the smaller species of *Orthotrichum*. Indeed the plane leaf margin is the main vegetative character in which it stands apart from them. Besides the absence of peristome, a smooth calyptra and an oblique lid to the capsule render this species distinct when in fruit.

Ecology. Its chief habitat is the trunk or branches of trees. It is, perhaps, particularly abundant on the elder. It grows less commonly on rocks or walls.

ADDITIONAL SPECIES

Zygodon conoideus (Dicks.) Hook. & Tayl. is a less common species closely resembling *Z. viridissimus*, but distinguished by its more slender habit and straighter leaves, rather larger leaf cells (up to 15μ), and the presence of a well-developed peristome in the capsule. It grows on trees.

ORTHOTRICHUM

General notes on the genus. Identification of the species of *Orthotrichum* often presents some difficulty owing to the close resemblance between many of the species in general habit, leaf shape and cell structure. Habitat, however, may be a useful guide, as some species always form their characteristic, neat, rounded cushions on rocks, others equally invariably on trees. Almost all the British species agree in having leaves which are held more or less straight and appressed when dry, but spread out very rapidly on moistening. The leaves of most species agree, too, in their elongate, mainly thin-walled (but locally thickened) basal areolation and the isodiametric, rather thick-walled, clearly defined cells of the upper part of the leaf.

Accordingly, special attention must be paid to the details of the fertile plants. Thus relevant points are: (i) the length of the seta and the extent to which the capsule in consequence emerges from the perichaetial leaves, (ii) the hairiness of the calyptra, (iii) the precise form of the ripe capsule when moist, (iv) the number of furrows on the dry capsule, (v) peristome details, and especially the exact appearance of the peristome teeth in the dry capsule. A particularly valuable character, however, lies in the stomata which are found scattered on the wall of the capsule. These may be observed quite easily by cutting a soaked capsule longitudinally in half, getting rid of the spores and mounting the halves in water, convex face upwards. On examination under the high power of the microscope it will be seen that the stomata are either fully exposed and hence superficial in position or are largely concealed by surrounding cells, and hence 'immersed' (fig. 98). This feature is absolutely constant for any given species. Spore size at times affords a valuable character.

98. ORTHOTRICHUM ANOMALUM Hedw.

Two distinct plants are included here: (i) typical *O. anomalum*, a not very common species of siliceous rocks, and (ii) var. *saxatile* (Wood) Milde, which is found on calcareous rocks where its neat, round, abundantly fertile cushions are often conspicuous; this is especially so in spring when each young, green, almost barrel-shaped capsule with its pale, bell-shaped calyptra stands out light against the dark green of the leaves that surround it.

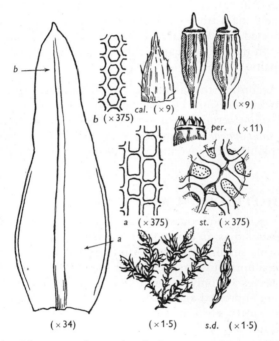

Fig. 98. *Orthotrichum anomalum*: *cal.* calyptra, *per.* peristome of typical form, *s.d.* shoot in a dry state, *st.* stoma from capsule wall; the right-hand capsule is of the typical form, the left-hand one is of the var. *saxatile*.

The typical form and the var. *saxatile* agree in size (erect stems 1–2 cm. tall), and in their dark green, widely ovate-lanceolate leaves which are spreading when moist but appressed (not curled or twisted) when dry. This is a feature, indeed, of all our common rock species of *Orthotrichum*.

The microscopic characters of the leaf, too, will confirm the genus rather than help very much to identify the species; for most species of *Orthotrichum* are alike in the nerve ceasing in the leaf apex, and the partially recurved leaf margins; also in the nearly round, thick-walled and well-defined (though papillose) cells of the upper part of the leaf.

O. anomalum is autoecious and usually very fertile. The bell-shaped, moderately hairy, pale calyptra and the erect capsule are typical of the genus *Orthotrichum*; a seta, 4–5 mm. long and thus lifting the ripe capsule just clear of the leaves, is a feature of this species; for in *O. rupestre* and *O. cupulatum* the seta is shorter and the ripe capsules scarcely emerge from among the leaves. In typical *O. anomalum* the capsule is ovoid-cylindrical, with 8 prominent ribs and 8 intervening fainter ridges; in var. *saxatile* it is narrowly cylindrical when ripe and has only the 8 main ribs. Under the high power the stomata are seen to be immersed (see Glossary, p. 30), whereas in *O. rupestre* they are superficial. In the old, dry capsule the 16 outer peristome teeth appear as 8 erect pairs.

Ecology. According to Dixon the typical form is usually found on siliceous rocks, but some continental authors regard it as indifferent. The var. *saxatile*, certainly, is found invariably on limestone rocks and walls. *Grimmia apocarpa* frequently grows with it, the separate cushions of the two plants often covering much of the surface of many limestone boulders.

99. **ORTHOTRICHUM AFFINE** Brid.

O. affine is the commonest British species of *Orthotrichum*, and is widely distributed as an epiphyte on the trunks and branches of trees, where it forms small, rather loose tufts of a dull, dark green colour. The branched upright stems are shorter than those of *O. lyellii*, and the lanceolate leaves are only 2–3 mm. long. As is usual in *Orthotrichum*, the leaves are appressed and scarcely twisted when dry, but spread out rapidly and widely when moistened. The leaf margin differs from that of *O. lyellii* in being widely recurved almost to the apex.

The elongated basal cells and the rounded, rather thick-walled slightly papillose cells higher up the leaf, do not furnish specific diagnostic characters. Indeed, although the gemmae found on the leaves of *O. lyellii* are hardly ever present here, the leaf characters by themselves rarely permit of the sure identification of *Orthotrichum* spp.; in most cases the capsules are needed for certainty.

O. affine is autoecious and may often be found fertile, bearing the capsules more or less concealed among the upper leaves, or raised just clear of them. The capsule is variable in this respect, as also in its precise shape. Normally it is broadly ellipsoidal to cylindrical, with a rather long neck, and when mature is marked by 8 longitudinal ridges and furrows. Useful additional characters are the slightly hairy calyptra, the superficial position of the stomata, and the fact that the 8 pairs of outer peristome teeth become reflexed outwards over the rim of the old dry capsule. The spores are only about 20μ in diameter, as against 30μ in *O. lyellii*.

Ecology. It appears to be catholic as regards choice of tree species on which to grow as an epiphyte. Often its cushions are associated with those of *Ulota crispa*. It grows at times on stones.

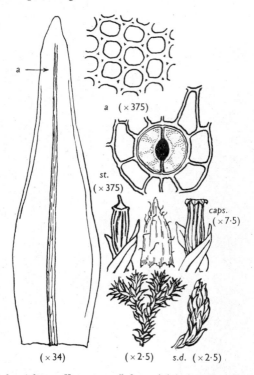

a (×375)

st. (×375)

caps. (×7·5)

(×34)   (×2·5)   s.d. (×2·5)

Fig. 99. *Orthotrichum affine*: *caps.* (left to right) capsule with lid, calyptra and old capsule showing reflexed peristome teeth; *st.* stoma from capsule wall, *s.d.* part of shoot in dry state.

100. ORTHOTRICHUM LYELLII Hook. & Tayl.

O. lyellii is our most robust species of the genus, the rather tall stems (2–6 cm.) bearing leaves which tend to be longer and taper to more acute points than in other species. Its loose-growing tufts are common on trees. It may be readily recognized as referable to *Orthotrichum* by the straight appressed character of the leaves in the dry state.

The individual leaf is linear-lanceolate, tapering and 3–4 mm. long.

The margin is lightly recurved near the base, but plane above. Usually the leaf is liberally covered with brown, multicellular gemmae which afford a good mark of identification (although such gemmae occur occasionally on the leaves of *O. affine*). Further useful microscopic

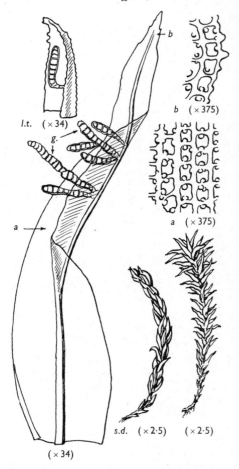

Fig. 100. *Orthotrichum lyellii*: *l.t.* tip of leaf with gemma, *s.d.* shoot in a dry state, *g.* gemmae.

characters are the prominently papillose cells and the faintly toothed leaf apex.

This species is dioecious and capsules are rarely seen. A pale-coloured inflated calyptra covers the young capsule. The mature structure is marked by its very short seta (0·5 mm.), long tapering neck and 8 longitudinal ridges. In the old dry capsule the 16 peristome teeth become recurved, and there are 16 inner teeth. The stomata here are superficial, their form thus not being obscured by surrounding cells when a portion of the capsule wall is examined under high power. The spores measure 25–35 μ across.

Ecology. It grows on a fair range of trees. Professor P. W. Richards suggests that it may have some preference for oaks, and Dr W. Watson, in his study of British woodland bryophytes, found *O. lyellii* to be particularly abundant on *Quercus petraea*. In Kent, however, Dr F. Rose has noted its preference for ash.

101. ORTHOTRICHUM DIAPHANUM Brid.

Growing usually on trees or wooden fences, more rarely on walls or on the ground, *O. diaphanum* is distinct from all other British species of *Orthotrichum* in the silvery white hair-points of the leaves. The neat rounded tufts seldom exceed 1 cm. in height, and the hyaline leaf points give the whole plant a rather greyish green appearance. One is reminded of *Grimmia* species. The leaves are held somewhat erect, both moist and dry.

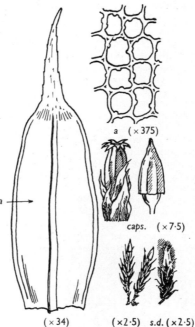

a (× 375)

caps. (× 7·5)

a

(× 34)   (×2·5)   s.d. (×2·5)

Fig. 101. *Orthotrichum diaphanum*: caps. ripe capsule showing peristome teeth (left), young capsule covered by calyptra (right); *s.d.* shoot in a dry state showing whitish brush formed by hair-points of the leaves.

Under the microscope the rather large (14–20 μ), rounded, thick-walled and clearly defined cells of the upper part of the leaf will distinguish this plant from any species of *Grimmia* or *Rhacomitrium* which it may resemble in the wide hyaline hair-point (but which are of course quite different in fruit). The hair-point is lightly toothed and has a very broad insertion at the leaf tip, much as in *R. heterostichum*. The leaf margin is recurved; the nerve disappears in or below the leaf apex.

Orthotrichum diaphanum is autoecious and, as in the last species, the seta (0·5 mm.) is so short as scarcely to raise the capsule above the leaves. In shape the capsule is more ovoid-ellipsoidal than in *O. affine*. It has only faint longitudinal ridges and furrows; further, the stomata in *O. diaphanum* are immersed (see Glossary, p. 30) and the peristome teeth are spreading, not reflexed as in *O. affine*, when the capsule is ripe and dry.

Ecology. It grows, most typically perhaps, on trees or wooden fences; but it also occurs on roadside pavements, wall-tops and other situations where it is growing on stone rather than wood. It flourishes in towns and even seems to have a preference for the neighbourhood of human habitations, perhaps because it is nitrophilous.

ADDITIONAL SPECIES

Of the thirteen remaining British species of this genus six are by no means rare.

O. rupestre Schleich. occurs on siliceous rocks; it may be known by its comparatively tall growth (2–5 cm.), immersed capsule and superficial stomata.

O. cupulatum Schwaegr. grows on calcareous rocks, often with *O. anomalum* var. *saxatile*, which it closely resembles, differing chiefly in its very sparsely hairy calyptra, and in the shape of its capsules; these are broadly barrel-shaped when ripe, and when old and dry are contracted just below the mouth.

O. striatum Hedw. (*O. leiocarpum* B. & S.), which grows on trees, is the only species without any trace of longitudinal furrows on the capsule.

O. rivulare Turn. is a plant of rocks and tree roots by water. It has broad, rounded leaf apices and resembles *Cinclidotus fontinaloides* (although smaller) rather than other species of *Orthotrichum*.

O. tenellum Bruch and *O. pulchellum* Brunton are small tree-growing species (neither of them common), both of which differ from *O. affine* in having the stomata on the capsule wall immersed, not superficial. *O. tenellum* is further remarkable for the bright orange-brown colour of its ripe capsules. *O. pulchellum* has the ripe capsule pale brown, with orange-red peristome; it is unlike all other British species of *Orthotrichum* in the evident twisting and curling of its leaves when dry.

102. **ULOTA PHYLLANTHA** Brid.

This species forms cushions, much as in *U. crispa*, on the branches of trees, but here the habit tends to be more robust and the plants are almost always barren. As in *U. crispa*, however, the leaves become much curled when dry, so that barren plants of these two epiphytic species look somewhat alike. *U. phyllantha* will grow at times on rocks, especially near the sea.

The gemmae at the tips of the leaves form the best mark of identification for this species. They are brownish, multicellular and just visible under a lens. The leaves themselves are oblong-lanceolate, and the nerve runs out into a short point on which the gemmae are borne. Each gemma consists of a short row of cells, narrowing into a stalk. Fortunately, these gemmae are almost always plentiful; for otherwise the leaf is like those of other members of the genus, the cells being short, nearly round in outline and thick-walled; but both here and in *U. crispa* the cells in the extreme leaf base are very narrow and much elongated.

Fertile material of this moss has been found only very rarely. The capsule differs from that of *Ulota crispa* in not being contracted below the mouth.

Ecology. Its habitat range is wider than that of most British species of *Ulota*, in that it includes both rocks and trees. In the latter habitat it is often found with *U. crispa*. It is particularly common near the sea, on the west coast of Britain, where it sometimes grows with *Grimmia maritima* in the spray zone.

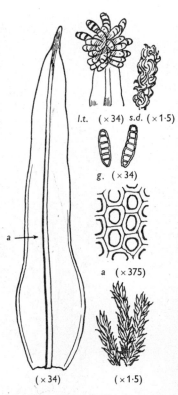

Fig. 102. *Ulota phyllantha*: *g.* gemmae, *l.t.* leaf tip bearing gemmae, *s.d.* shoot in a dry state.

103. **ULOTA CRISPA** (Hedw.) Brid.

This is the commonest British species of a genus that contains some of the most characteristic mosses to be found on trees in mountain districts. The neat round cushions of *U. crispa*, with leaves much curled and crisped when dry, are often conspicuous in great numbers on the branches of such trees, where they attract attention by their abundant capsules, each capsule with a strikingly hairy calyptra. The crowded stems in each cushion or tuft are commonly only 0·5–1 cm. tall, and the colour of the shoots is usually light yellowish green above, blackish below. The only mosses with which fertile *U. crispa* may be confused are other, less common species of *Ulota*, and certain species of the allied genus *Orthotrichum*, where the ribbed capsule and hairy calyptra also occur.

The narrow, tapering leaves, 2–3 mm. long, are distinctive under the microscope; for along each edge of the short expanded leaf base there run a few rows of empty hyaline cells, forming an obvious marginal

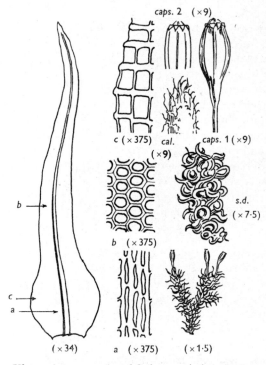

Fig. 103. *Ulota crispa*: *caps*. 1 and 2 show variations in the amount of contraction below the mouth in the ripe capsule; *cal.* calyptra, *s.d.* part of shoot in a dry state.

band. Above, the narrow part of the leaf is keeled and the nerve disappears in the acute leaf tip. The whole leaf thus has a rather characteristic form and outline, which, combined with the thick-walled, rounded cells, makes it easily identified.

Ulota crispa is autoecious and is often abundantly fertile. The hairy calyptra which covers the young capsule is conspicuous. The mature capsule is marked by 8 prominent ribs and is notable for its long tapering neck. The mouth is wide but there is a constriction just below it. With age the capsule becomes progressively narrower and more deeply furrowed. The dimensions of capsule and seta are the most certain means of separating this species from *U. bruchii*. Thus in *U. crispa* the capsule is 2 mm. and the seta 1–2 mm. in length, whilst in *U. bruchii* the capsule is 3–4 mm. and the seta 3–5 mm. The 16 peristome teeth are in pairs, and finally become recurved over the margin of the capsule mouth. There is also an inner peristome of 8 (occasionally 16) teeth.

Ecology. Characteristically epiphytic, it grows on living or decaying branches rather than on the trunks of trees. I am not aware that any species of tree is particularly favoured. It grows best in the wetter districts of north and west Britain and is comparatively uncommon in many of the drier types of woodland of south-east England.

ADDITIONAL SPECIES

U. bruchii Hornsch. is closely allied to *U. crispa*, and like it grows normally on trees. It differs in its larger size (leaves up to 4 mm., seta 3–5 mm., capsule 3–4 mm.). The capsule, when ripe, is markedly narrowed at the mouth itself, not just below it as in *U. crispa*.

U. hutchinsiae (Sm.) Hamm. is a dark brownish, straight-leaved plant which forms neat, usually abundantly fertile tufts on siliceous boulders (very rarely on trees) in many northern and western districts. It resembles *Orthotrichum* spp. in having leaves that are not curled when dry, but may be known by its colour and the very thick-walled, round cells in the upper part of the leaf.

104. FONTINALIS ANTIPYRETICA Hedw.

This aquatic species is well known even to non-bryologists, and its long, sparingly branched leafy shoots are a common feature of rivers and lake margins in all parts of Britain. Anchored to stones or tree roots, the lower parts of the submerged stems are often bare of leaves; but above, the dark green leafy branches may reach a length of 50–70 cm. In mountain streams more slender forms are frequent, and the colour is often a duller, brownish green, sometimes tinged with rusty red. The var. *gracilis* Schp. is a slender, often reddish form.

The leaves in *F. antipyretica* are highly characteristic, both in arrangement and in form. They are borne typically in 3 well-marked ranks and give the leafy shoots as a whole a 3-winged (or triquetrous) form. Each leaf is 4–7 mm. long and is folded so as to be boat-shaped with a sharp keel. Indeed, the leaves quite commonly split along the keel. The only other common British species, *F. squamosa*, lacks this keeled leaf.

The leaves are nerveless and their cells are very long; average measurements are $15 \times 150\mu$. The cells are thin-walled and tend to become much distorted in dried material. Sharply defined patches of specialized alar cells are lacking, but the leaf base as a whole is marked by somewhat shorter and wider cells, with slightly thicker, sometimes orange-coloured walls.

Although this is a variable plant, its robust habit and broad, keeled, nerveless leaves will readily distinguish it from other aquatic mosses such as *Eurhynchium riparioides* and species of *Drepanocladus*.

The species of *Fontinalis* are dioecious, and *F. antipyretica*, when growing submerged, is rare in fruit. The oblong-cylindrical capsules,

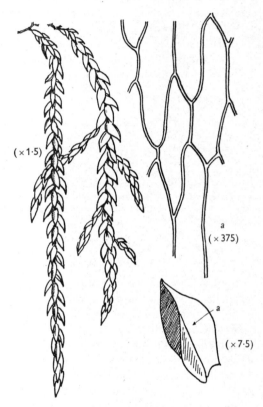

Fig. 104. *Fontinalis antipyretica*.

almost concealed among the perichaetial leaves, may be found more frequently, however, on plants exposed to the air in dry weather. The peristome is bright red with 16 outer teeth and 16 inner teeth united into a latticed cone.

Ecology. It will grow attached to wood or rock, and is mainly a plant of slow-flowing rivers and streams, also of lakes and ponds. Generally growing at no great depth, *F. antipyretica* has been noted as occurring in the *Isoetes lacustris* community in Esthwaite Water. Some forms (especially var. *gracilis*) occur in fast-flowing mountain streams. It appears to be in general equally tolerant of acid and calcareous waters, but the larger, typical forms have been found in some districts to prefer neutral or calcareous waters to those that are acid.

ADDITIONAL SPECIES

Fontinalis squamosa Hedw. is the only other species which is at all common. The absence of a keel to the rounded, concave leaf at once distinguishes it from slender forms of *F. antipyretica*. It is found in fast-flowing non-calcareous water.

105. CLIMACIUM DENDROIDES (Hedw.) Web. & Mohr

As indicated by the specific name, this moss is characterized by its dendroid habit. Indeed, here the upright secondary stems bear a more obvious resemblance to miniature trees than in any other British moss, the unbranched basal 2–3 cm. or more representing the 'tree-trunks', the crowded spreading branches above forming the leafy crown of the 'tree'. The blackish erect stems are stouter, and the branches thicker,

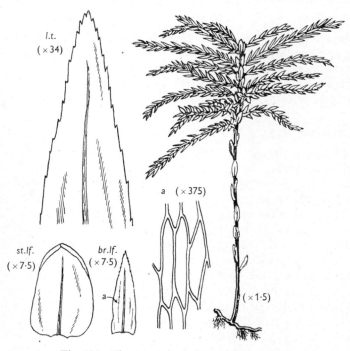

Fig. 105. *Climacium dendroides*: *st.lf.* stem leaf, *br.lf.* branch leaf, *l.t.* leaf tip.

with larger leaves than in *Thamnium*, so that the whole aspect of the plant is more bushy and less feathery. The prevailing colour, too, is here a lighter, more yellowish green. Connected below by a creeping primary stem resembling a rhizome, the 'miniature trees' occur in numbers together on moist ground in upland pastures or on coastal duneland. Growing as they do in grassy places, the tree-like character of the plants is not always apparent from a casual field inspection, but the dense clusters of bright green branches will attract attention.

The rather sparse stem leaves are broad and rounded at the apex, the crowded branch leaves (up to 3 mm. long) narrower, ovate-lanceolate and coarsely toothed towards the rather acute apex. Deep longitudinal folds are commonly present, and a single nerve extends to beyond mid-leaf. The cells are markedly elongated in both stem and branch leaves, a microscopic character which will at once separate *Climacium* from *Thamnium* in any doubtful case.

This species is dioecious and is almost always met with in the barren condition. The capsule is borne erect on a red seta about 2 cm. or rather more in length, several capsules usually springing from a single 'inflorescence'. In the rare fertile state *Climacium* thus becomes a most striking moss.

Ecology. It grows generally among grass, its most characteristic habitats being perhaps the margins of lakes in hill districts and moist hollows in sand dunes. It also occurs at times in woods, but less often on open moorland. It tolerates a soil reaction ranging from acid to moderately calcareous, growing, for instance, on the alkaline peat of fens.

106. HEDWIGIA CILIATA (Hedw.) P. Beauv.

This moss grows in stiff but straggling tufts on siliceous rocks in mountain districts. It suggests some species of *Grimmia* and *Rhacomitrium* in the white hair-points of its leaves, but the whole plant has a pale greyish green colour seen in few other mosses; this by itself is usually sufficient for ready identification in the field. The rather rigid stems are long (3–10 cm.) and repeatedly forked, with some additional side-branches, but owing to the small size of the leaves the shoots retain a slender, catkin-like form.

One certain microscopic feature which distinguishes *Hedwigia ciliata* from all species of *Grimmia* and *Rhacomitrium* lies in the absence of a nerve in the leaf. This feature, taken together with the broad-based, toothed, hyaline hair-point and the thick-walled, papillose cells, should identify this species without difficulty. The cells in the middle of the leaf base, however, are long, narrow and very thick-walled, reminding one of the leaf-base cells of *Rhacomitrium*.

In this autoecious moss the orange-brown capsule is borne immersed among the leaves. It is nearly globose and lacks a peristome, a fact which makes it hard to assess the precise affinities of the plant.

Ecology. Its habitat range agrees closely with that of *Ptychomitrium polyphyllum*, for it is never found on limestone, but is widely distributed on siliceous rocks and walls at moderate altitudes. It is fond of lake-side boulders, but occurs also on rocky outcrops on exposed mountain sides. *Hedwigia ciliata* is found chiefly in the north and west of Britain.

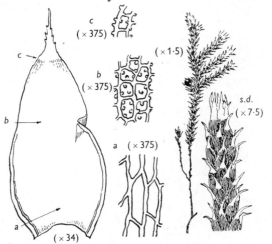

Fig. 106. *Hedwigia ciliata*: *s.d.* part of shoot in a dry state showing white brush at tip formed by hair-points of the leaves.

SPECIES OF RELATED FAMILIES

Cryphaea heteromalla (Hedw.) Mohr. is a rather slender, rigid, dark green plant of tree trunks, especially elder. The leaf has a single nerve and short rounded areolation. When in fruit it may be known by the rows of immersed capsules borne unilaterally on specialized branches.

Leucodon sciuroides (Hedw.) Schwaegr. is an uncommon species found chiefly on trees (rarely on rocks). On the erect, little-branched secondary stems the plicate nerveless leaves (2–3 mm. long) are crowded and overlapping, so that when dry the shoots are smoothly cylindrical and catkin-like. Elongated in the middle of the leaf base, the cells are oval and thick-walled above; the long cells are flanked by extensive patches of very small round cells.

Antitrichia curtipendula (Hedw.) Brid. is a robust plant which forms extensive mats on boulders or about tree bases in parts of west Britain. The leaf tip is unlike that of any other British moss, being narrow and studded with prominent hook-like teeth, the teeth at the extreme apex being in the form of a grapnel.

Pterogonium gracile (Hedw.) Sm. is a frequent plant on rocks and tree trunks, where the clustered shoots, rendered cylindrical by the closely appressed leaves, make it usually easy to recognize. The branches have a tendency to be curved rather uniformly in one direction. The non-plicate surface and toothed margins of the leaves distinguish it from *Leucodon sciuroides*.

Myurium hebridarum Schp. is a rare moss, mentioned here on account of its interesting distribution. Exceedingly abundant in suitable situations (moist ledges near the sea) in the Outer Hebrides, it has also been recorded from several of the Inner Isles and from a few stations on the mainland of north-west Scotland. Elsewhere it is known only from the Canaries and the Azores. It is a most distinctive plant, with extremely robust, swollen, golden green shoots and broad concave leaves which are suddenly contracted into long fine apices. The capsule is very rare; it has been found in the Azores, but never in this country. The plant looks most like a straight-leaved version of *Hypnum cupressiforme* var. *lacunosum*.

107. NECKERA CRISPA Hedw.

This is among the most robust and conspicuous mosses which grow on rocks and about the bases of trees in limestone districts. The conspicuous flattening of the curved, glossy shoots and the marked and very regular transverse undulations of the oblong leaves combine to give it a distinctive character. Indeed, *N. crispa* is a moss which, once it has been seen hanging in festoons on the face of some limestone rock, may be recognized again without any difficulty. It varies considerably

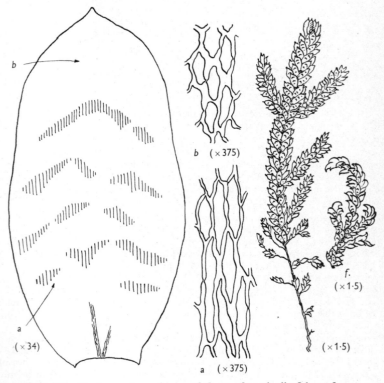

Fig. 107. *Neckera crispa*: *f.* part of shoot of markedly falcate form.

in the extent to which the tips of the branches become curved and curled, but the breadth of the flattened shoot (5–6 mm.) and the wrinkling of the glossy leaves through transverse undulations will prevent confusion with *N. complanata* or *Homalia trichomanoides*, both of which are smaller plants with much narrower shoots and unwrinkled leaves. Older parts of the plant are brown, but younger shoots are yellowish green.

Under the microscope the outline of the rounded leaf apex is seen to be broken by a distinct point or apiculus, and the leaf margin is finely toothed for some distance back from the apex. The nerve is weak or absent altogether. The cells are elongate below, shorter and rhomboid in shape above, and throughout are marked by thick walls with evident thin places, or 'pores', reminiscent of *Dicranum scoparium* and related species of *Dicranum*.

A dioecious plant, *Neckera crispa* is seldom seen in the fertile condition. The small ovoid orange-brown capsule is borne on a yellowish seta 6–10 mm. long.

Ecology. It grows typically on moist calcareous rock ledges, or in crevices in drier situations. It is locally plentiful in open grassland on the chalk. It occurs on Old Red Sandstone, but probably only in parts that are to some extent calcareous. In very moist places (e.g. near waterfalls) it will grow on the bases of trees.

108. NECKERA COMPLANATA (Hedw.) Hüben.

The yellowish green, shining patches formed by this species on trees or rocks bear a closer resemblance to *Homalia trichomanoides* than to

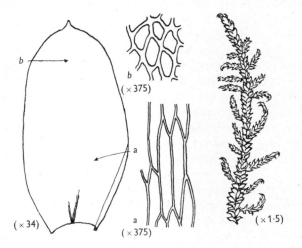

Fig. 108. *Neckera complanata.*

the much more robust *Neckera crispa*. At times the flattened, glossy character of the shoots may suggest a *Plagiothecium*, but the branching here is more widely angled and more regularly pinnate than in species of the latter genus. The leaves are curved downwards less markedly and the branching is more regular than in *Homalia*.

In its oblong shape and blunt, apiculate tip, as in its faintly toothed upper margins, the leaf agrees closely with that of *Neckera crispa*; but it is much smaller (scarcely 2 mm. long) and lacks the numerous fine transverse undulations. The cells come close to those of *N. crispa* in shape, but are not notably thick-walled. The cells in mid-leaf may be 10–15 times as long as broad, this longer, narrower type of cell and the much less sharply toothed leaf margin marking points of difference between this species and *Homalia trichomanoides*. Also, there is a very short, faint double nerve, whereas *Homalia* has a single nerve to mid-leaf.

Neckera complanata is dioecious and the capsules are not very common. The seta is pale yellow, 8–10 mm. long; the capsule itself narrowly ovoid, small, and pale orange-brown in colour.

Ecology. This species occurs on tree trunks much more commonly than does *N. crispa*. But it also grows on walls and rocks. It demands some shade and occurs in both hedgerows and woodlands. A marked preference is shown for neutral or basic conditions.

109. HOMALIA TRICHOMANOIDES (Hedw.) B. & S.

Among pleurocarpous mosses the genera *Homalia*, *Neckera*, *Hookeria* and *Plagiothecium* are remarkable for the conspicuous flattening of the shoot, which gives the leaves something of a 2-ranked appearance. All these genera, especially the first, have a superficial resemblance to leafy liverworts. *Homalia trichomanoides* forms robust and extensive mats on tree bases, on rocks and on the ground in woods. The flattened leafy shoots are glossy and the leaves tend to be curled downwards, so that the plant as a whole has a distinctive appearance that makes it generally easy to recognize. The shoots are wider than in *Neckera complanata*, and their branching is less regularly pinnate than is usual in that species. Indeed, at a quick glance *Homalia* is more likely to be mistaken for a liverwort such as *Plagiochila* sp. than for any other moss.

Under the microscope the leaf is seen to be of an unusual shape, curved, oblong and bluntly pointed. The margin is strongly toothed, especially towards the apex. The single nerve, though very faint and ill defined, is usually traceable to mid-leaf. One of the margins becomes strongly incurved in the basal third of the leaf, and this feature, taken in conjunction with the curved form of the leaf as a whole, imparts a lack of symmetry which is characteristic. The cells towards the leaf base are about six times as long as they are broad. Above they become very shortly rhomboid or diamond-shaped.

An autoecious species, *Homalia trichomanoides* is often abundantly fertile. The ovoid orange-brown capsule is borne on a reddish seta 1 cm. or rather more in length. In the young green capsule the long beak of the lid is notable, whilst in the mature state the well-developed pale yellowish double peristome is conspicuous.

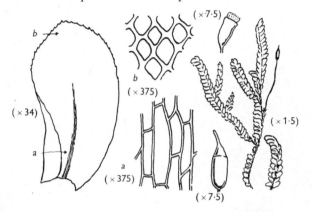

Fig. 109. *Homalia trichomanoides*: the lower of the two enlarged capsules shows the lid intact; in the upper it is removed to show the peristome.

Ecology. It grows most commonly on tree stumps or low on the trunk of a tree; more rarely it occurs on rock or stony ground in woodland or along a hedge bank. It is a plant chiefly of low altitudes and is particularly abundant in the calcareous districts of south-east Britain.

110. THAMNIUM ALOPECURUM (Hedw.) B. & S. (*Porotrichum alopecurum* (Hedw.) Mitt.). (Plate VIII)

This moss, which is not uncommon on the ground in woods and among shaded rocks, is one of the few British species with a 'tree-like' or dendroid habit. In *Thamnium alopecurum*, as in that other markedly 'tree-like' moss *Climacium dendroides*, the primary stems are horizontal and creeping, but from these primary stems there arise more or less upright shoots, which, being unbranched below but branched and leafy above, suggest miniature trees. In *Thamnium alopecurum* the 'trunk' of each tree-like shoot is longer (3–6 cm.) and the leaves smaller (1–1·5 mm.) than in *Climacium*; also the habit is less erect and the shoots are darker green and less glossy than in that plant. The two species are therefore not very likely to be confused. Both the horizontal and upright stems are of strong wiry texture, dark red to blackish in colour, and the upright ones are more or less covered with small, silvery, triangular scale-like leaves. The finer upper branches in some forms of this plant become tassel-like, with narrowly ovate leaves, the whole shoot being not uncommonly slightly flattened or turned to one side. At times it can resemble certain forms of *Eurhynchium striatum* and *Isothecium myosuroides*.

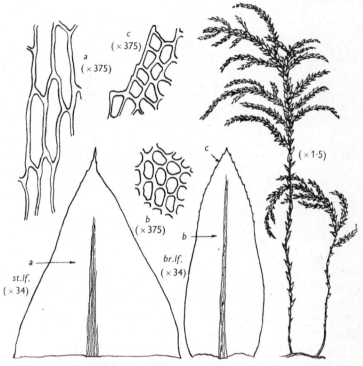

Fig. 110. *Thamnium alopecurum*: *st.lf.* stem leaf, *br.lf.* branch leaf.

Under the microscope the leaf is seen to be nerved nearly to the apex, which is itself prominently toothed. The cells in the stem leaves are variably elongated, but in the leaves of the finer branches they are uniformly short and rounded, a character by which the larger-leaved forms of the present plant are separated from *Climacium*.

A dioecious species, *Thamnium alopecurum* is not commonly fertile. Capsules arise, several together at the summits of the upright shoots, each on a red seta 1–1·5 cm. long. The lid is conspicuously beaked, the peristome large and reddish in colour.

Ecology. The plant has two distinct habitats: (i) very wet, shaded (and not necessarily calcareous) rocks by streams and waterfalls, (ii) comparatively dry calcareous woods, where it grows on tree roots, rock or soil, e.g. on calcareous boulder clay. It may be that two distinct ecotypes are involved. *Thamnium* sometimes becomes detached from the soil and forms balls (cf. *Leucobryum*).

HOOKERIALES

111. **HOOKERIA LUCENS** (Hedw.) Sm. (*Pterygophyllum lucens* (Hedw.) Brid.)

Hookeria lucens is one of the few British representatives of a mainly tropical family (Hookeriaceae). It is a plant of deeply shaded moist places and is remarkable for its flattened (complanate) shoots and large, translucent leaves. It presents very much the appearance of one of the more robust leafy liverworts, but close inspection will show that the leaves are not arranged strictly in two or three ranks as they are in that group. The plant tends to branch irregularly, the individual branches being 2–6 cm. in length.

The leaf of *H. lucens* is large (about 5 mm. long by 3 mm. broad), somewhat ovate in outline with a very broad, rounded apex, and of a peculiarly filmy, translucent quality. This is due in great part to the extremely large size of the rather thin-walled, approximately hexagonal cells which compose the delicate plate of tissue that forms the leaf. A cell of average size in this leaf attains 200–300μ in length and 50–80μ in breadth. The cells are thus detectable with the naked eye and are clearly seen with the weakest hand-lens. The leaf margin is entire and there is no nerve.

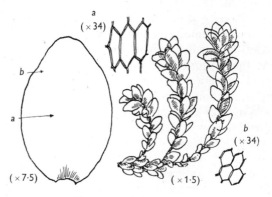

Fig. 111. *Hookeria lucens.*

H. lucens, though sometimes found barren, is an autoecious species, and the fruit is quite common. Raised on a stout, dark red seta about 2 cm. long, the capsule itself is shortly ovoid, and dark purplish red in colour. When ripe it is held horizontally or somewhat drooping.

Ecology. It occurs in habitats which are both deeply shaded and sheltered, growing on soil, or soil-covered ledges of rock. Given these conditions, it may be found alike in moist woodland, rocky moorland and on mountain slopes. Common in the north and west, it is rare in south-eastern England.

HYPNOBRYALES

112. LESKEA POLYCARPA Hedw.

This moss is rather common on and about the bases of trees by water, but being of slender habit and small dimensions it is apt to be overlooked. Numerous short, more or less upright branches arise from the prostrate primary stem, so that the moss forms rather dense low patches of growth.

The leaves, which are regularly arranged and rather crowded, are minute (hardly 1 mm. in length), ovate-lanceolate, tapering to short, not very acute points. The single nerve ends well below the leaf tip. The most useful microscopic character, however, is the short, rounded-hexagonal areolation, which is almost uniform throughout the leaf. Such uniformly short cell structure is comparatively unusual among pleuro-carpous mosses, and will automatically exclude the majority of the larger 'hypnoid' genera. *Amblystegium serpens* is a common species which somewhat resembles the present plant (and often grows with it); but it is of still more slender habit, the cells are longer, and the capsules are quite different.

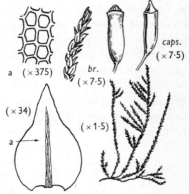

Fig. 112. *Leskea polycarpa*: *caps.* capsules, (left) showing peristome, (right) with lid intact; *br.* one of the finer branches; the leaf at bottom left is from a main stem.

Leskea polycarpa is autoecious and is commonly very fertile. The capsule, which is borne erect on a reddish seta scarcely 1 cm. long, is remarkably long and narrow. The seta arises from the primary stem—

a generic character. A small white calyptra is present on the young capsule—a further point of resemblance to *Amblystegium serpens*.

Ecology. This is a plant of rather restricted range ecologically. It grows usually on trees or wooden fences (less commonly on the ground), and almost always near water (and places liable to flooding). The neighbourhood of eutrophic lowland streams is especially favoured, and in this it agrees closely with *Amblystegium serpens*.

RELATED SPECIES

Heterocladium heteropterum (Bruch) B. & S., which occurs on wet, shaded rock ledges, has freely branched thread-like stems that bear ovate, pointed, finely toothed leaves up to 0·5 mm. in length. On the finer branches the leaves are extremely minute and narrower. The nerve is short and ill defined, and the cells average twice as long as broad, but the best distinction from species of *Amblystegium* and *Amblystegiella* (rare) lies in the prominent papillae that are seen when the back of a folded leaf is viewed under high power.

113. ANOMODON VITICULOSUS (Hedw.) Hook. & Tayl.

This robust, irregularly branched moss, which is often a conspicuous feature of calcareous banks and walls, is remarkable for the great change which it undergoes between wet and dry states. When moist, the wide loose patches which it forms are notable for the vivid green colour of the shoots and the evenly spreading character of the leaves;

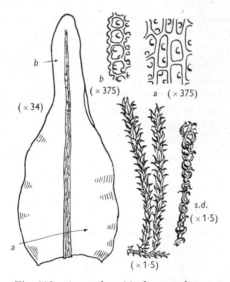

Fig. 113. *Anomodon viticulosus*: *s.d.* part of shoot in a dry state.

when dry, the leaves become crisped and appressed to the stem, and the colour is changed to a singularly dull green. Owing to the length (3–8 cm.) and sparse branching of the secondary shoots and their tendency often to grow somewhat upright, this species may be taken for an acrocarpous moss, but closer examination will show that the long leafy branches are connected below with a horizontal runner-like primary stem. A common, and sometimes abundant species, *A. viticulosus* can usually be recognized without much difficulty in the field.

The leaf is of unusual shape, broad at the base, then narrowing rather abruptly, and terminating in a blunt apex. It is 1·5–3 mm. long. Under the microscope the rather broad single nerve is seen to extend almost to the leaf tip. The shortly hexagonal cells are highly papillose.

This species is dioecious, and the capsule—erect and cylindrical on a pale yellow seta, 2 cm. long—is not common.

Ecology. It is a markedly calcicole species, growing on rock, soil or wood in limestone and chalk districts; it demands some shade, and appears to do best on steep hedge banks and limestone walls, where it is a conspicuous member of a bryophyte community that also includes *Ctenidium molluscum* and the liverwort *Porella platyphylla*.

114. THUIDIUM TAMARISCINUM (Hedw.) B. & S. (Plate IX)

T. tamariscinum is among the first mosses which the beginner learns to recognize, so distinct is it in its bright green colour, robust habit and intricate 'frond-like' manner of branching. This latter character is due to the fact that the principal secondary stems, which arise from a prostrate primary stem system, are usually themselves pinnately branched three times. This tripinnate branching and flattened frond-like form give the plant a totally different appearance from that of most other pleurocarpous mosses, where the branching is usually only simply pinnate, or at most bipinnate. Apart from a few considerably less common species of *Thuidium*, the only British moss which at all resembles the present plant is *Hylocomium splendens*, and this may be known at once by its red stems. In woodland and on fairly shaded banks generally, *Thuidium tamariscinum* is one of the most plentiful (as well as most noticeable) British mosses.

The blackish or green, stiff stems are only sparsely leafy near the base. These stem leaves are very broad and heart-shaped to triangular in form. The branch leaves are much narrower. Both types of leaf have finely acute, toothed apices. The microscope will show the cells to be short, though somewhat irregular both in size and shape. They have prominent papillae, and if a complete shoot tip be mounted, this feature will be very evident on the backs of the leaves. The simple, acute apical

Plate IX. *Thuidium tamariscinum* (natural size)

Plate X. *Pseudoscleropodium purum* (natural size)

cell of the branch leaf will distinguish *T. tamariscinum* from less common species of the genus.

In addition to the leaves, the stem of this plant is covered with a felt of variously branched structures termed paraphyllia.

T. tamariscinum is dioecious, and the large curved capsule, on its long red seta, is seldom seen. It has a well-developed double peristome.

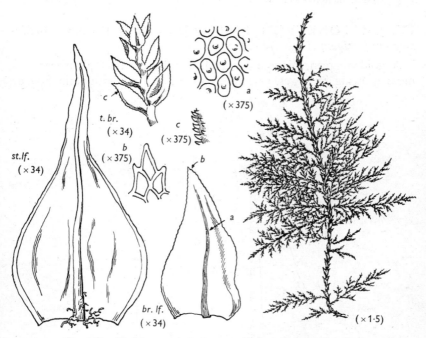

Fig. 114. *Thuidium tamariscinum*: *st.lf.* stem leaf, *br.lf.* leaf of primary branch, *t.br.* tip of fine branchlet.

Ecology. While its most characteristic habitats are woods and hedge banks, *T. tamariscinum* will grow among grass in open situations, especially in north and west Britain. Then it occurs with other mosses, such as species of *Rhytidiadelphus* and *Hylocomium*, in close association with flowering plants. It is perhaps less abundant in highly calcareous conditions, and on chalk grassland is sometimes replaced by *Thuidium philiberti*.

ADDITIONAL SPECIES

T. abietinum (Brid.) B. & S. is a plant of very different appearance from *T. tamariscinum*. The branching is simply pinnate only, but the leaves are those of a *Thuidium*. It grows among grass in calcareous places.

The remaining two species considered here are scarcely distinguishable from *T. tamariscinum* in the field, though in both the tripinnate branching is commonly less luxuriantly developed than in that plant. Under the microscope the apical cells of the branch leaves, which are crowned with 2 or 3 papillae, at once distinguish either of these species from *T. tamariscinum*.

T. philiberti Limpr., which appears to be widespread in chalk grassland, is further known by the long fine points of its stem leaves.

T. delicatulum (Hedw.) Mitt. is a plant of western distribution; it has stem leaves which are not drawn out to fine points, and in habit it is usually rather slender. It is found in sheltered wooded valleys, by waterfalls, etc.

115. CRATONEURON FILICINUM (Hedw.) Roth (*Amblystegium filicinum* (Hedw.) De Not.)

A golden green, pinnately branched, falcate-leaved pleurocarpous moss growing in a wet, calcareous situation, be it mountain bog or

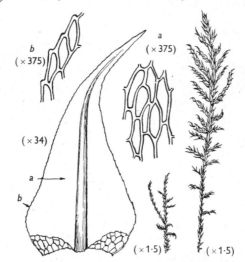

Fig. 115. *Cratoneuron filicinum*: the left-hand shoot represents a dwarf form, the right-hand shoot a fairly robust form of the species.

duneland 'slacks', will most likely be this species or the next. Both are highly variable plants, and luxuriant forms of the present species can appear very like some of the smaller and less richly branched forms of the next. In general, however, the pinnately branched shoots of *Cratoneuron filicinum* are shorter and stiffer in growth, with the leaves less strongly falcate than in *C. commutatum*. Indeed, some forms of the present plant are so slender as to approach *Amblystegium serpens*. The stems have a dense covering of paraphyllia and interwoven reddish brown radicles, a further point of agreement with *Cratoneuron commutatum*, and one that will separate *Cratoneuron* as a whole from most other related or superficially similar 'hypnoid' mosses.

The stem leaves are nearly heart-shaped, with well-developed auricles; the branch leaves are considerably narrower. There is a useful combination of microscopic leaf characters which will usually identify this species with certainty: broad, strong nerve to the apex; rather short, approximately hexagonal cells (3–5 times as long as broad); toothed leaf margin; and very well-defined, often orange-coloured patches of enlarged alar cells. The leaf of *C. commutatum* is at once distinct in its longer, narrower areolation.

C. filicinum is dioecious, and the curved capsule, on a seta 4–5 cm. long, is not at all common.

Ecology. It occurs in a rather wide range of wet habitats, but is found most typically where high water content is combined with calcareous conditions; as, for instance, on damp roadsides and marshy ground by streams in limestone districts, and in wet hollows in calcareous sand dunes. A frequently associated species is *Campylium stellatum*.

116. CRATONEURON COMMUTATUM (Hedw.) Roth (*Hypnum commutatum* Hedw.)

This species agrees with *Cratoneuron filicinum* in the pinnate branching of the principal shoots, and in its prevailing golden green colour, often grading to a rich orange-brown. It is, however, usually much larger. Moreover, the falcate curving of the leaves is more marked here than in *C. filicinum*, and the leaves taper to much longer, finer points. Reddish brown radicles are often conspicuous on the lower parts of the stems. Two rather distinct forms (which have nevertheless been found to intergrade) are included in this species: (i) a plant especially characteristic of calcareous bogs and waterfalls ('*Hypnum commutatum* Hedw.'), with small leaves and plumose branching; (ii) a plant of bogs, chiefly on high moorland ('*Hypnum falcatum* Brid.'), which has larger leaves that give the less regularly pinnate shoots a much stouter, more robust appearance. These have often been regarded as distinct species (e.g. in Dixon's *Handbook*) but are probably mere habitat modifications.

As there are several other marsh and bog mosses with strikingly falcate leaves (members of the old section *Harpidium* of the genus *Hypnum*, and now placed in the genus *Drepanocladus*), the microscope is needed to confirm the identification of *Cratoneuron commutatum*. This should show the following characters: (i) the stout nerve (70–90μ at base) characteristic of *Cratoneuron*; (ii) well-marked alar cells (variable in extent, colour and wall thickness); (iii) a leaf margin toothed to a variable extent; (iv) markedly elongated cells in the leaf as a whole (50–80μ long and 8–10 times as long as wide). This last feature will distinguish *C. commutatum* with certainty from robust states of *C. fili-*

cinum, which has much shorter cells. The presence of paraphyllia among the true leaves will serve as a useful confirmatory character, distinguishing *C. commutatum* from any species of *Drepanocladus* that may resemble it. The stem leaves in the present plant are strongly plicate.

This species is dioecious, and the large curved, cylindrical capsule, much resembling that of *Cratoneuron filicinum*, is not very common.

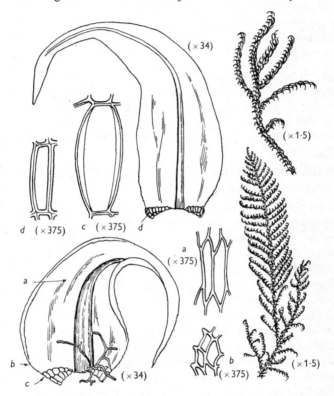

Fig. 116. *Cratoneuron commutatum*: the upper shoot and leaf are drawn from the variety formerly known as *Hypnum falcatum*; the lower shoot and leaf from typical *C. commutatum* (*Hypnum commutatum*).

Ecology. Habitats where the typical plumose form may be found include streamside ledges, springs, waterfalls and fenland. Its presence seems always to be correlated with calcareous conditions and a nearly neutral or slightly basic soil reaction. The irregularly branched, larger-leaved varieties* ('*Hypnum falcatum*') would appear to be less exacting in

* *Cratoneuron commutatum* var. *falcatum* (Brid.) Moenk. and var. *virescens* (Schp.) Richards & Wallace in an Annotated List of British Mosses by P. W. Richards and E. C. Wallace, *Trans. Brit. Bryol. Soc.* (1950), pt. 4.

requirements, occurring more generally over wet moorland and in mountain springs, but almost certainly avoiding the most acid habitats. Further study of the ecological limits of these different forms is desirable, especially in view of the light that it might throw on the wisdom of including them under a single species.

117. **CAMPYLIUM STELLATUM** (Hedw.) Lange & C. Jens. (*Hypnum stellatum* Hedw.)

Campylium stellatum is aptly named, the specific epithet alluding to the star-like manner in which the leaves stand out from the branches. This is especially evident at the tips of the shoots and gives the plant a distinct appearance by which it may generally be known from other species with which it grows in marshy places. It is commonly rather robust, and irregularly branched; but the principal shoots grow more or less upright, are glossy in texture and of a bright golden-green colour.

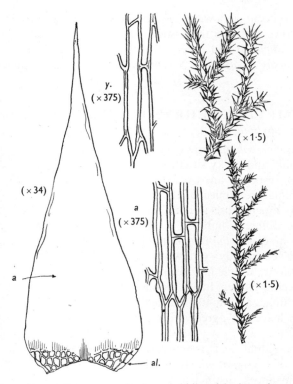

Fig. 117. *Campylium stellatum*: *y.* cells from middle of leaf much younger than the leaf drawn on left; the lower shoot is drawn from a slender form of the species; *al.* alar cells.

With a little experience it is not generally a difficult plant to recognize, although some of its slender forms at times may prove puzzling.

A leaf from one of the larger stems is 2–3 mm. long, with a very broad clasping base and a long fine straight point standing out at a wide angle with the stem. Finer branches, and slender forms of this moss, will be found to have much smaller leaves. Under the microscope the leaf is seen to be practically nerveless (occasionally with a faint nerve, single or double), with an entire margin. The elongated cells which occur throughout most of the leaf are not particularly characteristic, but the clasping auricles at the leaf base are marked by short, wide, greatly enlarged cells, their rather thick walls colourless or orange. These patches of alar cells are usually very distinct and well defined in *C. stellatum*, but, like the nerve, are subject to a certain amount of variation.

This moss is dioecious, and fruit is rather rare. The capsule is curved, cylindrical, and is borne on a seta 2·5–5 cm. long.

Ecology. It occurs in a wide range of moist habitats. Thus, in calcareous dune hollows and fens the associated species are *Cratoneuron commutatum*, *C. filicinum* and others, whilst on wet moorland it grows with *Drepanocladus revolvens* or *Scorpidium scorpioides*. It appears to avoid shade.

118. CAMPYLIUM CHRYSOPHYLLUM (Brid.) Bryhn (*Hypnum chrysophyllum* Brid.)

This species, which grows on calcareous hills and rocky situations, is in many respects an exceedingly slender miniature replica of the last. Besides its much more slender form and very different habitat, it also differs from *Campylium stellatum* in the more constantly yellowish or golden colour, which has given rise to the specific name. In its manner of branching, as in its markedly squarrose leaves, it agrees with the larger species.

The individual leaf is narrower than that of *C. stellatum*. It is also, in most instances, distinct in several microscopic characters. Thus, in *C. chrysophyllum* a single rather faint

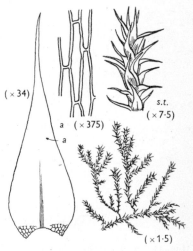

Fig. 118. *Campylium chrysophyllum*: *s.t.* tip of shoot.

nerve extends to about mid-leaf, the areolation is on the average shorter than in the last species (cells 5–10 times as long as broad); and the alar cells, whilst short and forming distinct patches, are not usually so swollen or so transparent as are those of *C. stellatum*.

The capsule here is even rarer than that of the last species, which it resembles. The small size and characteristic colouring of the plant, however, taken in conjunction with the habitat, will usually make *C. chrysophyllum* an easy species to recognize. Nevertheless, puzzling forms occur at times, in some respects intermediate between these two species of *Campylium*, forms which have suggested to some authorities that the present plant should be ranked as a subspecies of the last.

Ecology. Favouring considerably drier habitats than *C. stellatum*, and exclusively calcicole, this species occurs chiefly in chalk quarries, and among grass or on rock ledges in limestone districts. It is also found on sand dunes. A frequently associated species is *Camptothecium lutescens*.

ADDITIONAL SPECIES

Campylium polygamum (B. & S.) Lange & C. Jens. (*Hypnum polygamum* (B. & S.) Wils.), a plant of marshy ground, is the only other species in this genus which is at all common. Being autoecious (or synoecious) it is usually very fertile, and differs in this from *Campylium stellatum*, which it otherwise resembles. It is also distinct in its less widely spreading leaves and in the nerve that extends at least to mid-leaf.

119. LEPTODICTYUM RIPARIUM (Hedw.) Warnst. (*Hypnum riparium* Hedw.)

This is a rather common moss in lowland habitats near water. It forms loose, untidy patches on soil, wooden palings, rotting stumps or fallen branches on the margins of lakes or pools. In its bright green colour and glossy texture it sometimes looks like *Brachythecium rutabulum*; at other times it may resemble some forms of *Drepanocladus aduncus*. *Leptodictyum riparium* is, in fact, a notably variable plant, and identification should be confirmed under the microscope. Perhaps the most useful field characters lie in the rather soft texture of the plant as a whole, and the widely spreading character of the fairly narrow, finely pointed leaves, both wet and dry.

The leaf is 2–4 mm. long, straight and tapering from a not very broad base to a rather long fine point. Under the microscope the long single nerve ($\frac{3}{4}$ of total length of leaf) separates it at once from *Campylium stellatum*, or from those species of *Plagiothecium* which it sometimes resembles in its markedly flattened branches. The absence of longitudinal folds and the lack of teeth along the leaf margins readily separate it from *Brachythecium rutabulum*. The long narrow cells become shorter and wider in the basal angles of the leaf, but well-defined patches of alar cells are not formed.

Leptodictyum riparium is autoecious and capsules are common. The orange-red seta is 1·5–2·5 cm. long, the capsule narrowly ovoid to cylindrical, curved, with a conspicuously large peristome.

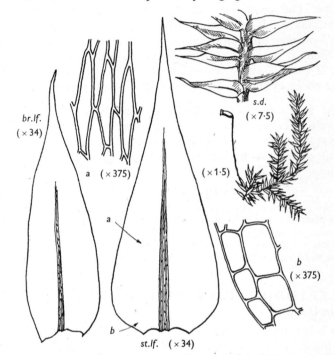

Fig. 119. *Leptodictyum riparium*: *st.lf.* stem leaf, *br.lf.* branch leaf, *s.d.* fragment of shoot in a dry state.

Ecology. The various substrata—rock, wood and soil—on which this species grows are mentioned above, its presence almost always being associated with river bank or pond margin. It is mainly a lowland species and is most common in calcareous districts. I have found it to be the sole species responsible for the extensive growths of moss on the metallurgical coke of the bacterial filter beds of the Reading Sewage Farm.

120. AMBLYSTEGIUM SERPENS (Hedw.) B. & S.

This species is the most slender of our common pleurocarpous mosses. Indeed, in the low, soft, yellowish or mid-green patches which it forms on decaying tree stumps or other shaded sites it is hard to distinguish clearly the leafy nature of the branches, for the individual leaf is barely 0·5 mm. in length, almost too minute for the naked eye to see. Despite its small size, *A. serpens* is a moderately conspicuous moss, for the prostrate stem system throws up very numerous, more or less erect or spreading branches, so that quite extensive patches may be formed. These patches lack any trace of glossiness.

The leaf is ovate-lanceolate and tapers to a fairly fine point. The microscope will reveal the rather faint nerve which seldom extends much above mid-leaf. The margin is entire or very slightly toothed. The short cells, approximately hexagonal or rhomboid in outline, afford a useful character; they are only about four times as long as broad, not greatly elongated as in so many related mosses.

This species is autoecious, and is nearly always abundantly fertile. The curved, narrowly cylindrical capsule is borne on a red seta about 1·5 cm. long; it is relatively large for the size of the plant, and in the young stage the whitish calyptra is conspicuous.

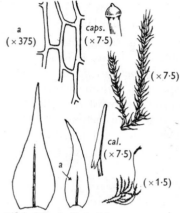

st.lf. (× 34)   br.lf. (× 34)

Fig. 120. *Amblystegium serpens*: *cal.* young capsule covered by calyptra, *caps.* part of older capsule showing lid, *st.lf.* stem leaf (robust state), *br.lf.* branch leaf.

Ecology. Growing equally readily on wood, rock and soil, this species is most characteristic of moist woodlands and hedge banks in southern England, but is widespread generally. Rotting tree stumps and decaying logs are, perhaps, especially favoured. It appears to avoid very acid conditions.

ADDITIONAL SPECIES

A. varium (Hedw.) Lindb. grows in the same habitats as *A. serpens*, from which it differs in the longer nerve, extending well up into the leaf tip. It is also a rather larger plant, and the leaf cells are relatively wide and thick-walled.

RELATED SPECIES

Mention should be made of two species belonging to the closely related genus *Hygroamblystegium*. These are both aquatic plants that grow on stones submerged in streams.

H. tenax (Hedw.) Jennings (*Amblystegium irriguum* (Wils.) B. & S.) has rather rigid leaves with faintly toothed margins and a stout nerve that vanishes in the extreme leaf tip.

Hygroamblystegium fluviatile (Hedw.) Loeske (*Amblystegium fluviatile* (Hedw.) B. & S.) is very deep green, and has numerous long, nearly parallel branches. The leaves are of softer texture, with entire margins and a very stout nerve distinct to the extreme apex. *Hygroamblystegium fluviatile* is most typically a plant of fast-flowing streams in limestone country.

121. **DREPANOCLADUS ADUNCUS** (Hedw.) Warnst. (*Hypnum aduncum* Hedw.)

This is among the common 'hypnoid' mosses of lowland pools and ditches. In its loose mode of growth and soft texture it resembles *Drepanocladus fluitans*, but the curving of the shoot tips seen in that species is seldom well developed here, and the colour is most often green (rarely yellowish or dark brown). It is extremely variable in size, and in extent of branching. At times it is minute, pinnately branched, with curved leaves 1 mm. long; yet forms occur in which the stems extend

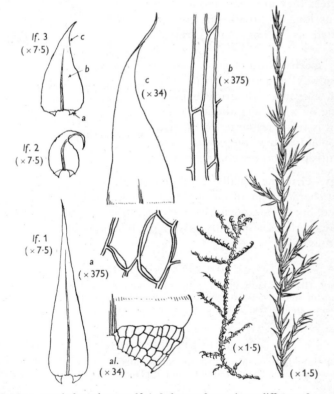

Fig. 121. *Drepanocladus aduncus*: *lf*. 1–3, leaves from three different forms of the plant; *lf*. 1 is from the shoot on the extreme right, *lf*. 2 is from the shoot next on the left, *lf*. 3 is from a third form, of intermediate habit; *al*. alar cells.

unbranched for 25–30 cm., bearing straight, widely spaced leaves
5–6 mm. in length.

In so variable a plant it is essential to examine leaves carefully under
the microscope. The leaf is fairly broad just above the base, and tapers
to a long fine tip; the nerve usually extends well above mid-leaf but is
not so long as that of *D. fluitans*; it may be quite short and faint. The
cells are long and narrow, with patches of enlarged transparent cells
forming the well-defined basal auricles. The entire margin will separate
this leaf from that of *D. fluitans*, and the absence of longitudinal folds
on the leaf surface will distinguish it from any form of *Cratoneuron
commutatum*. The base of the leaf and the auricles are 'hollowed out'
in a rather characteristic manner.

Drepanocladus aduncus is dioecious, and in most forms of the plant
capsules are rare. The oblong, curved capsule is borne on a seta 2–5 cm.
long.

Ecology. It occurs in a fairly wide range of marshland and semi-aquatic
habitats. It is one of the numerous species of *Drepanocladus* and allied
plants which form a characteristic association in moist hollows of sand
dunes (such as the 'slacks' of the Lancashire coast). It is less tolerant
of acid conditions than is *D. fluitans*, being found chiefly in neutral
to slightly basic waters.

122. DREPANOCLADUS FLUITANS (Hedw.) Warnst. (*Hypnum fluitans* Hedw.)

This species is locally common in bogs and marshy places, at low to
moderate altitudes. Although varying in colour from green, through
yellowish, to chestnut brown, and variable also in habit, it may generally
be known by the hooked ends of the shoots, and soft, rather glossy
texture of the plant as a whole. At the tips of the shoots the long narrow
leaves become strongly curved to one side, otherwise the falcato-secund
character is not so well marked as in many related mosses; in fact in
some of the lax, attenuated forms of *Drepanocladus fluitans* the leaves—
except at the extreme shoot tips—spread evenly in all directions. These
curved shoot tips are often considerably paler in colour than the lower
parts. In habit *D. fluitans* varies from the long, almost simple stems
(up to 30 cm. in length) of some submerged forms, to the short stems
with closely pinnate branching met with at times in terrestrial habitats.
Many named varieties exist, but these are by no means well defined.

The individual leaf is 3–4 mm. in length; it is curved, widening above
the narrow base, then narrowing gradually into the long fine apex.
There is a long, strong nerve; other microscopic characters to be noted
are: (i) margin lightly toothed, most distinctly so near the apex, (ii) longi-

tudinal folds (plicae) lacking, (iii) cells very long and narrow except at the basal angles where well-defined patches of short, hyaline or coloured cells form auricles. The toothed margin separates it from *D. aduncus*, the absence of folds in the leaf surface from forms of *Cratoneuron commutatum*, whilst the presence of auricles will usually distinguish it from

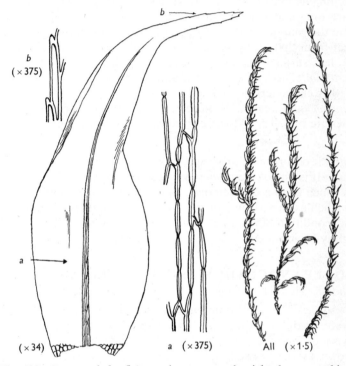

Fig. 122. *Drepanocladus fluitans*: the stems on the right show something of the range of variation met with; only the terminal part of each shoot is shown.

Drepanocladus revolvens. Most forms of the last two species, however, are very different in colour and habit.

D. *fluitans* is normally autoecious and the capsule is not rare. It is curved, cylindrical, and is borne on a very long seta (4–8 cm.).

Ecology. It occurs chiefly, if not invariably, in relatively acid pools, bogs and marshes. It will tolerate habitats ranging from aquatic to almost wholly terrestrial; thus it grows in deep pools and runnels but may also be met with on lowland heaths in places which are only periodically wet. *D. fluitans* is rarer in calcareous districts.

123. **DREPANOCLADUS REVOLVENS** (Sm.) Warnst. (*Hypnum revolvens* Turn.)

This is a plant of bog pools and wet peaty places, where it tends to be conspicuous owing to its robust habit and the evident tinge of orange, crimson or purple which is usually present, with yellowish green, in the glossy tufts. The branching is irregular, the plant often forming a number of long secondary shoots which are themselves only sparsely branched. The crowded leaves are uniformly and strongly curved to one side, the

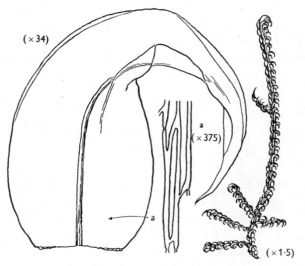

Fig. 123. *Drepanocladus revolvens.*

individual leaf being sometimes so curled as almost to form a circle (a feature in which it resembles that of *Drepanocladus uncinatus*). The leaves are always more regularly and more markedly curved than in *D. fluitans* or *D. aduncus*, but robust forms of the present species may bear some resemblance to certain states of *Scorpidium scorpioides*.

The leaf under the microscope shows a single nerve, extending about $\frac{3}{4}$ of the leaf length, a margin without teeth, and a leaf surface not scored by longitudinal folds or furrows. The cells are extremely long and narrow (4–6×70–120μ), thick-walled and porose, especially towards the leaf base, where only very small, ill-defined patches of short cells will be found. Definite auricles are never formed, nor is the leaf base concave as it is in *Drepanocladus aduncus*.

D. revolvens may be dioecious or autoecious, but the fertile condition is rare. The curved, cylindrical capsule is large, and is borne on a seta 2–5 cm. long.

Ecology. Most typically it is a plant of lake margins and peat bogs in the north and west, where a frequently associated species is *Scorpidium scorpioides*. There would seem to be some doubt as to its status in calcareous bogs and fens, most forms which occur there being perhaps best referred to var. *intermedius* (Lindb.) Richards & Wallace (*Hypnum intermedium* Lindb.), a plant of more slender habit and shorter leaf cells.

124. DREPANOCLADUS UNCINATUS (Hedw.) Warnst. (*Hypnum uncinatum* Hedw.)

This species occurs on damp rocks and in boggy places (less often on walls) in hilly districts; it is distinct in its slender but compact habit, pale golden green or yellowish olive colour, and in the strong and regular manner in which the sickle-shaped leaves are curved to one side. In this

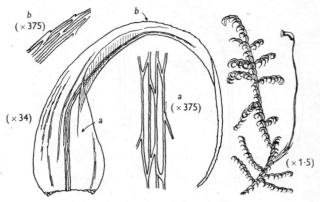

Fig. 124. *Drepanocladus uncinatus.*

feature it resembles *Drepanocladus revolvens*, but in colour that species is very different. The stems of *D. uncinatus* are commonly 4–10 cm. long, and regularly branched in a pinnate manner. It forms more or less prostrate, spreading patches, the tips of the curved shoots being conspicuous for their pale colour and glossy character. This feature, taken in conjunction with the long fine points of the leaves, may suggest a resemblance to *Ctenidium molluscum*, but the leaves on the branches are much larger than in that plant, and have their tips developed into obvious 'hooks', whence the specific epithet *uncinatus* derives.

The leaf is 2–3 mm. long, narrow in form, and tapers to an exceedingly fine, distinctly toothed apex. The nerve is weak and narrow, but extends far up the leaf. Strongly developed longitudinal folds occur. The cells are long and narrow, as in others of this group, giving place at the basal angles to rather ill-defined patches of thin-walled short cells, which form very small auricles.

Drepanocladus uncinatus is autoecious, and is quite commonly fertile. The long plicate perichaetial leaves afford a useful diagnostic character. The orange-brown seta attains about 2·5 cm. in length. The ripe capsule is of a similar colour, and of the curved, cylindrical form common in the Hypnaceae.

Ecology. It grows about the edges of moist upland woods and on wet ground near streams on moorland. It is most commonly found on a rock or gravelly substratum.

ADDITIONAL SPECIES

D. lycopodioides (Brid.) Warnst. (*Hypnum lycopodioides* Brid.), which occurs at times in bogs (it is abundant, for example, in the duneland 'slacks' near Southport, Lancs), is known by its very robust habit and rich golden colour. The strong, single nerve of the leaf distinguishes it from *Scorpidium scorpioides*, which it resembles in its long stout branches and crowded concave leaves.

Drepanocladus exannulatus (B. & S.) Warnst. (*Hypnum exannulatum* B. & S.), is closely allied to *Drepanocladus fluitans*, but may be known generally by its more compact and stiffer habit, more regularly pinnate branching and, often too, by a deep purplish colour not found in *D. fluitans*. The leaves are more strongly curved and the patches of inflated alar cells tend to be more extensive than in *D. fluitans*. It occurs chiefly in upland bogs, and in some districts is commoner than the latter species.

125. HYGROHYPNUM LURIDUM (Hedw.) Jennings (*Hypnum palustre* Huds.)

This is a common moss on rocks in mountain streams, where it forms loose, untidy patches, usually of a dull yellowish or brownish green colour. The whole texture of the plant is soft and limp, with irregular leafy branches arising from a prostrate primary stem which is commonly

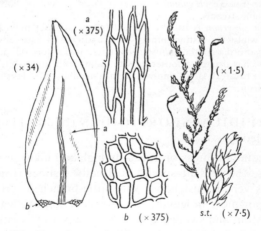

Fig. 125. *Hygrohypnum luridum*: *s.t.* terminal part of shoot.

18-2

more or less bare of leaves. Often the branches are quite short and markedly curved to one side at their tips. At other times the habit is laxer, the branches long and weak, and the falcate tips not well shown.

If sometimes there is a slight resemblance to some forms of *Drepanocladus fluitans* in the soft texture and falcate branch tips of this plant, all such resemblance disappears when the leaf is examined closely. This has a rather concave form, and the tip is short and blunt compared with species of *Drepanocladus*. The leaf is also quite small, attaining only 1·5 mm. in length.

Under the microscope the leaf shows a variable nerve—either long and single or quite short and double—an entire margin and long, narrow cell structure. The edges of the leaf are often characteristically incurved near the apex. Specialized alar cells occur, but they are opaque and granular, and rather thick-walled (not transparent and thin-walled as in *Hygrohypnum ochraceum*); nor do they form very distinct or well-defined auricles.

H. luridum is autoecious and is often abundantly fertile. The capsule is rather short (2 mm.) and ovoid in shape. It is held in an inclined or horizontal position on an orange-red seta, 1–2 cm. long.

Ecology. Although perhaps most plentiful in fast-flowing mountain streams, it will grow, on rock or wooden substrata, in the slowly flowing water of large rivers, e.g. the middle reaches of the Thames. Base-rich habitats are favoured.

ADDITIONAL SPECIES

The only other species of this genus which are not rare are *Hygrohypnum eugyrium* (B. & S.) Loeske (*Hypnum eugyrium* (B. & S.) Schp.) and *Hygrohypnum ochraceum* (Turn. ex Wils.) Loeske (*Hypnum ochraceum* Turn. ex Wils.); both are plants of rocks in mountain streams.

Hygrohypnum eugyrium is usually marked by its compact habit of growth and glossy texture, whilst the distinct orange-brown patches of alar cells in all but the youngest leaves afford a good diagnostic character. It is autoecious and fruits freely.

H. ochraceum has usually a looser growth habit, with long, soft, dull yellowish green shoots. It has very large, thin-walled, hyaline alar cells. It is dioecious—and is very rare in fruit.

126. SCORPIDIUM SCORPIOIDES (Hedw.) Limpr. (*Hypnum scorpioides* Hedw.)

A rather common moss on rocks by highland lakes or streams, and in peat bogs, the present species is usually distinct from all related mosses in its very robust habit, with little-branched, prostrate stems often 12 cm. or more in length, and crowded, concave leaves that give the shoots a stout and swollen appearance. It is usually either yellowish green at the curved shoot tips, becoming dark brown or blackish below, or tinged with orange or crimson at the tips and deep blackish purple in the lower parts. Smaller forms occur at times, but *Scorpidium* is a moss that, once known, is usually recognized without difficulty. In the one-sided curvature of the leaves, especially at the tips of the branches, it resembles species of *Drepanocladus*, and small forms may occasionally be mistaken in the field for *D. revolvens*.

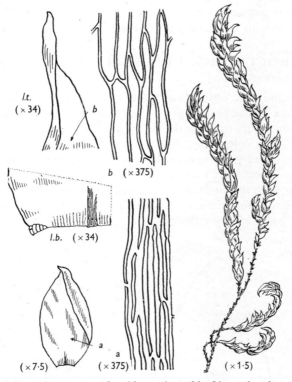

Fig. 126. *Scorpidium scorpioides*: *l.b.* portion of leaf base showing alar cells, *l.t.* leaf tip; the cells at *b* are taken from a much younger leaf than those at *a*. *Note.* The leafy shoot drawn is from a robust state of the species.

The leaf is 2–4 mm. long, very wide and concave in the middle, narrowed at the base, and abruptly contracted at the tip into a short and normally not very acute point. The surface of the leaf is variously wrinkled, with irregular transverse and longitudinal folds, which are accentuated by the concave shape of the leaf as a whole. The margin is entire, and the nerve usually very short and forked. The cells are long and very narrow, thick-walled and porose, with a few enlarged and colourless alar cells forming the small auricles. From *Drepanocladus* it is immediately distinct in the feebly developed nerve.

Scorpidium scorpioides is dioecious, and the large, curved capsule, on a very long seta, is not common.

Ecology. It occurs most commonly on wet peat or peat-covered rocks in the mountain country of the north-west; but it will also grow in fens, so that it cannot be regarded as confined to acid habitats; indeed some authorities regard it as calciphile. I have frequently found it associated with *Drepanocladus revolvens*.

127. **ACROCLADIUM CUSPIDATUM** (Hedw.) Lindb. (*Hypnum cuspidatum* Hedw.)

The specific name of this plant alludes to one of its most characteristic features, the 'spear-head' tips of the branches, which result from the leaves being tightly rolled in the bud. This feature, taken with the bright or yellow-green colour (occasionally orange-brown), glossy texture and vigorous mode of growth, will usually serve to identify this very common marshland species. At times it becomes a conspicuous element in the vegetation, with its numerous pinnately branched upright shoots arising from the prostrate primary stem which creeps on the wet mud or between the bases of larger plants.

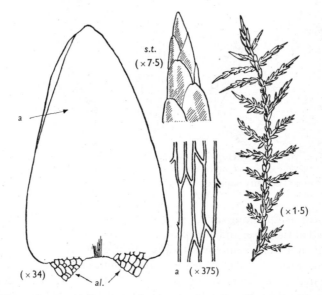

Fig. 127. *Acrocladium cuspidatum*: *s.t.* shoot tip showing cuspidate form, *al.* alar cells forming auricles.

The leaves on the main stems are large (2·5 mm. long), but on the finer branches considerably smaller; in the former broadly ovate in outline, in the latter narrower; they taper from a wide base to a blunt, rounded apex. There is never any trace of the falcate curving of the leaf which is found in so many 'hypnoid' mosses.

The nerve is short and double, or wanting, the leaf margin entire and the areolation long and narrow. Very distinct patches of large, thin-walled, colourless or orange cells at the basal angles of the leaf form well-marked auricles. These characters will help in separating the present species from *Pleurozium schreberi*, which it resembles in some respects. The latter plant, however, has a different mode of growth and is found in a very different habitat.

Acrocladium cuspidatum is dioecious, and whilst barren plants are abundant, the fruit is not very common. The large, curved capsule, however, is distinctive, and conspicuous on its long (4–7 cm.) red seta.

Ecology. As a species of wet ground *A. cuspidatum* is common in moist clay pastures, in calcareous bogs, and in wet hollows of sand dunes. It also has another, quite different habitat, namely, relatively dry chalk grassland, especially on sheltered, north-facing slopes.

ADDITIONAL SPECIES

The four remaining species of *Acrocladium* that occur fairly commonly in Britain all differ from *A. cuspidatum* in having a strong single nerve running for most of the length of the leaf.

A. cordifolium (Hedw.) Richards & Wallace (*Hypnum cordifolium* Hedw.) occurs in pools and marshes, and is distinct in its little-branched, often nearly erect stems 8–20 cm. in length, and large (2–5 mm. long), widely spaced, spreading leaves.

Acrocladium sarmentosum (Wahl.) Richards & Wallace (*Hypnum sarmentosum* Wahl.), a plant of mountain bogs, has a distinctive dark purplish or crimson colour and lacks the regularly pinnate branching of *Acrocladium cuspidatum*.

A. giganteum (Schp.) Richards & Wallace (*Hypnum giganteum* Schp.), a local plant of marshes, may be known by its robust, bushy habit, with the crowded branches spreading in all directions.

Acrocladium stramineum (Brid.) Richards & Wallace (*Hypnum stramineum* Brid.), which may be found among other mosses in moorland bogs, looks a little like a weak and slender variant of *Acrocladium cuspidatum*. Its stems are very long (8–20 cm.), weak and scarcely branched; the leaves (1·5–2 mm.) are held more or less erect and some of them commonly bear tufts of rhizoids.

128. ISOTHECIUM MYURUM (Brid.) Brid. (*Eurhynchium myurum* (Brid.) Dix.)

This species grows in extensive yellowish green mats which, together with those of certain other species, largely cover the bases of trees, rotting stumps and rocks in some moist woods. From all except *Isothecium myosuroides* it is distinguished by its sub-dendroid habit, the more or less erect secondary stems being unbranched for 1–3 cm., but freely branched above to give each shoot a distinctive tassel-like appearance. The crowded branches tend, moreover, to be curved and turned to one side, a tendency which enhances the tassel-like effect. Each branchlet appears swollen and smoothly cylindrical when dry, owing to the concave form of the crowded, overlapping leaves. Thus it has a robust appearance quite different from that of *I. myosuroides*.

The concave leaf is ovate-oblong in outline; and, although the shape of the leaf tip varies, it never tapers to a long fine point as in *I. myo-*

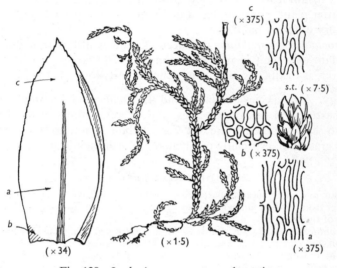

Fig. 128. *Isothecium myurum*: *s.t.* shoot tip.

suroides. Often the rounded outline is broken by a short apiculus. The margin of the leaf is less strongly toothed than in *I. myosuroides*. The cells throughout much of the leaf are long, narrow and rather thick-walled. The best microscopic character, however, lies in the very short, almost isodiametric cells that extend a considerable way up the margins of the leaf, above the well-defined patches of opaque (green or orange) alar cells. It has sometimes been stated that the alar cells are uniformly shorter than in *I. myosuroides*, but this is not constantly so.

Although dioecious, *I. myurum* is not uncommonly fertile. The erect, symmetrically ovoid capsule, with long-beaked lid, will identify the plant with certainty.

Ecology. It is commoner than *I. myosuroides* in many parts of the midlands and south of England, where it grows chiefly on tree bases; but it is often the less common plant in the west, where it is found on rocks and tree bases alike. Although in many places, especially where a stream flows through a mountain wood, the two species grow together, *I. myurum* is absent from most of the more markedly acid habitats tolerated by *I. myosuroides*.

129. ISOTHECIUM MYOSUROIDES Brid. (*Eurhynchium myosuroides* (Brid.) Schp.)

Isothecium myosuroides is a common and widely distributed moss, forming rather loose yellowish or mid-green mats or patches on tree stumps or rocks in woodland. Small forms occur on exposed boulders

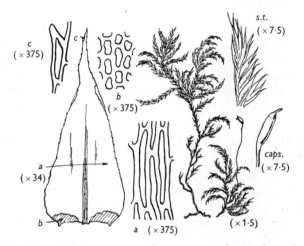

Fig. 129. *Isothecium myosuroides*: *caps.* capsule, *s.t.* shoot tip.

and a luxuriant variety, differing markedly from the typical form, is found occasionally on mountain ledges. Usually the numerous curved or upright shoots have something of the tassel-like or sub-dendroid form met with in the last species, and this feature, coupled with the finely pointed shoot tips, will generally make *I. myosuroides* an easy plant to recognize. Typically, it forms massive cushions on some boulder, or loosely festooning the base of a tree, and the relatively bare base and feathery summit of each shoot conveys something of the 'mouse's tail' effect whence it gets its specific name. A lens will show the leaves to be narrowly ovate-lanceolate, and drawn out into fine points in contrast with the blunt and apiculate leaf tips of *I. myurum*. On the main secondary stems the leaves are shorter and wider, becoming heart-shaped to triangular in outline, and often suddenly narrowed above the middle.

A useful microscopic character lies in the sharply but unevenly toothed leaf margins, especially of the branch leaves. Even near the leaf apex cells will be found up to 40–50×6–8μ (longer than the corresponding cells of *I. myurum*). The alar cells form well-defined, opaque (green or orange-brown) auricles, composed of oval or nearly isodiametric cells. The very short cells that occur just above the auricles do not extend upwards along the leaf margin as they do in *I. myurum*. The nerve is extremely variable; at times it is faint or lacking.

This species is dioecious, but capsules are not rare. Raised on an orange-red seta 1–2 cm. long, the capsule is slightly curved and ovoid in form; it is held inclined at an angle, not erect as in the last species. The lid is only shortly pointed and the whole capsule (2–2·5 mm. long) appears small for the size of the plant.

Ecology. This species is very characteristic of, but by no means confined to, *Quercus petraea* (*Q. sessiliflora*) woods of western Britain. Like *Isothecium myurum*, it grows on trees, stumps and boulders—on the latter often in fully exposed situations in moist western districts. It extends to many acid types of habitat where *I. myurum* is not found.

130. CAMPTOTHECIUM SERICEUM (Hedw.) Kindb.

This common plant of walls and the bases of trees is likely to attract attention by the bright silky sheen on the pale green tips of the shoots. The glossy character is especially noticeable in the dry state, when the numerous short branches will appear somewhat curved. This species differs from the closely allied *C. lutescens* in its prostrate, creeping habit, each of the principal secondary stems giving rise to a mass of short shoots. The latter tend to grow nearly upright, and as the plant is of fairly robust habit, quite extensive and deep cushions are formed. It will often spread over a wall-top in a thick felted mat.

The leaf is narrowly lanceolate, straight and tapers to a long fine point. It is longer and narrower than in any of the more robust species of *Brachythecium*, plants which *Camptothecium sericeum* resembles in habit. Notable microscopic characters of the leaf are the long single nerve, extending nearly to the apex, the deeply plicate character of the leaf surface, and the long, extremely narrow cells (approx. 4–6×90–130μ). No species of *Brachythecium* has such narrow cells, nor the leaf surface so deeply and regularly plicate. The margin may be faintly toothed above, whilst at the base of the leaf are found small but well-defined toothed auricles.

Camptothecium sericeum is dioecious, but capsules are fairly common. The capsule is borne erect on an orange-red seta about 2 cm. in length. It is symmetrically ovoid-cylindrical, 2–3 mm. long.

Ecology. It will grow on a wide range of substrata, including wall-tops, the surfaces of boulders, the bases of living trees and roof-tops. It is most typical and most abundant, however, on walls and boulders in limestone districts.

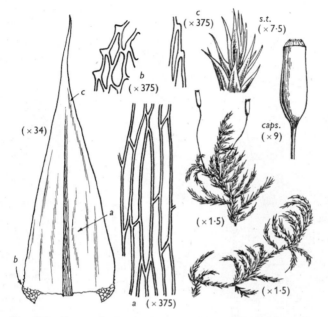

Fig. 130. *Camptothecium sericeum*: *caps.* capsule, *s.t.* shoot tip.

131. CAMPTOTHECIUM LUTESCENS (Hedw.) Brid.

The specific name of this moss refers to the yellowish colour which is normally an evident feature of the loose tufts that it forms on chalk hills or calcareous duneland. It lacks the creeping habit of the last species, and the tufts fall apart readily when gathered. The principal ascending stems bear short erect branches, but these are not so densely massed as in *C. sericeum*. Only exceptional forms, in fact, can be confused with that species. The pale yellowish colour and the long, straight, finely pointed leaves will readily separate *C. lutescens* from other bryophytes with which it grows intermixed; indeed, it is usually an easy plant to recognize. Like the last species, it is narrower in the leaf than any species of *Brachythecium*.

In microscopic leaf characters it bears a close resemblance to *Camptothecium sericeum*. As in that plant, the leaf has a long single nerve, a deeply plicate surface, and rather well-developed basal auricles. The cells are long and very narrow (3–6μ wide); in the leaf base they tend to have thicker, more obviously porose walls than those of *C. sericeum*, and the auricles are less distinctly toothed than in that species.

This species is dioecious and is usually barren. In fertile plants the short, slightly curved form of the capsule (1·5–2 mm. long) will help to confirm the identification.

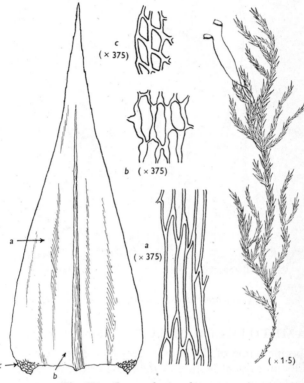

Fig. 131. *Camptothecium lutescens*.

Ecology. It grows on the ground—seldom on rocks—in calcareous places. Thus it is one of the few mosses which are frequent in chalk grassland. Where dunes are composed largely of white shell sand it becomes an important moss of the semi-fixed dune community. In more acid conditions its place may be taken by *Brachythecium albicans*.

132. BRACHYTHECIUM ALBICANS (Hedw.) B. & S.

B. albicans, though quite common, can easily be overlooked. It is of slender habit, the numerous ascending or upright secondary stems arising irregularly, often, in fact, growing up between the branches of other mosses which share the habitat of bare earth, sand or gravel. It may be distinguished by its pale yellowish green colour and silky texture. The slender character of the sparingly branched erect stems is accentuated by the tendency of the leaves, when dry, to lie appressed to the stem with only their long fine tips diverging, thus giving the shoots a somewhat string-like appearance. One is reminded of a diminutive *Camptothecium lutescens*, but the present plant is not only more slender, but also is much less freely branched than that species.

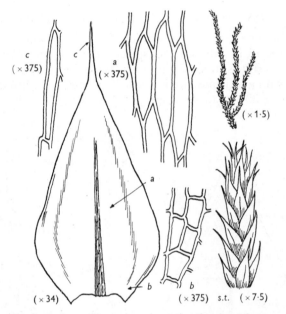

Fig. 132. *Brachythecium albicans*: *s.t.* shoot tip; the shoot fragment at top right shows a comparatively slender state of the species.

The leaf is 1·5–2 mm. long, broadly ovate in general outline (a feature distinguishing *Brachythecium* from *Camptothecium*), but rather suddenly contracted near the tip into a fairly long, fine point. The leaf is slightly smaller, and the apex longer and finer than in most other common species of the genus. The leaf surface has well-marked longitudinal folds or furrows, but this feature is less pronounced than in *Camptothecium*. The leaf margin is plane and entire; the single nerve extends to just beyond mid-leaf. Clearly defined auricles are not formed, but short cells extend some way up the margins from the basal angles before giving place to the relatively long narrow cells ($50–80 \times 10–15\,\mu$) that mark the leaf as a whole.

Brachythecium albicans is dioecious, and the shortly oval capsule, on a smooth seta, is comparatively rare.

Ecology. In its more slender forms it often occurs as a colonist of bare patches of sand or gravel; more luxuriantly, and very commonly, it grows among grass on sandy soils. It is equally a plant of coastal dunes and of inland sandy heaths and gravelly ground. Though widely distributed, it is perhaps commonest in the south.

133. BRACHYTHECIUM RUTABULUM (Hedw.) B. & S.

This is among the first mosses which the beginner will come to know, since it is robust in habit and conspicuously abundant in many places, especially on stones, tree stumps and decaying branches in woodland. Although with a little practice it is usually recognized without difficulty, *B. rutabulum* is not an easy species to define. The numerous short,

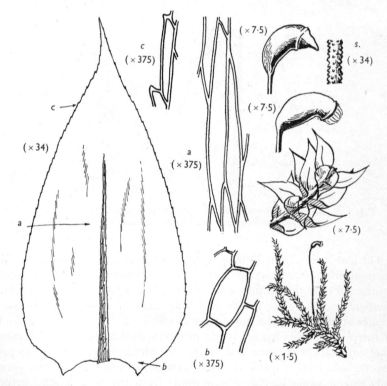

Fig. 133. *Brachythecium rutabulum*: *s.* part of seta; capsules are shown with and without lids; the shoot at bottom right was drawn from a plant in the dry state, the fragment just above it is from a moistened plant.

irregularly placed secondary stems (with nothing of the regularly pinnate character seen in so many 'hypnoid' mosses), the bright green or yellowish colour and glossy texture of the plant, and the widely spreading leaves, both wet and dry, are a combination of characters by which it may generally be known at sight. Certain non-typical states may, however, present more difficulty, notably some forms which occur on stream banks or in marshes and somewhat resemble *B. rivulare* or *Leptodictyum riparium*.

The larger leaves attain 2–3 mm. in length and are broadly ovate in outline, tapering to acute tips. The leaf, however, is without the long fine point of *Brachythecium albicans* and *Cirriphyllum piliferum*. The nerve extends just beyond mid-leaf. The leaf surface is not strongly plicate. The margin is marked by minute but distinct teeth, and this denticulate leaf margin will separate marsh forms of the present plant from *Leptodictyum riparium*. Shorter, wider cells occur at the basal angles of the leaf, but clearly defined patches of alar cells are lacking; elsewhere the cells are long and narrow (80–150 × 6–14 μ).

In this autoecious moss the capsules are commonly plentiful; the absence of a beak to the lid of the capsule indicates *Brachythecium*, and the rough seta (due to papillae that are just visible with a lens) affords a useful specific character. The seta is longer (1·5–3 cm.), and the capsule larger (2·5–3·5 mm.) than in *B. velutinum*, another common species with a rough seta.

Ecology. Except in some markedly acid habitats, this species is almost ubiquitous in lowland districts, growing on rocks, soil and decaying wood. It is often found on lawns in gardens. Well-grown, abundantly fruiting plants may be found in almost any type of moist woodland, including fen carr; certain less typical forms occur frequently in grass on river banks and in marshes. Dr E. W. Jones considers that it is probably nitrophilous.

134. BRACHYTHECIUM RIVULARE (Bruch) B. & S.

This species, although structurally akin to *B. rutabulum*, usually occurs in moister habitats (e.g. in and about the margins of streams) and is distinct in its bright or golden green colour and less spreading leaves. The shoot tips have, indeed, something of the 'spear-head' form met with in *Acrocladium cuspidatum*. Typically it is a robust plant with crowded secondary shoots, and its bright colour and glossy character render it conspicuous. It is, however, very variable.

The leaf is of similar size and general shape to that of *Brachythecium rutabulum*, but is less finely pointed at the tip and more strongly decurrent at the base. It is also more strongly and regularly plicate. Under the microscope these deep longitudinal folds are an evident feature, but the most useful confirmatory character for separation from *B. rutabulum* lies in the cells at the basal angles of the leaf, which are swollen, hyaline or orange, forming clearly defined auricles. In the single nerve to mid-leaf and the toothed margin the leaf resembles that of *B. rutabulum*.

B. rivulare is dioecious and capsules are not common. The seta resembles that of the last species in being rough with papillae. The capsule is rather shorter and more swollen in form.

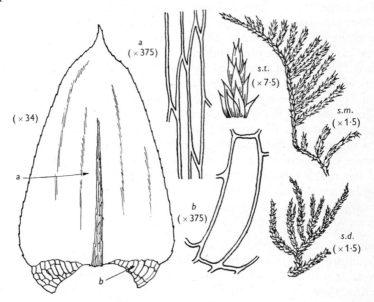

Fig. 134. *Brachythecium rivulare*: *s.d.* and *s.m.*, portions of shoots in dry and moist conditions respectively; *s.t.* shoot tip. The leaf on the left is from a main stem and shows the enlarged alar cells forming prominent auricles.

Ecology. It is a characteristic member of the specialized flora which develops about small springs on mountains, where almost constantly associated species are *Philonotis fontana* and *Bryum pseudotriquetrum*; but it is common on wet rocks in and by streams generally, and by waterfalls. It is then a member of a rich flora that may include many pleurocarpous species such as *Hyocomium flagellare* and *Cratoneuron commutatum*. A less usual habitat is on the ground in marshy woods.

135. BRACHYTHECIUM VELUTINUM (Hedw.) B. & S.

This is by far the commonest of the smaller British species of *Brachythecium*, and is among the most likely mosses to be found covering rocks, walls or tree roots in gardens and similar places. It requires to be distinguished carefully from certain rarer species of *Brachythecium* and from one other common plant of similar size, *Eurhynchium confertum*. It agrees with *Brachythecium rutabulum* in mode of branching and in the loose patches which result; indeed, it may be pictured as a miniature replica of that very common plant. When dry, the leaves become rather widely spreading, and the whole plant has a bright green colour and glossy texture, just as in the much larger *B. rutabulum*.

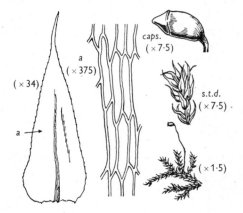

Fig. 135. *Brachythecium velutinum*: *s.t.d.* tip of shoot in a dry state, *caps.* capsule.

The leaf is 1–1·5 mm. in length, narrowly ovate-lanceolate, tapering to a rather long fine point. It is distinct from the leaf of *Eurhynchium confertum* only in its narrower outline and more finely drawn-out apex. The microscopic characters—long narrow cells, single nerve to mid-leaf, minutely toothed margin and absence of well-defined auricles—are without special diagnostic value.

An autoecious species, *Brachythecium velutinum* is usually very fertile. The seta, which is 1–1·5 cm. long, is rough with papillae; this feature and the absence of a long beak to the lid of the capsule will separate it from *Eurhynchium confertum*.

Ecology. It will grow on living and dead wood, and on stony ground. In many districts it is extremely common on banks, on garden paths and rocks, and on tree stumps in woodland. Specially abundant on calcareous soils in the south-east, *Brachythecium velutinum* is, however, uncommon in some parts of west and north-west Britain.

136. BRACHYTHECIUM PLUMOSUM (Hedw.) B. & S.

This robust moss usually grows on rocks by water, where it is often conspicuous on account of the wide thick mats which it forms. Characteristically it is marked by a strong tinge of golden brown in the green of the crowded secondary shoots, by a shining silky texture and by a tendency for the leaves, at least at the shoot tips, to become turned to one side. In this last character *B. plumosum* differs from all other British species of this genus except certain states of *B. albicans* and *B. rutabulum*. Thus, with its somewhat specialized habitat, and distinctive form, colour and texture, *B. plumosum* is not usually hard to recognize.

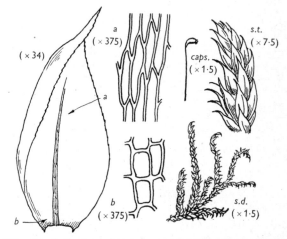

Fig. 136. *Brachythecium plumosum*: *caps.* detached capsule and seta, *s.d.* shoot in a dry state, *s.t.* tip of moistened shoot.

The leaf, which in size and general shape comes near to that of *B. rivulare*, is distinct in its markedly concave form, asymmetrical base and obliquely set, not very finely pointed apex. The concave, closely overlapping leaves give the shoot as a whole a very different appearance from that of *B. rutabulum*, where the leaves are widely spreading.

In most microscopic leaf characters—elongate cells, finely toothed margin, nerve to just above mid-leaf—this species agrees with *B. rivulare*, but it lacks the clearly defined auricles of that plant. Shorter cells occur at the basal angles, but they are not swollen to form well-defined areas of hyaline or coloured tissue.

The most useful fruiting character in this autoecious moss is the seta, which is smooth for much of its length, becoming somewhat rough with papillae only in the upper part.

Ecology. Its most characteristic habitat is on siliceous boulders by lakes and fast-flowing streams in upland regions. A frequently associated species is *Rhacomitrium aciculare*. It will also grow on wood by water, and on wet rock ledges on mountains. It seldom occurs by slow-flowing mature rivers and is absent from many south-eastern counties.

ADDITIONAL SPECIES

Brachythecium populeum (Hedw.) B. & S., which is fairly common in some districts on rocks in woods, and on wall-tops, is readily known by its narrow leaves, with the nerve extending into the extreme leaf tip.

B. glareosum (Bruch) B. & S. is uncommon but widely distributed in calcareous places. It has the robust habit of *B. rutabulum* but is known by its long, fine, twisted leaf apex; it is distinguished from *Camptothecium lutescens* by its wider cells, especially towards the base of the leaf (cells in basal angles about $20 \times 40\mu$, thin-walled, against $10 \times 20\mu$, thick-walled, in *C. lutescens*); and from *Cirriphyllum piliferum* it differs in the less abruptly contracted leaf tip.

RELATED SPECIES

The genus *Scleropodium*, with two British species, is closely related to *Brachythecium*. Both are uncommon plants.

Scleropodium caespitosum (Wils.) B. & S. (*Brachythecium caespitosum* (Wils.) Dix.) occurs on stony ground, etc. In size it is near *B. velutinum*, but its crowded, concave leaves give the curved shoots a distinct appearance, and the leaf apex is not long and fine as in that species; the seta is rough.

Scleropodium illecebrum (Hedw.) B. & S. (*Brachythecium illecebrum* (Hedw.) De Not.) has similar curved shoots, crowded concave leaves and rough seta; but the leaf shape here is that of *Pseudoscleropodium purum*. It occurs chiefly near the sea, being most frequent in the south.

137. CIRRIPHYLLUM PILIFERUM (Hedw.) Grout (*Eurhynchium piliferum* (Hedw.) B. & S.)

A rather common moss of grassy banks and woodland clearings, this species resembles *Pseudoscleropodium purum* in its pinnately branched secondary stems and light green colour; but it is less robust in habit and is quite distinct in the long fine points of the leaves. The specific epithet '*piliferum*' alludes to these hair-like leaf points, although they do not catch the eye as do the hyaline hair-points on the leaves of many acrocarpous mosses. They have the effect, however, of giving the shoot tips a slender, finely drawn-out appearance which contrasts strongly with the stout, notably blunt shoot tips of *P. purum*.

The form of the leaf provides the best character for the recognition of this species. No other common British pleurocarpous moss has a leaf of this precise shape—concave and broadly ovate in general outline, but abruptly narrowed near the tip to form a greenish hair of variable length. The nerve extends to just beyond mid-leaf, and the leaf margin is lightly toothed. The cells are long and narrow except for fairly well-defined groups of short cells at the decurrent basal angles.

Cirriphyllum piliferum is dioecious and capsules are rarely seen. The reddish, papillose seta is 2–4 cm. long, the capsule itself curved, and nearly horizontal when ripe. The lid has a long beak, as in the genus *Eurhynchium*.

Ecology. It shows a preference for shady and moderately calcareous conditions. Thus it occurs on the ground in woods where the soil is not markedly acid. It is often common on heavy clay soil.

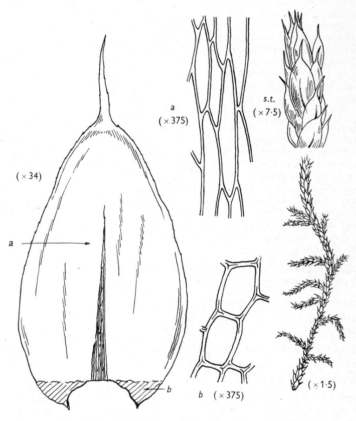

Fig. 137. *Cirriphyllum piliferum*: *s.t.* tip of shoot.

ADDITIONAL SPECIES

Cirriphyllum crassinervium (Tayl.) Loeske & Fleisch. (*Eurhynchium crassinervium* (Tayl.) B. & S.) is normally a robust woodland moss of a bright glossy green; it can be distinguished from any other species which it may resemble in a general way, such as some of the larger species of *Brachythecium*, by the shortly pointed, very concave leaves and stout nerve; this is about 80μ wide at the leaf base, and extends to just above mid-leaf, often forking near its extremity.

138. EURHYNCHIUM STRIATUM (Hedw.) Schp.

This species grows in large, loose masses on banks or in woodland, and is usually easy to recognize, for it is robust in habit and the long, arched secondary stems are freely branched and notably rigid, so that the plant has a characteristic bushy appearance. This is enhanced by the widely spreading character of the leaves which alter little when dry. Indeed, *E. striatum* is superficially more like *Hylocomium brevirostre* than most other species of its own genus, but it differs from that plant

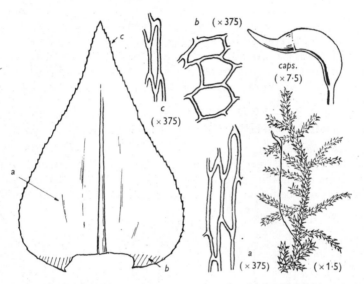

Fig. 138. *Eurhynchium striatum*: *caps.* capsule with beaked lid intact.

in its green (not red) stem and lack of paraphyllia. From *Brachythecium rutabulum*, which it resembles in some states, it may be known even when barren by its more rigid character and differently shaped leaves. The plant as a whole is mid- or yellowish green in colour. It can be confused (though not under the microscope) with some forms of *Thamnium alopecurum*.

A typical stem leaf is 1·5–2 mm. long, triangular to heart-shaped in outline, with a narrow insertion and pronounced auricles. It tapers to a relatively short and broad but acute apex. The leaves on the finer branches are smaller but do not differ greatly from the stem leaves in shape. All the leaves have strong longitudinal folds when dry (hence the name '*striatum*'), but these are much less conspicuous when the plant is wet.

An important microscopic character is the sharply toothed margin of the leaf. The long narrow cells (40–80 × 5–10μ) become thick-walled and porose towards the base of the leaf, and patches of short, wide cells occur in the auricles.

Eurhynchium striatum is normally dioecious, but the bright reddish brown capsules are not very rare. The smooth seta is 2–3 cm. long. The long beak to the lid of the curved, cylindrical capsule will distinguish the plant with certainty from any species of *Brachythecium*.

Ecology. It is common on the ground in woods and along hedge banks where the soil is calcareous or at least moderately rich in mineral nutrients. *Rhytidiadelphus triquetrus* often grows with it in these habitats and when (more rarely) it occurs in sheltered chalk grassland. Professor P. W. Richards has found it to be often by far the commonest moss on the ground in the calcareous boulder clay woods of East Anglia. More rarely it occurs on grassy sea cliffs and elsewhere, but always on neutral to basic soils.

139. EURHYNCHIUM PRAELONGUM (Hedw.) Hobk.

E. praelongum is an extremely common moss on shaded banks or about the bases of trees in woodland, often growing in wide, untidy, deep green patches in places where light is deficient and few other bryophytes will be found. It is generally an easy plant to recognize, on

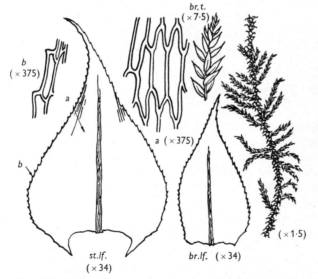

Fig. 139. *Eurhynchium praelongum*: *br.t.* tip of fine branch, *br.lf.* branch leaf, *st.lf.* stem leaf.

Note. The cells at *a* do not represent the full length attained by leaf cells in this species.

account of its rather regularly pinnate branching,* the main stems being comparatively thick, with broad widely spreading leaves, and the finer branches very slender, with much smaller, narrowly lanceolate leaves. Not many mosses show this dimorphic character of shoot and leaf so plainly, and in this, as in the frond-like form of the shoots, this species reminds one of *Thuidium tamariscinum*. The stems, however, are much weaker, the colour a duller green and the branching less elaborate than in *Thuidium*. *Eurhynchium praelongum* is closely related to *E. swartzii*; the differences between them are considered under that species.

A leaf from one of the principal stems has an unusual and characteristic shape, broadly heart-shaped, but sharply narrowed and decurrent at the base, and suddenly contracted to a narrow point at the apex. The single nerve to mid-leaf, rather long narrow cells and sharply toothed margin will be evident under the microscope. The much smaller, narrower branch leaves agree in microscopic details.

E. praelongum is dioecious and capsules are not very common. The seta is 2–3 cm. long and rough with papillae; the capsule itself is curved in form and the lid long-beaked.

Ecology. Perhaps the most marked ecological feature of *E. praelongum* is its shade tolerance. On many earthy banks in very deep shade it is the only moss present. It favours clay soil but will also grow on stones and rotting tree stumps, rarely, however, on living wood. It appears to be indifferent as regards soil reaction. Though most abundant in woods and hedgerows in the south, it is not confined to lowland habitats and may be found in mountain woodlands. It also occurs in marshes and fields. The var. *stokesii* is most common in mountain woods in the west.

140. EURHYNCHIUM SWARTZII (Turn.) Curn.

This rather common plant of calcareous ground is structurally closely allied to the last. Usually, however, it is of a yellowish green general colour, with branching less regularly pinnate, and with the lateral branchlets much less fine than in *E. praelongum*. Also, there is in *E. swartzii* by no means so strong a contrast between the stem leaves and the branch leaves as regards size and form. In their straggling habit and semi-prostrate manner of growth the two plants are alike, but in practice *E. swartzii* is generally recognized without great difficulty by its combination of structural and colour differences; it is more glossy when dry, and it shows a distinct preference for a more open type of habitat.

---

* The most regularly pinnate and bipinnate branching is seen in the var. *stokesii* (Turn.) Hobk., which has sometimes been regarded as a distinct species.

The leaves, both on the main shoots and on the branches, are broadly ovate to cordate, somewhat narrowed and decurrent at the base, and acute at the apex, but not drawn out into fine points as in *E. praelongum*. The single nerve extends just beyond mid-leaf, and the leaf margin is sharply toothed. The cells, though elongated, are shorter than in *E. praelongum*. An average measurement is about 6×40–50μ.

A dioecious species, *E. swartzii* is rare in the fertile state. The capsule resembles that of the last species, but the seta is shorter.

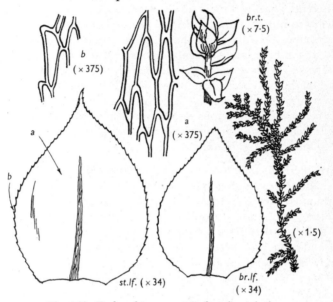

Fig. 140 *Eurhynchium swartzii*: *br.t.* branch tip,
br.lf. branch leaf, *st.lf.* stem leaf.

Ecology. It occurs, perhaps most characteristically, in the moss flora of bare patches in chalk grassland and often associated with *Camptothecium lutescens* and *Campylium chrysophyllum*. Detailed investigation would probably show that it was restricted to the more calcareous types of habitat. This species and *Eurhynchium praelongum* often grow together; *E. swartzii*, however, is relatively rare on wood. On the other hand, it is sometimes an important moss in fallow fields on limestone.

141. EURHYNCHIUM RIPARIOIDES (Hedw.) Jennings
(*Eurhynchium rusciforme* (Neck.) Milde)

Robust in habit, bright or deep green in colour, often with a fine metallic sheen when dry, *E. riparioides* is one of our commonest and most conspicuous aquatic mosses. It forms extensive patches on boulders, wood or stonework, low on the banks of streams or sometimes actually submerged. The stems are often very long and only sparingly branched above, whilst below they become bare of leaves and discoloured. The way in which the leaves stand out from the stems reminds one of *Brachythecium rutabulum*, but the present species is normally a plant of stronger, more rigid growth, with more densely crowded leaves and more flattened shoots than *B. rutabulum*. Further, *Eurhynchium riparioides* when growing submerged in fast-running water is usually

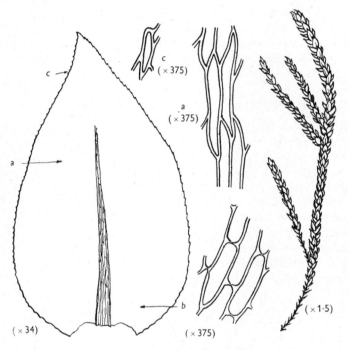

Fig. 141. *Eurhynchium riparioides*.

distinct in its long shoots which may extend for 8–15 cm. with scarcely a branch, the lower parts wiry and rough with the persistent bases of eroded leaves.

The leaf is 1·5–2·5 mm. long, broad with a narrow insertion; it is approximately ovate, somewhat concave in form and shortly acute at the apex. The single nerve extends for about three-quarters of the leaf length; another important microscopic character is the closely and rather strongly toothed leaf margin. Except at the basal angles (where there are wide, but ill-defined patches of short oval cells) the cells are very long and narrow.

E. riparioides is autoecious, and the capsule not uncommon. Borne on a smooth seta 1–2 cm. long, it is nearly oval in form, and is held horizontally. The long beak to the lid is a notable feature.

Ecology. It demands running water and is thus confined to the banks of streams and rivers. Preferring swiftly flowing water, in lowland districts it is found chiefly about mill-races and waterfalls. It will grow attached to wood or rocky substrata, and will tolerate both acid and calcareous conditions. It appears to demand at least periodic submersion in water, and its limitations in this respect might be worth investigating. I have noticed that, when the moss flora of boulders in shaded mountain streams shows zonation, this species (often almost pure) occupies the lowest zone, and is thus fully exposed only in time of drought.

142. EURHYNCHIUM CONFERTUM (Dicks.) Milde

This common plant of garden stones, walls and the bases of trees is of similar slender proportions to *Brachythecium velutinum*, which it closely resembles in its rather small, moderately glossy patches of growth and in the general appearance of the irregularly branched stems and spreading leaves. The more slender habit will at once distinguish it from other common members of related genera, such as *B. rutabulum*; but, on the other hand, it is very much more robust than *Amblystegium serpens*, a plant with which it often grows.

The ovate-lanceolate, toothed leaves are about 1 mm. long; nerved to mid-leaf or rather beyond, they will furnish scarcely any microscopic character of real help in separating this species from *Brachythecium*

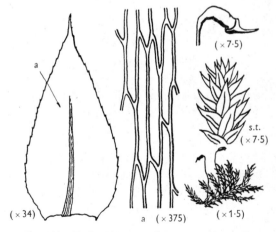

Fig. 142. *Eurhynchium confertum*: *s.t.* shoot tip.

velutinum. The leaf of *Eurhynchium confertum,* however, is less finely drawn-out at the apex and the plant has a less silky texture than that of *Brachythecium velutinum.*

Eurhynchium confertum is autoecious, and is commonly very fertile. Only then are the characters available which can separate it clearly from the smaller species of *Brachythecium.* These characters are: (i) the evident beak to the lid of the capsule and (ii) the smooth seta, which is without a trace of the papillae seen in *B. velutinum.*

Ecology. Its habitat range overlaps that of *B. velutinum,* but it is more shade-demanding. Although most common on walls, rocks and stones in gardens and woodlands, it also grows on trees. It is mainly a lowland species and in many districts it is among the mosses most likely to be found on stone walls. It becomes rarer, however, in much of the mountain country of the north and west.

ADDITIONAL SPECIES

Eurhynchium murale (Hedw.) Milde is a closely related species which is common on rocky banks and walls in some districts, especially on limestone. It is easily recognized by its low dense habit, glossy character and very concave, obtusely rounded or shortly pointed leaves. It is usually very fertile.

RELATED SPECIES

In the allied genus *Rhynchostegiella* are placed several plants of very slender habit, none of them very common.

R. pallidirostra (A.Br.) Loeske (*Eurhynchium pumilum* (Wils.) Schp.) bears some resemblance to *Amblystegium serpens,* but has more shortly pointed and more strongly toothed leaves, and a short ovoid capsule. It grows in neat tufts on deeply shaded stones, rocks or soil.

Rhynchostegiella tenella (Dicks.) Limpr. (*Eurhynchium tenellum* (Dicks.) Milde), which is quite common on rocky banks and walls in calcareous districts, is distinct from all equally slender plants in its very narrow, finely pointed, entire leaves. It is also notable for its silky texture and bright yellowish green colour, and the casual observer might mistake it in the field for a species of *Dicranella.*

143. **PSEUDOSCLEROPODIUM PURUM** (Hedw.) Fleisch. (*Brachythecium purum* (Hedw.) Dixon). (Plate X)

This will be among the first species which the beginner will find on grassy banks, on heaths or in open woodland, for it is one of the most abundant British mosses. It is an easy plant to recognize, on account of its robust habit and pinnately branched blunt-tipped shoots, rendered stout by the crowded concave leaves; the prevailing colour is a very light shade of green. The bluntly rounded shoot apices are highly characteristic, being approached most closely perhaps by *Pleurozium schreberi,* but in that plant the stems are red, not greenish as in this species.

The leaf is about 2 mm. long, very wide and concave in form, with its broad, rounded apex interrupted by a short protruding point, or apiculus. This unusual shape of leaf, coupled with microscopic characters recalling species of *Brachythecium* (single nerve to mid-leaf, slightly plicate leaf surface, lightly toothed margin and cells elongate except at basal angles) will serve to confirm the identification. The cells in the extreme leaf base have their thick walls interrupted by obvious thin places; in few related mosses is this porose type of cell so well seen.

Pseudoscleropodium purum is dioecious and capsules are rather rare. The seta is 2–5 cm. long, and smooth. The capsule is broadly ovoid-cylindrical, somewhat curved, and held horizontally.

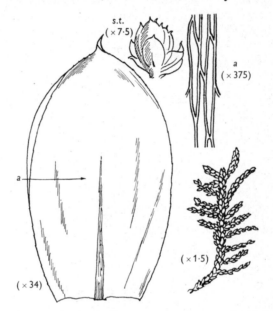

Fig. 143. *Pseudoscleropodium purum*: *s.t.* extreme tip of shoot.

Ecology. With *Thuidium tamariscinum, Pleurozium schreberi, Rhytidiadelphus squarrosus* and others it forms a bryophyte community which grows in close association with flowering plants on heathland and in woodland clearings. *Pseudoscleropodium purum* is tolerant as regards soil reaction and grows equally well in grassy places on the chalk.

144. PLEUROZIUM SCHREBERI (Brid.) Mitt. (*Hypnum Schreberi* Brid.)

This species, which is among the most abundant mosses of heathy places, has something of the light green colour and robust, pinnately branched shoots seen in *Pseudoscleropodium purum*, but it is at once distinct in its bright red stems, which in the moist state will show clearly through the leaves. Also, although the stem leaves are bluntly rounded at their tips, the ends of the finer branches are drawn out into much more slender points than in *P. purum*. Indeed, the form of the shoot tip here often approaches the 'spear-head' point seen in *Acrocladium cuspidatum*.

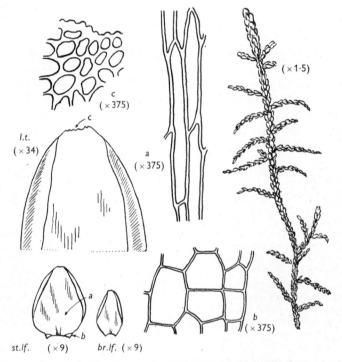

Fig. 144. *Pleurozium schreberi*: *br.lf.* leaf from slender branch, *st.lf.* stem leaf, *l.t.* leaf tip.

The leaf attains a length of 2–2·5 mm. and is very broadly ovate in outline; its rounded form, with no trace of falcate curvature and with the apex notably blunt, will prevent confusion with robust forms of the variable *Hypnum cupressiforme*, or indeed with any other common heathland moss. *Acrocladium cuspidatum* seldom grows in the same habitat, and *Pseudoscleropodium purum* has a very different appearance as a rule, with its stouter, blunter shoots and greenish stem colour.

Under the microscope the leaf has the almost entire margin, slightly indented hooded apex, short double nerve and elongate cells of *Acrocladium cuspidatum*. It differs, however, in one important respect; here the basal angles of the leaf are occupied by patches of short, rather thick-walled, somewhat opaque or orange cells, which do not form obviously decurrent auricles; whereas in *A. cuspidatum* the prominent auricles project downwards well beyond the line of the leaf base, and are composed of greatly swollen, thin-walled, hyaline or orange cells.

Pleurozium schreberi is dioecious, and capsules—at least in Britain— are rare.

Ecology. Strongly calcifuge, this plant is an 'indicator' of acid soil conditions. It is commonly the principal moss in Callunetum and on the ground in pine woods. In grassy lowland heaths it often grows with *Pseudoscleropodium purum*, *Hypnum cupressiforme*, *Rhytidiadelphus squarrosus* and others; whilst on peaty ledges on the slopes of mountains it occurs with *Hylocomium splendens* and *Rhytidiadelphus loreus*. It grows luxuriantly amid the wealth of other bryophytes in the damp, open type of highland oakwood (Quercetum petraeae).

RELATED SPECIES

Orthothecium intricatum (Hartm.) B. & S. is an uncommon though widely distributed mountain plant which stands out from other members of the damp calcareous rock-ledge community on account of its narrow silky leaves and reddish colour. In size and form it resembles *Hypnum cupressiforme* var. *resupinatum*, but the nerveless leaf lacks the distinct alar cells found in that species.

Orthothecium rufescens (Brid.) B. & S., although rather rare, is one of the most striking mosses of moist base-rich rock clefts on mountains. It is a much larger plant, notably glossy, and strongly tinged with salmon pink; the leaves are plicate and three times as long as the smooth leaves of *O. intricatum* (3 mm. against 1 mm.).

Entodon orthocarpus (La Pyl.) Lindb. (*Cylindrothecium concinnum* (De Not.) Schp.) resembles *Pleurozium schreberi* in general appearance and in its nerveless, concave, glossy leaves. The stems, however, are never red, and the leaves lack obvious auricles. Moreover, the habitat is quite different, *Entodon orthocarpus* being a rather rare plant of calcareous sand dunes and old chalk grassland.

145. ISOPTERYGIUM ELEGANS (Hook.) Lindb. (*Plagiothecium elegans* (Hook.) Sull.)

In its prostrate, silky patches and flattened leafy shoots this species is like a miniature of *Plagiothecium denticulatum*. *Isopterygium elegans* is, however, rather more an upland species, although it is often abundant on sands and gravel in lowland districts. As in *Plagiothecium denticulatum* secondary shoots are numerous, somewhat crowded, and little-branched.

The rather narrowly ovate-oblong leaf of *Isopterygium elegans* is about 1 mm. long, slightly longer on some of the main shoots, but much shorter on the finer branchlets. It tapers to a finer point than that of *Plagiothecium denticulatum*. It also differs from the leaf of that species in its less decurrent base, i.e. there are no wings extending down the stem.

The microscope shows that the leaf is practically nerveless, has elongate cells and a few faint teeth on the plane leaf margin. The one character, however, which separates it conclusively from small forms of *P. denticulatum* is the narrowness of the cells, which are only about 5μ wide, as against 12μ in *P. denticulatum*.

Isopterygium elegans is dioecious and capsules are exceedingly rare. Bunches of thread-like shoots, however, are frequently formed in the axils of the leaves. These shoots bear only rudimentary leaves and serve for vegetative propagation.

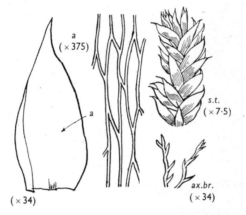

Fig. 145. *Isopterygium elegans*: *s.t.* shoot tip, *ax.br.* fragments of thread-like branches from leaf axil serving for vegetative propagation.

Ecology. A point of difference from *Plagiothecium denticulatum* is that this species does not grow on wood. It occurs, however, on sandy and gravelly banks, peat and siliceous rock ledges. I do not regard it as a primary colonist of rock and suspect that it demands some capping or 'nidus' of humus. It is a calcifuge.

ADDITIONAL SPECIES

Isopterygium depressum (Bruch) Mitt. (*Plagiothecium depressum* (Bruch) Dixon) is a rather rare plant of shaded limestone rocks and tree bases. It differs from *Isopterygium elegans* in its shorter leaf point and wider cells ($7–10\mu$). In fresh specimens the cells can be seen each to bear a row of minute papillae, a feature not found in any other British species of the genus.

I. pulchellum (Hedw.) Jaeg. & Sauerb. (*Plagiothecium pulchellum* (Hedw.) B. & S.) occurs on mountain ledges. It differs from *Isopterygium elegans* in its even more slender and less markedly flattened shoots, and leaves with entire margins. The very delicate leaves are often curved and turned to one side. In cell size it agrees with *I. elegans*.

146. PLAGIOTHECIUM DENTICULATUM (Hedw.) B. & S.

This common moss of woods and hedge banks is distinguished from most other 'hypnoid' genera by the way in which the shoots and leaves are flattened in one plane. This flattening, together with the rather distant leaf arrangement, at least on the finer branches, to some extent

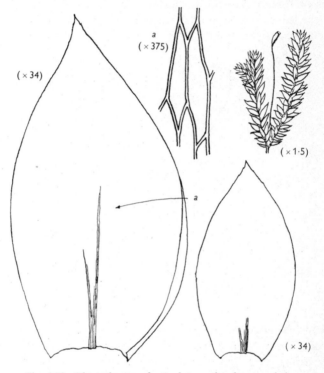

Fig. 146. *Plagiothecium denticulatum*: the shoot and the large leaf (left) are of the var. *majus*.

Note. Many of the leaf cells are much longer than those shown at *a*.

recalls the 2-ranked leafy shoots of *Fissidens* or a foliose hepatic, but closer inspection shows that the leaves are not strictly 2-ranked here. From some more closely related plants *Plagiothecium denticulatum* is separated with more difficulty. It is, however, more robust than *Isopterygium elegans*, but considerably smaller than *Plagiothecium undulatum*. It is normally marked by its bright green colour and glossy texture—features which set it apart from *P. silvaticum*, a species which forms somewhat similar flattened patches of growth in the same habitats, but is usually of a noticeably dull colour and texture.

The leaf is 1·5–2·5 mm. long, ovate to oblong in shape, with a rather asymmetrical base and an acute, but not finely drawn-out apex. The very short double nerve is in general a useful character, but forms are known in which one branch of the nerve extends almost to mid-leaf. The margin is plane and entire except for a few faint teeth towards the apex. The long narrow cells (up to 150×10–15μ) give place to shorter ones in the leaf base, but there are no well-defined patches of distinctive cells in the basal angles.

P. denticulatum is autoecious and rather commonly fertile. The nearly upright, cylindrical, light brown capsule is borne on an orange-red seta 1–3 cm. in length.

Ecology. It will grow on rotten wood, humus or stones; also on soil-capped walls and peaty banks. It demands some shade and prefers a slightly acid soil. It is perhaps most characteristically a plant of woodland banks but, given shade, is not confined to the neighbourhood of trees.

147. PLAGIOTHECIUM UNDULATUM (Hedw.) B. & S. (Plate XI)

This fine moss is seen at its best in woods in northern and western Britain, where the robust, greatly flattened, pale whitish green shoots spread over the ground in a characteristic manner. The flattened shoots and the transverse wrinkling of the leaves remind one of *Neckera crispa*, but that moss occurs in quite different habitats and lacks, too, the curiously pale colouring and soft texture of *Plagiothecium undulatum*. The principal secondary shoots are little branched, and in this *P. undulatum* differs from many of the robust 'hypnoid' mosses that grow with it. Once known, it is an easy plant to recognize. *Pseudoscleropodium purum* sometimes resembles it in colour, but the shoots there are less noticeably flattened.

The leaf is about 3 mm. long, with the rather broad, slightly asymmetrical base (narrowed at insertion) and shortly acute apex characteristic of most members of the genus. The undulations of the leaf surface, due to transverse wrinkling, are very evident both wet and dry. In microscopic characters this leaf, with its short double nerve, nearly entire margin and long narrow cells, resembles that of *Plagiothecium denticulatum*.

P. undulatum is dioecious, and fertile plants are not very common. The long, curved cylindrical capsule is borne on an orange-red seta 2–4 cm. in length.

Ecology. This species is comparatively uncommon in south-east England, but is a typical member of the moss flora of upland oakwoods (Quercetum petraeae). Here it is associated with *Rhytidiadelphus loreus*, *Dicranum majus* and other mosses, and the liverwort *Bazzania trilobata*. It will grow in the open on wet moorland and on mountain ledges in north and west Britain, in such places often being associated with *Pleurozium schreberi*.

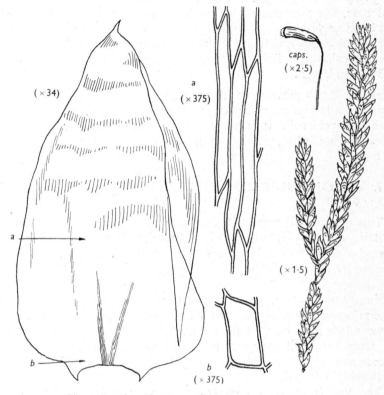

Fig. 147. *Plagiothecium undulatum*: *caps.* old capsule.

ADDITIONAL SPECIES

Plagiothecium silvaticum (Brid.) B. & S. is the only one of the four remaining species which is at all common. It is a fairly robust plant of hedge banks and woods. It differs from *P. denticulatum* in its normally much duller, olive-green colour, less glossy texture and leaves that become more shrivelled on drying and thus give the shoots a different appearance from those of *P. denticulatum*; also it is dioecious and capsules are rather rare.

Plate XI. *Plagiothecium undulatum* (natural size)

Plate XII. *Hypnum cupressiforme* (natural size)

148. HYPNUM CUPRESSIFORME Hedw. (Plate XII)

H. cupressiforme is at once our most abundant and most variable British moss. Several of the varieties are extremely well marked and constant, so that they merit separate treatment here. In one or other of its forms this species occurs in almost all types of terrestrial habitat. The special preferences of the principal varieties will be considered under 'Ecology', p. 309.

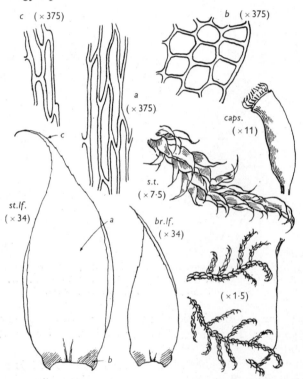

Fig. 148. *Hypnum cupressiforme* (typical form): *br.lf.* branch leaf, *st.lf.* stem leaf, *s.t.* shoot tip, *caps.* capsule showing peristome.

In its most typical form *H. cupressiforme* is a prostrate-growing, pinnately branched pleurocarpous moss of moderately robust habit, forming flat patches or extended carpets on wall-tops, boulders and decaying logs, and on the ground. The name *cupressiforme* would seem to refer to the very symmetrical way in which the rather concave, curved leaves overlap one another somewhat as on a cypress branchlet. This gives the plant an appearance by which it may, with practice, be recognized quite readily—at least in its more typical states. The leaves, too, are curved and turned regularly downwards, especially at the tips of the branches. In this falcato-secund character it resembles species of *Drepanocladus*, but it is at once distinguished from that genus by its nerveless leaves. Each leaf is ovate-lanceolate in shape, tapering to a fine point which is either entire or very lightly toothed. The leaf cells are long and narrow (40–80×3–6μ) except at the basal angles, where there are characteristic well-defined patches of small greenish cells.

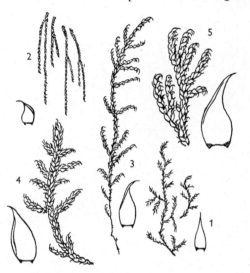

Fig. 148a. *Hypnum cupressiforme* (varieties): 1, var. *resupinatum*;
2, var. *filiforme*; 3, var. *ericetorum*; 4, var. *tectorum*; 5, var.
lacunosum. All shoots × 1·5, all leaves × 7·5.

Hypnum cupressiforme is dioecious, but capsules are quite common, except in some of the varieties. The capsule is curved, cylindrical, with a beaked lid.

The following are the most distinct varietal forms of this moss:

Var. *resupinatum* (Wils.) Schp. Of slender habit and very silky texture, this variety differs markedly from the typical form in that its leaves are almost straight or slightly bent to one side (not downwards). It is quite without the 'cypress-branch' character of typical *H. cupressiforme* and bears a strong superficial resemblance to a small species of *Brachythecium* or *Eurhynchium*, but differs, of course, in its nerveless leaves. The capsule is long-beaked.

Var. *filiforme* Brid. This is another quite distinct variety, differing from typical forms in its extremely slender habit and in its manner of branching. It is only distantly and irregularly pinnate, and produces a number of nearly parallel, long thread-like branches. The individual branch, however, is merely a miniature of the typical form with its regularly overlapping rows of curved leaves. Capsules are very rare.

Var. *ericetorum* B. & S. Again a more slender plant than the typical form, this variety is marked by its more or less erect habit and long regularly pinnate branches. It is pale whitish green in colour. The leaves, less close-set than in the typical form, are very strongly curved, giving a hooked effect to the silky shoot tips.

Var. *tectorum* Brid. This variety is notably robust, and in its numerous more or less erect, swollen branches it has lost the character of the typical form, for the leaves are very densely crowded, concave and only lightly curved. In colour it is dark or olive green.

Var. *lacunosum* Brid. (*H. cupressiforme* Hedw. var. *elatum* B. & S.). This has many of the characters of the last variety, but is even more robust. The branches are few, short and swollen with the crowded, very concave leaves. It is the largest of all the forms of *H. cupressiforme*, and its stout erect branches display nothing of the flattened character of the typical plant. Its colour is a rich golden green, often tinged with bronze, and the leaf is curved and shortly pointed.

Ecology. It grows on wood, rock and soil in gardens and woodland; it is a common moss of grassy heathland, where associated species are *Pleurozium schreberi* and species of *Hylocomium* and *Rhytidiadelphus*. Equally, it occurs on calcareous soil, and is commonly found on sand dunes. *Hypnum cupressiforme* is not rare on mountain ledges, and poorly grown plants may be met with almost anywhere on open moorland.

Several of the varieties have well-marked ecological preferences; thus, var. *filiforme* is practically confined to the trunks and larger branches of trees; var. *ericetorum* occurs on heaths and in upland woods (especially among *Calluna* and always on acid soil); var. *tectorum* is a plant of grassland (acid and calcareous), also of walls and roofs; and var. *lacunosum* grows mainly on the ground in limestone regions. Var. *resupinatum* is common generally, but especially on tree bases.

ADDITIONAL SPECIES

Hypnum patientiae Lindb. is a rather uncommon plant which grows on gravelly or clayey soil among thin grass, often on the edges of woods or on damp roadsides in hilly districts. It is distinguished from *H. cupressiforme* by the patches of large, thin-walled cells which form conspicuous auricles at the base of each leaf (in contrast with the small granular-looking alar cells of *H. cupressiforme*). It is robust in habit and of a characteristic yellowish green colour, with the shoot tips very glossy and strongly hooked.

RELATED SPECIES

Pylaisia polyantha (Hedw.) B. & S. strongly resembles *Hypnum cupressiforme* var. *resupinatum*, but the leaf differs in its less well-defined patches of alar cells. Moreover, *Pylaisia polyantha* is autoecious and the erect capsule is distinct in its shortly conical lid. It is a rather rare epiphytic species.

149. CTENIDIUM MOLLUSCUM (Hedw.) Mitt. (*Hypnum molluscum* Hedw.). (Plate XIII)

This is one of the most abundant and readily recognized mosses of calcareous habitats, whether chalky banks in lowland districts or base-rich rock ledges on mountains. It will be readily identified by the pale or yellowish green colour, silky texture, and pinnately branched shoots with crowded, strongly curled leaves. Although it forms wide and

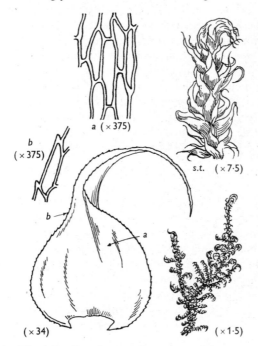

a (× 375)

b (× 375)

s.t. (× 7·5)

b

a

(× 34)

(× 1·5)

Fig. 149. *Ctenidium molluscum*: *s.t.* shoot tip.

intricate patches, the individual shoots are delicate and each branchlet bears quite small, sickle-shaped leaves at its tip. These leaves are so strongly curled as to form almost a complete circle; it is to these curled leaf apices, viewed in the mass, that *Ctenidium molluscum* owes much of its distinctive appearance. With a little experience it can usually be recognized at sight, although poorly marked forms occur, with more distant branchlets, and these bear some resemblance to *Cratoneuron commutatum* or other mosses.

The stem leaves, which are usually broadly heart-shaped, are about 1 mm. long; the branch leaves are smaller and much narrower. The absence of nerve separates this leaf from those of *Drepanocladus* and related genera; the curved form and wavy outline of the apex will distinguish it from that of *Hyocomium flagellare*, which it resembles in its toothed margin. The long narrow cells give place to shorter ones in the leaf base, where clasping auricles are formed.

Ctenidium molluscum is dioecious and is most commonly barren. The small, more or less ovoid capsule is held horizontally, on a dark purple seta.

Ecology. It is strongly calcicole, its chief habitats including chalk grassland, banks in limestone districts and base-rich rock ledges on mountains. Sometimes it grows very luxuriantly along the bases of walls in limestone country, with *Camptothecium sericeum* and other mosses. It occurs, less typically, in fens.

RELATED SPECIES

Ptilium crista-castrensis (Hedw.) De Not. (*Hypnum crista-castrensis* Hedw.) is a striking species, which is occasionally plentiful on the ground beneath conifers in mountain districts. It resembles the most robust and regularly plumose forms of *Ctenidium molluscum*, but is more erect in habit; it is easily identified by the form of the large stem leaves. These are rather over 2 mm. in length, and although hooked at their tips as in *C. molluscum*, are square-based (not auricled) and deeply plicate.

150. HYOCOMIUM FLAGELLARE B. & S.

H. flagellare is a rather common moss in the hill districts of north and west Britain, where its robust festoons, usually of a bright golden green colour, or deep green with pale tips, are conspicuous around waterfalls or on the banks of fast-flowing streams. The principal shoots are often very long (8–20 cm.), variously divided to form branches which may be either pinnate or, more typically, only sparingly branched themselves. Thus its habit is notably loose and free, lacking the regularly and closely pinnate character of *Cratoneuron commutatum* and the still more densely branched and tightly curled appearance of *Ctenidium molluscum*. Besides, the leaves here are nearly straight in form, and this gives the shoot as a whole a different aspect from that of the above-mentioned plants, where the leaves are strongly curved and turned to one side.

The leaf is about 1 mm. long, and is approximately heart-shaped, with clasping auricles at the base and a shortly acute apex. Broad below, it narrows rather suddenly towards the tip.

Under the microscope the very coarsely toothed margin of the leaf is a notable feature. A nerve is lacking, or appears only as a short forked structure. These features, combined with the long narrow cells, and characters of leaf form and outline, separate *Hyocomium flagellare* from all other mosses with the possible exception of *Ctenidium mollus-*

cum. The latter usually has a very different and much denser growth habit, and is also distinct in its falcate leaves with long fine apices of characteristically wavy outline.

Hyocomium flagellare is dioecious and fruit is rare. The rather large capsule is borne on a very rough seta 2–2·5 cm. long.

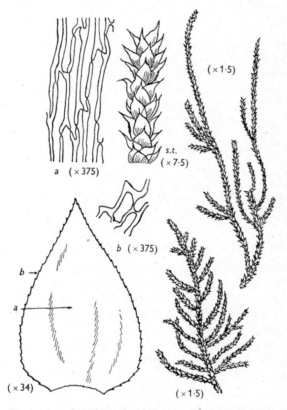

Fig. 150. *Hyocomium flagellare*: the shoots on the right show two different types of branching, both of them common; *s.t.* shoot tip.

Note. The leaf (bottom left) is a branch leaf, which lacks the pronounced auricles and abruptly contracted apex of the stem leaves.

Ecology. It grows by waterfalls and on the rocky banks of streams in the mountainous parts of north and west Britain, where it is one of the most abundant and characteristic species. By contrast, it is rare in most parts of the south-east. *H. flagellare* is almost certainly calcifuge on the whole, to some extent replacing *Cratoneuron commutatum* which is found in comparable habitats on limestone.

151. RHYTIDIADELPHUS TRIQUETRUS (Hedw.) Warnst.
(*Hylocomium triquetrum* (Hedw.) B. & S.)

The robust growth and bushy habit of this plant make it a conspicuous moss as it grows in wide patches on some roadside bank or in a woodland clearing. It is usually of a noticeably pale, somewhat yellowish green colour. The bushy habit is due to the nearly erect growth and short, irregular or pinnate branches which do not lie in

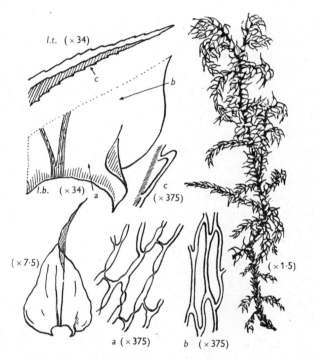

Fig. 151. *Rhytidiadelphus triquetrus*: *l.b.* part of
leaf base, *l.t.* leaf tip.

one plane as happens in so many of the pinnately branched 'hypnoid' mosses. The red stems are remarkably rigid and tough; they bear large, widely spreading leaves, which are, however, without the long recurved tips found in *Rhytidiadelphus squarrosus* and *R. loreus*.

The length of a stem leaf varies from 3 to 6 mm. It becomes very wide immediately above the narrow base, so that the outline of the leaf as a whole is almost triangular. There is a double nerve of variable length and the leaf surface is markedly plicate. The margin is toothed. There is nothing very distinctive in the long narrow cell structure which, as in other species of this genus, becomes thick-walled and porose in the leaf base. Spiny papillae occur on the back of the leaf, towards the tip.

R. *triquetrus*, like the other British species of the genus, is dioecious, and the capsule is not common. It is borne on a seta 2·5–3·5 cm. long.

Ecology. Its most typical habitat, perhaps, is on the ground in woodland clearings, especially on calcareous clay soils; but it also occurs in many types of open situation, including chalk downs, sand dunes, heaths and sheep-grazed moors.

152. **RHYTIDIADELPHUS SQUARROSUS** (Hedw.) Warnst. (*Hylocomium squarrosum* (Hedw.) B. & S.)

This plant, which somewhat resembles *Rhytidiadelphus loreus* but is more slender, is one of the commonest mosses in moist grassy places where it either forms local patches or occurs as scattered stems in the turf. It has the red stem, irregularly pinnate branching and widely spreading leaves of *R. loreus*, but the stems are commonly only 5–12 cm. in length, and are much weaker than those of *R. loreus*. Moreover, as the name suggests, the leaves are squarrose, arranged evenly all round the stem with their long points spreading outwards or reflexed, but not curved to one side as in *R. loreus*. The star-like spreading character of the leaves is especially evident at the shoot apex.

A leaf from one of the main stems is 2–3 mm. in length. It has the

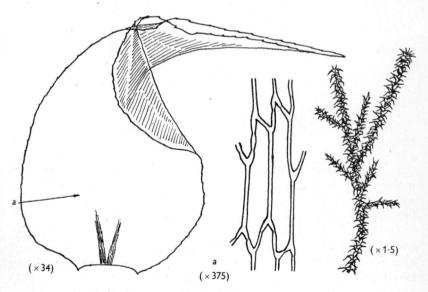

Fig. 152. *Rhytidiadelphus squarrosus.*

broad base and suddenly contracted apex of *R. loreus*. The long channelled leaf tip, however, is straighter in *R. squarrosus*, not sickle-shaped.

The leaf of *R. squarrosus* is only faintly plicate, and the cells are somewhat wider than in *R. loreus*. It has the bluntly toothed margin and short forked nerve of its immediate allies. The cells in the basal angles of the leaf usually become rather strongly thickened and orange in colour, a feature also seen in *R. loreus*. If there is any doubt as to its identity, the best characters for the recognition of *R. squarrosus* will be the absence of one-sided curvature of the leaves, and the only faintly plicate leaf surface.

R. squarrosus is dioecious and capsules are rather rare. The capsule is reddish and shortly oval, with an acutely conical lid. The seta is 2–3 cm. long.

Ecology. It is indifferent as regards soil reaction, and grows equally in chalk grassland, among grass on lawns, in neutral pastures and on acid heaths. Grassy rides in woods are also a favourite habitat and some authorities regard it as a shade-loving species. My own experience has been to find it in a great variety of fairly exposed habitats, including calcareous sand dunes and acid mountain ledges—always, however, in close association with flowering plants. Thus its place in a succession must be that of a late-comer, not an early colonist.

153. **RHYTIDIADELPHUS LOREUS** (Hedw.) Warnst. (*Hylocomium loreum* (Hedw.) B. & S.)

This robust and striking species is likely to be among the first to attract attention in the ground flora of the 'highland' type of oak wood that occurs so widely in the north and west. The stems are red, prostrate, ascending or variously arched, and are either pinnate or rather sparingly and irregularly branched. The branches and main stems alike bear a dense covering of large, curved, widely spreading leaves. It is distinct from the less common *Hylocomium brevirostre* in the absence of paraphyllia among the leaves. A single branching shoot of *Rhytidiadelphus loreus* commonly attains 15–20 cm. in length; indeed, it differs notably from *R. squarrosus* in its typically much more robust habit.

A single leaf from a well-grown main stem is about 3 mm. long. It is broad at the base, and narrows above rather suddenly into the long, strongly curved apex. Often all the leaves along a branch will show a tendency to be curved in one direction, much as in the regularly falcato-secund shoots of a species of *Drepanocladus*; but this is not always so.

The microscope reveals a strongly plicate leaf surface, a short faint double nerve, and a toothed leaf margin. In its elongate, thick-walled cell structure *Rhytidiadelphus loreus* is like the last species. On the whole, however, *R. loreus* can be identified so readily by its striking habit that close attention to microscopic detail will be unnecessary.

R. loreus is dioecious. The short curved capsule is borne on a red seta about 3 cm. long; fertile plants are not uncommon in moist mountain woods.

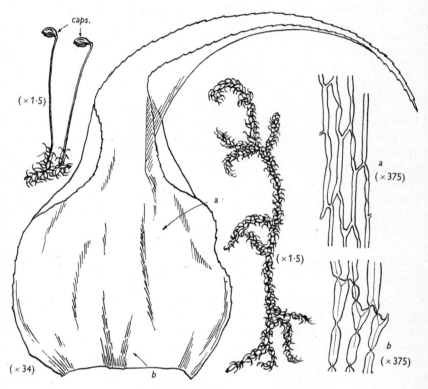

Fig. 153. *Rhytidiadelphus loreus*: *caps.* old capsule.

Ecology. This moss is rather rare in the south-east, but is extremely common on acid humus in the upland woods of the north and west. Its ecological range to some extent parallels that of *Plagiothecium undulatum*. Like that species, it comes out into the open on mountains in western Britain, occurring commonly with *Pleurozium schreberi* and others on the acid humus of turfy ledges at moderate to high altitudes.

154. **HYLOCOMIUM SPLENDENS** (Hedw.) B. & S.

In its somewhat flattened, bipinnate frond-like shoots this robust and common plant of heaths, moors and upland woods reminds one of *Thuidium tamariscinum*. Here, however, the stems are red, not greenish to black as in that species. Also there is a bright silky gloss

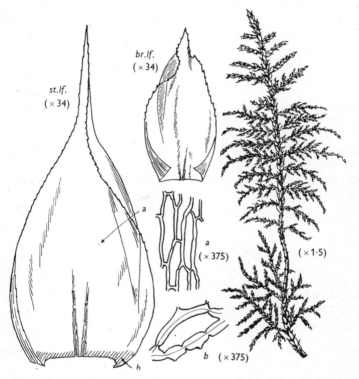

Fig. 154. *Hylocomium splendens*: *st.lf.* stem leaf, *br.lf.* branch leaf.

on the shoots, a feature that is quite lacking in the much more vivid green fronds of *Thuidium*. Sometimes, in its less regularly or less obviously bipinnate forms, *Hylocomium splendens* may bear a superficial resemblance to *Pleurozium schreberi*, a plant with which it very commonly grows. Closer examination, however, reveals obvious differences: for example, the stems in *Hylocomium splendens* are clothed with minute scale-like structures, the paraphyllia, in addition to the normal leaves; paraphyllia do not occur in *Pleurozium schreberi*.

The leaves from a main stem are nearly 3 mm. long, and broadly ovate in general outline, but abruptly narrowed above into rather long straight tips. Branch leaves are very much smaller, and shorter in the point. The leaves are often somewhat concave. Good microscopic characters, on stem- and branch-leaves alike, are the short double nerve, sharply toothed margin and elongate cell structure, the cells at the extreme leaf base becoming very thick-walled and often orange in colour. With several related mosses this species shares the plicate or longitudinally 'pleated' leaf surface, but reference to such detail should seldom be necessary when identifying a plant so distinct as this; for in its frond-like leafy shoots it differs from all its common allies.

Hylocomium splendens is dioecious, and the rather short, broad, curved capsule is not common. It has a long-beaked lid, and is borne on a seta about 2·5 cm. long.

Ecology. It grows nearly always in close association with flowering plants, typically amongst grass and heather; but it occurs in a wide range of plant communities, including those of acid woodland, sheltered chalk grassland and fixed dunes. In upland woods of north and west Britain it often grows luxuriantly with other mosses such as *Rhytidia-delphus loreus*, at times carpeting the ground under conifers where little else is growing. Very often it is associated with *Pleurozium schreberi*, and with this species and others it ascends on turfy ledges to the highest altitudes.

ADDITIONAL SPECIES

Hylocomium brevirostre (P. Beauv.) B. & S. is a plant of mountain woods. It has the red stem of its immediate allies, but in habit and general appearance it somewhat resembles robust forms of *Brachythecium rutabulum* or *Eurhynchium striatum*. It is known from *Rhytidiadelphus loreus* by the presence of numerous paraphyllia on the stem and by the much straighter leaf tips.

RELATED SPECIES

Rhytidium rugosum (Hedw.) Kindb. (*Hylocomium rugosum* (Hedw.) De Not.) is a striking, but rather rare moss of calcareous grassland and limestone rocks. The yellow-green to brown colour of the robust leafy shoots and the wide, concave, transversely undulate leaves—curved and turned to one side at the shoot tips—may suggest *Drepanocladus lycopodioides*; but the latter is found in much wetter habitats, and is known at once by its entire leaf margins.

HEPATICAE (*Liverworts*)

JUNGERMANNIALES

JUNGERMANNINEAE

(JUNGERMANNIALES ACROGYNAE)

1. ANTHELIA JULACEA (L.) Dum.

In his remarks on this species in the *Student's Handbook of British Hepatics*, MacVicar justifiably claims that it is 'the most conspicuous hepatic on our highland mountains'. On those moist rock ledges where it occurs, by mountain streams especially, its dense tufts or wide cushions, dull bluish green when wet and pale grey when dry, can hardly fail to attract attention; it is unknown, however, in many parts of the country. Whilst the tufts are remarkably dense and robust, the individual stems are very slender, 1–3 cm. long with numerous nearly erect branches that are covered with minute appressed leaves.

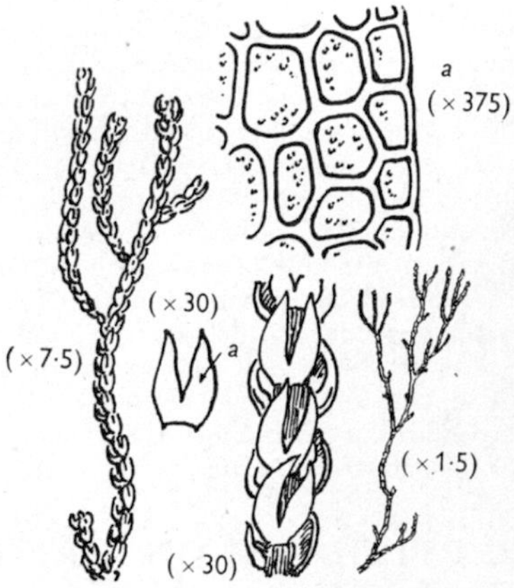

The microscope shows three ranks of leaves, the underleaves here resembling the lateral ones. Each leaf is deeply cleft to near its base. In no other genus of British liverworts with appressed leaves deeply cleft into two does the 3-ranked arrangement occur.

Fig. 155. *Anthelia julacea.*

Thus, *Gymnomitrium concinnatum*, which small forms of this plant may resemble superficially in the bluish grey tips of the shoots, has the leaves in two ranks only. Each leaf is only about 0·5 mm. long.

Since the leaves are fairly closely appressed to the stem and densely overlapping, little detail of cell structure can be seen without dissecting off an individual leaf. The cells are then seen to be notably unequal in size, those near the middle of the leaf lobe being oblong and about $25\,\mu$ long, the marginal cells shorter and smaller. The apical cell is often rather large and nearly colourless. The walls of all the cells are fairly strongly and evenly thickened.

Anthelia julacea is dioecious. Fertile plants are not rare, however, especially where (as in parts of the north-west) this species descends to near sea-level. The bracts are larger than the leaves, and the grooved, tubular perianth, which projects beyond them, has a toothed mouth. The capsule is nearly globose.

Ecology. Its chief habitats are constantly wet ledges and wet gravelly or peaty ground by streams, from about 1500 ft. upwards. Often it is associated with *Andreaea alpina*, sometimes with *Scapania undulata* and *Philonotis fontana*. In Britain *Anthelia julacea* is rare or absent except in North Wales, the Lake District and the Scottish Highlands.

ADDITIONAL SPECIES

A. juratzkana (Limpr.) Trevis. is the only other British species. It occurs on moist bare soil from 1900 ft. upwards. The stems are usually only 2–4 mm. long, but the best distinctions from *A. julacea* lie in the ovoid, scarcely exserted perianth, and larger spores (measuring 17–20μ against 14–15μ in the common plant). It is rare, but is of interest in being a plant of 'late snow patches'.

RELATED SPECIES

Herberta, a genus which has two British species, agrees with *Anthelia* in its three ranks of deeply cleft leaves, all of almost equal size. However, the species of *Herberta* look very different from *Anthelia* spp. on account of their orange- or red-brown to olive colour, robust habit, and spreading leaves.

Herberta adunca (Dicks.) Gray varies from olive-green to dark yellow-brown, grows to a length of 3–7 cm., and forms erect rigid tufts on exposed ground on mountains. The leaves are divided for $\frac{2}{3}$ of their length into two acute lobes. Superficially it resembles a moss such as *Rhacomitrium aquaticum*.

Herberta hutchinsiae (Gottsche) Evans is a local plant but is abundant on some mountain ledges in the west. Forming erect tufts of a characteristic red-brown or yellowish colour and with stems up to 12 cm. long, it is very conspicuous and an easy plant to identify. The two leaf lobes are drawn out to long fine points and the leaves are curved and turned to one side—features which suggest, at first sight, a moss of the genus *Dicranum*.

2. PTILIDIUM CILIARE (L.) Nees.

The moderately robust, much-branched leafy shoots and olive green to reddish brown colour of *P. ciliare* make it usually a conspicuous and easily recognized liverwort. It will often be found among mosses in grass or heather in hill districts, or in heathy places in the lowlands, but is seldom very abundant. The stems are 2–6 cm. long and are covered with leaves that are unlike those of any other common British liverwort.

Each leaf is about 1·5 mm. long and equally broad, and is irregularly divided into four unequal lobes. Each lobe is fringed in a striking way with outgrowths termed cilia, and these give the leaf its distinctive appearance. The leaves appear to be in three ranks, since the underleaves resemble the others except in their somewhat smaller size. The cell structure is notable for the large size of the corner thickenings. Each cell is roundish, about 30μ in diameter.

This species is dioecious and the conspicuous plants are female, the male plants being very small. Although the swollen, pear-shaped perianths are not uncommon, capsules are very rare in Britain.

Ecology. MacVicar mentions wet moors as one of its chief habitats, but, while certain forms occur on very wet ground, it is in general more a plant of *Calluna* and grass heath. In these habitats it often grows closely intermixed with the heather or grasses. Its distribution suggests that it is mildly calcifuge. It sometimes occurs in the turf on the tops of mountains, and is known to ascend to over 4000 ft. in Britain. *Ptilidium ciliare* is rare in the south-east.

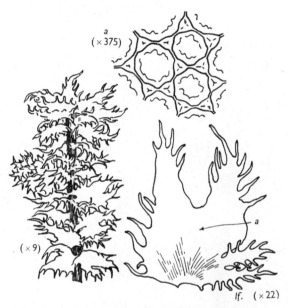

Fig. 156. *Ptilidium ciliare*: the left-hand sketch shows only a small part of the plant; *lf.* lateral leaf.

ADDITIONAL SPECIES

P. pulcherrimum (Web.) Hampe, which is relatively rare, is the only other British species. It grows on the trunks of trees, being locally frequent also on little-decayed fallen trunks and logs; it occurs more rarely on rocks. Thus it differs from *P. ciliare* in habitat, as also in its smaller size. The bases of the broadest leaf segments here are only 6–10 cells wide, whereas in *P. ciliare* they reach a breadth of 15–20 cells.

3. BLEPHAROSTOMA TRICHOPHYLLUM (L.) Dum.

This small liverwort is not conspicuous in the field, but its appearance under the microscope is unlike that of any other British species. It grows in small, yellowish green patches on moist shaded banks, or as scattered stems among mosses. Each stem is 0·5–2 cm. long, with frequent lateral branches.

The leaves are arranged in three ranks on the stem, the underleaves differing little from the lateral ones. Each leaf is divided to its base into three or four thread-like segments, each of which is composed of a chain of 8–14 narrowly oblong cells. These filaments are not matched by any other common British hepatic and bear a superficial resemblance to the branches of some filamentous alga; only the finer leaf segments of *Trichocolea tomentella* and of some species of *Lepidozia* and *Ptilidium* at all resemble them. Each cell in these leaf segments appears 4-sided, with walls thickened, especially at the angles. The larger cells attain $50 \times 20\mu$.

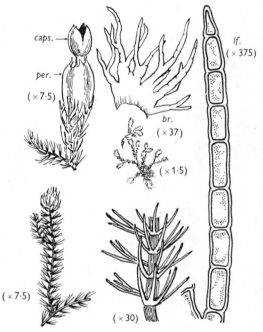

Fig. 157. *Blepharostoma trichophyllum*: top left, fertile shoot; *caps.* capsule, *per.* perianth; bottom left, sterile shoot; bottom centre, part of sterile shoot; *br.* bract, *lf.* part of single leaf.

Blepharostoma trichophyllum is normally paroecious. The bracts immediately below the perianth are divided into forked segments which commonly differ from the leaves in being two to three cells broad at the base. The perianth extends far beyond the bracts and is fringed with 'cilia' at its mouth. Each of these thread-like structures is composed of a row of about seven cells.

Ecology. Its chief habitats are sheltered rock crevices and ledges. It is perhaps calcifuge. Principally a mountain species, it is rare in the southern counties, where it may be found occasionally on sheltered banks, sandstone rocks and decaying wood.

4. TRICHOCOLEA TOMENTELLA (Ehrh.) Dum.

This plant forms extensive, pale whitish green patches in very damp shaded situations. Its light colour should attract attention. The stems are twice or thrice pinnately branched and attain 5–12 cm. in length. Superficially it resembles a species of *Thuidium* rather than a liverwort.

Under the microscope the appearance of the leafy shoot is very remarkable. The stem itself is relatively broad and solid-looking, but it bears a dense mass of leaves and paraphyllia, the former deeply divided into thread-like segments, the latter mere thread-like outgrowths, so that the cylindrical stem appears as if densely clothed in some

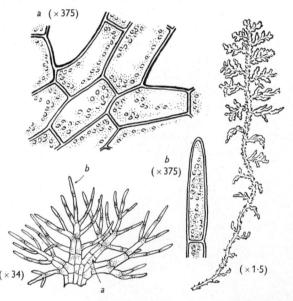

Fig. 158. *Trichocolea tomentella.* The drawing at bottom left is of a lateral leaf.

filamentous algal growth. On careful separation each typical leaf is seen to be deeply divided into about four main segments. Each segment, which is about three cells wide at its base, bears a number of pinnately arranged, curved filamentous outgrowths, which are only one cell wide. The leaf is thus divided much as in *Ptilidium,* but has thread-like chains of cells reminding one of *Blepharostoma.* The underleaves differ from the lateral ones chiefly in being somewhat smaller and more finely divided. The cells composing the filaments are long and narrow (up to 70μ); those in the leaf base are shorter and wider.

Trichocolea tomentella is dioecious and is unusual in that there is no perianth. After fertilization a cylindrical sac is formed for the protection of the developing sporogonium by an upgrowth of the stem tissue immediately surrounding the archegonia. Fertile plants are very rare in Britain.

Ecology. It is confined to wet habitats, and is perhaps most characteristic of shaded swampy ground in alder thickets by streams. It also grows occasionally on wet shaded rocks. It is widely distributed but nowhere very common.

5. BAZZANIA TRILOBATA (L.) Gray

This species is unlike any other British liverwort. *B. tricrenata*, which most resembles it, is quite distinct in its much more slender habit. *B. trilobata* is one of our largest leafy liverworts and grows in wide patches or cushions of dull or yellowish green colour and firm texture, on rocky banks and ledges in woods on acid soils. The stems are 3–10 cm. long and the flattened or convex leafy shoots are notably broad (4–5 mm.), the leaves being crowded and overlapping. These shoots normally show forked branching very clearly, whilst from their undersides arise conspicuous small-leaved 'flagella'.

The leaves in *Bazzania* spp. overlap in the incubous manner, itself a comparatively unusual character among British hepatics. But by far

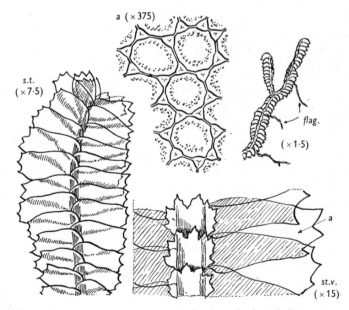

Fig. 159. *Bazzania trilobata*: *s.t.* part of a shoot in dorsal view, *st.v.* small part of the same shoot in ventral view, *flag.* flagella.

the most striking feature of the present plant (shared only by *B. tricrenata*) is the 3-toothed or shallowly 3-lobed apex of the strongly asymmetrical leaves. The leaf is ovate-oblong, but appears as if sheared off abruptly (truncate), to give this highly characteristic 3-toothed apex. Large, roundish, irregularly toothed underleaves are present.

Under the microscope, the cell cavities are seen to be rounded in outline, the walls being thin save for the conspicuous corner thickenings. An average cell measures 30μ across, and a triangular corner thickening $6-7\mu$.

B. trilobata is dioecious, and in Britain female plants are much commoner than male. Fertile plants are rare.

Ecology. A calcifuge, *B. trilobata* is one of the characteristic bryophytes of the *Quercus petraea* woods of the north and west, associated species often including the mosses *Dicranum majus* and *Rhytidiadelphus loreus*. It will grow on acid soil, boulders, or tree bases. It is rare in the south-east.

ADDITIONAL SPECIES

Bazzania tricrenata (Wahl.) Trev. occurs on rocky and grassy ledges on mountains, being frequent in the West Highlands of Scotland, in Wales and the Lake District. It is distinct from *B. trilobata* in its yellowish or red-brown colour, slender habit and narrower leaf apex. It often grows with *Herberta hutchinsiae*.

6. LEPIDOZIA REPTANS (L.) Dum.

The first impression of this common liverwort is of exceedingly fine, pinnately branched stems, matted together to form wide patches of dull or deep green colour, on banks or rotten wood. Its proportions are those of one of the most slender 'hypnoid' mosses, such as *Amblystegium serpens*. Individual stems are commonly 0·5–2 cm. long, and bear three ranks of minute leaves. It is, however, very variable in size.

The details of leaf shape and arrangement are clearly seen only under the microscope. *Lepidozia* is then seen to differ from all other British genera, in that the leaves are deeply 3- to 4-lobed to one-third of their length, the underleaves closely resembling the lateral leaves. Only in such genera as *Barbilophozia* is this 4-lobed leaf character met with again, and there the whole plant is robust and the underleaves are minute. Thus *Lepidozia reptans* can be confused only with certain rarer species of *Lepidozia*.

The leaf-cells in *L. reptans* are rather small, about 25μ across, with slightly thickened walls but without obvious corner thickenings.

The monoecious 'inflorescence' here is a feature of distinction from *L. pearsoni*, which is dioecious. The long cylindrical perianth is narrowed at its furrowed, toothed apex. The bracts which surround its base are much larger than the ordinary leaves and are 3- to 6-lobed. The capsule is oblong-cylindrical, and yellow-brown in colour.

Ecology. Its chief habitat is rotten wood; but it will also grow on moist banks, and on peat or humus on the ground in woods. It occurs less commonly on sandstone rocks and in shady situations on moors.

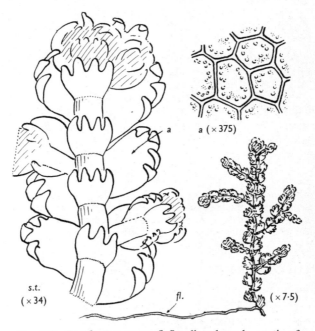

Fig. 160. *Lepidozia reptans*: *fl.* flagellum branch, *s.t.* tip of shoot in ventral view.

7. LEPIDOZIA SETACEA (Web.) Mitt.

This species varies much in habit, sometimes forming compact, dark olive green patches, at other times occurring as scattered, nearly erect stems among *Sphagnum* or other mosses. The stems are extremely slender and thread-like, 1–3 cm. long. Whilst agreeing with *Lepidozia reptans* in its three ranks of 3- or 4-lobed leaves, it is very different from that plant in the details of its leaf structure, and in being most typically a plant of bogs and moorlands. In a general way it resembles *Blepharostoma trichophyllum*, but that species is a conspicuously light yellowish green in colour, and quite different in habit.

The leaf is in most cases divided almost to its base into three segments, each segment being very narrow, two cells wide for most of its length and one cell wide at its tip. In *Blepharostoma* the leaf segments are much longer, and only one cell wide throughout. The underleaves are like the lateral leaves, but smaller. The leaf cells are mostly oblong, and measure about $20 \times 30\mu$. The walls are not markedly thickened. Under a high power the cuticle is seen to be rough with minute papillae.

Lepidozia setacea is dioecious, and fertile material is rare. The bracts are much larger than the vegetative leaves and are fringed above with long fine teeth, or 'cilia'. The perianth is cylindrical and is itself fringed

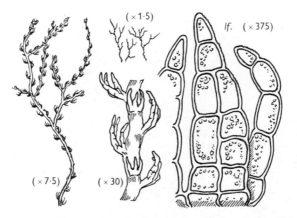

Fig. 161. *Lepidozia setacea*: bottom centre, small portion of shoot in ventral view; *lf.* part of single lateral leaf.

with 'cilia' at its mouth. On the male plants antheridia occur on specialized short branches which consist of 4–5 pairs of bracts, each protecting a single globose stalked antheridium.

Ecology. By far its most characteristic habitat is in acid bogs, where it occurs commonly in *Sphagnum* patches, a frequently associated species being *Odontoschisma sphagni*. It also grows on wet moorland slopes.

ADDITIONAL SPECIES

Lepidozia pinnata (Hook.) Dum. is a rather rare western species allied to *L. reptans*, but differing in being dioecious and in its more close-set underleaves, with longer, more acute lobes. It forms characteristic large swollen patches of pale colour.

L. trichoclados K. Müll., another western species, closely resembles *L. setacea*, the only certain distinction lying in the bracts and perianth, which in *L. trichoclados* are fringed merely with shortly projecting cells, not with fine multicellular threads. Usually, however, *L. trichoclados* grows in swollen cushions and occurs on moist peat banks rather than on wet moors.

L. pearsoni Spruce is frequent in some western districts. It resembles the more elongated forms of *L. reptans*, but may be distinguished with certainty by its dioecious 'inflorescence'. The long slender stems (3–6 cm.) and very distant leaves are also characteristic.

8. CALYPOGEIA TRICHOMANIS (L.) Corda

This and the closely allied *C. fissa* are both common on rather dry
sandy and peaty banks, and on wet moorland. Both species are variable,
but typical plants of each show well-marked distinguishing characters.
C. trichomanis forms flat patches of mid- or bluish green colour, often
with tips made conspicuous by the light green clusters of gemmae. Individual stems are 1–4 cm. long, and the two ranks of obliquely inserted,
widely spreading leaves give the shoots a flattened appearance.

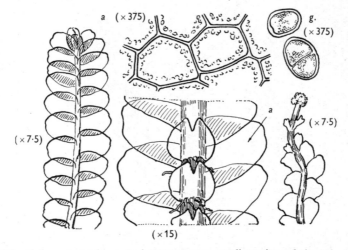

Fig. 162. *Calypogeia trichomanis*: bottom centre, small portion of shoot in ventral
view, showing underleaves; other shoots are seen in dorsal view, showing incubous
leaf arrangement; *g.* gemmae from tip of stem on extreme right.

The leaves are crowded, and differ from those of the majority of
British genera in their incubous arrangement. The individual leaf is
broadly ovate, normally entire, but sometimes showing two small teeth
at its apex. Relatively large underleaves are present. These are nearly
round in outline, but are bilobed to about $\frac{1}{4}$ of their length, the margins
of the lobes being entire (cf. the much broader, often toothed underleaves of *C. fissa*).

The leaf cells are 5- to 7-sided in surface view, and uniformly thin-
walled. They are 30–40μ wide, but may be 50–70μ long. The gemmae
which are so often a feature of the shoot tips in this plant are 1- or
2-celled structures.

C. trichomanis has antheridia and archegonia on the same specialized
branches, which arise in the axils of the underleaves. The development
of the sporophyte is unusual throughout the genus *Calypogeia* in that

there is no perianth, but instead there is a nearly cylindrical bulb-like structure termed a *perigynium* or *marsupium* (cf. *Saccogyna*), the growth of which keeps pace with the developing embryo and helps in its nutrition. This is found in comparatively few British hepatics. Capsules, however, are rare.

Ecology. A calcifuge, it is most common on sandy banks and sandstone rocks, in shade. It also occurs on peaty banks, the sides of ditches and on moorland. According to MacVicar it ascends to 3000 ft.—higher than does *Calypogeia fissa.*

9. CALYPOGEIA FISSA (L.) Raddi

This species forms delicate, pale green patches in the same range of habitats as *C. trichomanis* but usually in slightly damper situations. Sandy or peaty banks and moorland bogs are thus likely places to find it. It is occasionally difficult to distinguish from *C. trichomanis* and by some authors is not considered a distinct species. Well-developed plants of the two, however, are separated with ease. *C. fissa* is usually more slender and of more fragile texture; but in length of stems, and in the oblique, incubous leaf insertion, the two species are alike.

The leaves of *C. fissa* differ from those of *C. trichomanis* in being narrowed at their tips, which are regularly notched. The underleaves

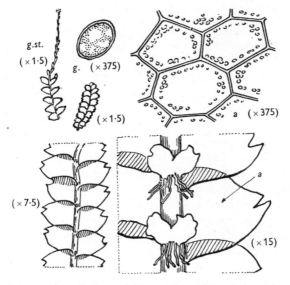

Fig. 163. *Calypogeia fissa*: bottom right, small portion of shoot in ventral view showing underleaves; other shoots are seen in dorsal view showing incubous leaf arrangement; *g.st.* stem bearing gemmae at apex, *g.* gemma.

are proportionately shorter and wider than those of *C. trichomanis*, and much more deeply cleft into two lobes which are themselves notched. Only fairly typical, well-developed plants, however, show these characters clearly.

The leaf cells are thin-walled, as in *C. trichomanis*. On the whole they tend to be larger (up to 55μ wide), but cell size is not in itself a very reliable character for separating these two species. The gemmae resemble those of the last species. Their clusters are often borne on the tips of special attenuated shoots, when they appear to the naked eye like pale glistening pin-heads, but under the microscope the individual gemma is seen to be 1- or 2-celled.

C. fissa is monoecious, with the antheridia on separate branches. It is, however, rarely fertile. In the details of the female branch and capsule it closely resembles *C. trichomanis*.

Ecology. A calcifuge, like *C. trichomanis*, its chief habitats are sandy banks and sandstone rocks, in shade; but it also grows on peat and is the commoner of the two species in *Sphagnum* bogs. MacVicar states that it is rare above 1000 ft.

ADDITIONAL SPECIES

Calypogeia arguta Nees & Mont. is the only other species which is at all common. It grows in delicate whitish green patches on moist banks in very sheltered places (e.g. the vertical sides of drainage ditches in bogs). It is most frequent near the west coast. Microscopically, it may be identified at once by the two acute, divergent teeth at the apex of each leaf, and by the underleaves which are divided into narrow segments. Species of *Lophocolea* resemble it superficially, but they occur in different habitats and bear leaves that overlap succubously.

10. CEPHALOZIA BICUSPIDATA (L.) Dum.

This is a very common liverwort on moist soil in a wide range of habitats, but it is rather easily overlooked on account of its small size. The stems are only 0·5–2 cm. long and often grow intermixed with mosses. Usually, however, the pale, whitish green colour of the plant makes the flattened tufts or carpets stand out against a background of dark soil, but at times the plant is a dull mid-green and is then inconspicuous.

The leaves vary from very oblique to nearly transverse in their insertion, and those towards the tip of a shoot are often more nearly erect and overlapping. Each leaf is deeply divided into two pointed segments. Certain long shoots with very small leaves may be seen; these are the 'flagella', a notable feature of some species of *Cephalozia*.

Under a low power of the microscope it will be seen that the surface of the stem is marked by large translucent cells, of which about four comprise the width of the stem. The segments of the leaf normally diverge, but in some forms of the plant they may converge like pincers, as they do even more markedly in *C. connivens*. The cells in a typical leaf measure about 40μ across, and are thus nearly three times as broad as those of all our common species of *Cephaloziella*; but leaves with smaller cells will commonly be found in poorly developed specimens. There are no underleaves.

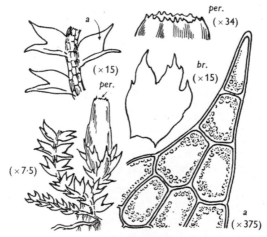

Fig. 164. *Cephalozia bicuspidata*: *br.* bract, *per.* mouth of perianth.

Cephalozia bicuspidata is monoecious and is commonly fertile. Perianths are conspicuous tubular structures on special short lateral branches. The perianth is freely exposed for $\frac{3}{4}$ of its length, and is narrowed at the shortly toothed mouth. Surrounding the base of the perianth are the bracts, which are of diagnostic value in this genus. In *C. bicuspidata* they are 2-pronged, like the vegetative leaves. A third row of leaves (the bracteoles) distinguishes these short reproductive branches. Leaves of more concave form enclose the antheridia; these may be at the middle or the ends of ordinary branches.

Ecology. An acid substratum is always favoured. In other respects it is catholic in choice of habitat; for whilst most common on sandy banks, it also occurs on loamy banks and paths in woodland, in *Sphagnum* bogs and occasionally on rotten wood.

ADDITIONAL SPECIES

Two other species may be mentioned; both are plants of acid habitats.

Cephalozia connivens (Dicks.) Lindb., which occurs in bogs and on wet banks, has the leaves nearly longitudinally inserted, and the tips of the leaf lobes turned in pincer-wise. It may be known by its very large leaf cells ($40–60\mu$ wide).

C. media Lindb. grows in various habitats, including peaty and sandy banks and *Sphagnum* bogs. This species has the leaf lobes turned inwards as in *Cephalozia connivens*, and is best distinguished from that species by the much smaller leaf cells (23–35 μ). Both these plants lack 'flagella'.

RELATED SPECIES

Cladopodiella francisci (Hook.) Joerg. (*Cephalozia Francisci* (Hook.) Dum.) is a small species which forms yellowish green to reddish brown patches on sandy or peaty soil on heaths. The leaves are markedly concave, and merely notched at the apex, and the leaf cells are small (about 20 μ).

Cladopodiella fluitans (Nees) Joerg. (*Cephalozia fluitans* (Nees) Spruce) occurs in bogs and submerged in moorland pools. In leaf form and insertion it resembles *Gymnocolea inflata*, but is distinguished by the presence of 'flagella' and by its large leaf cells (35–42 μ).

Nowellia curvifolia (Dicks.) Mitt. is an uncommon plant of decaying tree trunks in the north and west. It has the 2-pronged leaf of *Cephalozia bicuspidata*, but each leaf is very concave and is rendered asymmetrical by a conspicuous sac-like swelling on one side. The cells are small (about 20 μ) and tend to be square in outline. The older parts of the plant are almost always strongly tinged with rose red.

11. ODONTOSCHISMA SPHAGNI (Dicks.) Dum.

As the name suggests, this species may be found in patches of *Sphagnum* and other bog or wet moorland vegetation, where it occurs as scattered stems or loose tufts. It is green, often variegated with reddish brown. Individual stems are 2–8 cm. long. Branches always arise from the underside of the stem, and frequently take the form of slender, nearly leafless 'flagella'. Rhizoids are numerous along the undersides of the stems and 'flagella'. The two ranks of leaves are very obliquely (nearly longitudinally) inserted, and often appear to form a double row along one side of the stem, though elsewhere on the same

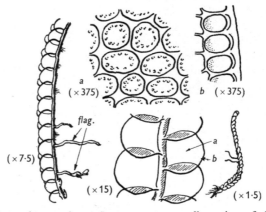

Fig. 165. *Odontoschisma sphagni*: bottom centre, small portion of shoot in dorsal view, flattened out to show succubous leaf arrangement; other shoots are seen in lateral view; *flag.* flagella.

shoot they may be widely spreading and not obviously turned to one side. Underleaves are absent from mature stems, but may be found near the tips of young branches.

The broadly ovate or nearly orbicular leaves (0·7 mm. long) are slightly concave and have entire margins. Where crowded, they overlap one another in a succubous manner. In this, as in shape, they resemble some species of *Jungermannia*. These, however, are plants of very different habit from *Odontoschisma*; moreover, they grow in quite different habitats from *O. sphagni*. From *Mylia anomala*, which often grows with it, the present species is distinguished by the absence of gemmae (and by much smaller cells).

Under the microscope the leaf shows two well-marked characters: (i) the evident margin formed by oblong cells with nearly equally thickened walls, and (ii) elsewhere in the leaf rounded cells about 20μ wide, thin-walled except for obvious (but variable) corner thickenings. Under a high power the cuticle of the leaf is seen to be rough with papillae.

Odontoschisma sphagni is dioecious and fertile plants are rare. Perianths are conspicuous when present, since they extend far beyond the lobed bracts. They occur on special short branches which arise on the underside of the stem.

Ecology. It is usually but not invariably associated with *Sphagnum* spp., most often growing closely intermixed with those mosses. *Odontoschisma sphagni* is of interest in being chiefly a plant of 'raised'* bogs and valley bogs, rather than of 'blanket'* bog. Although this perhaps implies that it is most typical of lowland areas of bog and wet heath, it is also widely distributed and not rare in the great areas of wet hill moorland in the north and west.

ADDITIONAL SPECIES

O. denudatum (Nees) Dum. is a rather uncommon plant of peaty banks, sandstone rocks and decaying stumps. Unlike *O. sphagni* it commonly bears gemmae on the tips of the apical leaves. It may always be known from the commoner plant by the very large corner thickenings of the cells; often they are almost as large as the cell cavities. *O. denudatum* is, moreover, a smaller plant (stems 1–2 cm. long) than *O. sphagni*.

CEPHALOZIELLA

General notes on the genus. All species of *Cephaloziella* are exceedingly minute plants which are notably difficult to determine. The deeply 2-cleft leaves are less than 0·3 mm. long and are very small in relation

* See Sir A. G. Tansley, *The British Islands and their Vegetation*, for definitions of these terms.

to the width of the stem. Thus the width of the stem is commonly $70-80\mu$ and that of the transversely inserted leaves only $80-120\mu$. In all the common species the stem is composed of denser tissues than the stems of *Cephalozia* spp. This gives the stem a dark, opaque appearance not found in *Cephalozia*. This is an important character, for small specimens of *Cephalozia* spp. (with cell dimensions below the average) are often mistaken by beginners for *Cephaloziella*. 'Flagella' are lacking. The perianth is tubular, with 3–6 longitudinal folds near its tip.

The species of this genus commonly form dark green, red-brown or blackish patches on the ground, often looking like some terrestrial alga, but very frequently they grow intermixed with mosses such as species of *Campylopus*, *Dicranum*, *Sphagnum* and *Leucobryum glaucum*. Such scattered stems may be recognized as *Cephaloziella*, but determination of the species may be impossible (indeed, is often out of the question even for the expert) in the absence of 'inflorescences'. Dr E. W. Jones informs me, from long experience of this difficult genus, that good fertile material may usually be found by careful searching, and that it is often easier to ascertain, by careful collecting *in the field*, whether male and female plants are in different patches (and the species therefore probably dioecious), than it is to do so from herbarium specimens. In order to see whether an 'inflorescence' is paroecious it is essential to examine it while it is very young, since by the time the perianths are obvious all traces of antheridia have often vanished.

12. CEPHALOZIELLA BYSSACEA (Roth) Warnst. (*C. Starkii* (Funck) Schiffn.)

C. byssacea is the most robust of the three species that are fairly common in Britain, but even so the stems are only 3–10 mm. long and the leaves are microscopically small. Although it has sometimes been described as the only common species in this country, this is not so; for in some habitats, and perhaps over the country as a whole, *C. hampeana* and *C. rubella* are by far the commoner plants. The colour of *C. byssacea* is usually dark green to blackish. In habit it is prostrate or ascending, with numerous thread-like, densely massed but relatively little-branched stems.

The leaves are transversely inserted and relatively distantly placed on the stems, except near the shoot tips where they become crowded. Each leaf is divided to beyond the middle into two segments. These segments are divergent and usually fairly acute. In addition to the two ranks of deeply divided leaves, there are narrow, undivided (lanceolate-subulate) underleaves. They stand out from the stem and should not be difficult to distinguish from the rhizoids which may also be numerous along the under surface. The underleaves afford a useful diagnostic character, for *C. rubella* lacks them and in *C. hampeana* they are absent except near the 'inflorescence'.

A typical leaf, which is about 0·2 mm. long, is composed of relatively few cells. Each segment of the leaf is but 5–12 cells broad at its base, and the largest cells are only 12–15μ wide, half the width of the leaf cells of *Gymnocolea inflata* and one-third the width of those found in well-grown plants of *Cephalozia bicuspidata*. Two-celled gemmae may occur at the tips of the sterile stems.

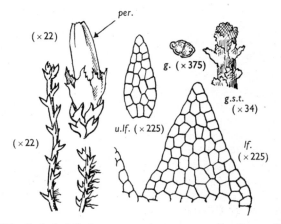

Fig. 166. *Cephaloziella byssacea*: *g.* gemma, *g.s.t.* tip of gemmiferous shoot, *lf.* part of leaf, *per.* perianth, *u.lf.* underleaf.

Cephaloziella byssacea is dioecious. Male 'inflorescences' may be found at the tips or in the middle of branches, their concave bracts being in 6–12 closely overlapping pairs. The involucral bracts are united at their bases and their segments have toothed margins. Three-quarters of the long tubular perianth projects beyond the bracts.

Ecology. Early statements concerning the ecological requirements of this species should be accepted with caution, owing to the frequency with which other species of *Cephaloziella* have been mistaken for it. Dr E. W. Jones regards this species as characteristic of relatively dry habitats, where it will grow on earth, rock or on the boles of trees. It agrees with all the other common species in being calcifuge.

ADDITIONAL SPECIES

Cephaloziella hampeana (Nees) Schiffn. is a green or brownish green plant common on raw humus (e.g. amongst *Calluna*), clayey and sandy soil, and in marshes. The leaves on well-developed shoots are relatively large, with lobes 4–10 cells wide at base, and cells in mid-leaf $11–18\mu$. There are no underleaves except near 'inflorescences'. It is monoecious, with separate male 'inflorescences'.

Cephaloziella rubella (Nees) Warnst. is marked by its prevailing dark brown or reddish brown colour. It is especially common on raw humus, and often grows intermixed with mosses on heathland. The leaf lobes are usually 4–5 cells wide at the base, the cells most commonly about 12μ. There are no underleaves. *C. rubella* is monoecious, the same plant having some paroecious, and some separate male and female, 'inflorescences'.

13. **LOPHOCOLEA BIDENTATA** (L.) Dum.

The nearly prostrate, pale whitish green leafy shoots of this plant will be found among mosses, or forming pure patches, on moist lawns, banks and similar grassy places. Fairly robust (leafy shoots 2–3 mm. across) it will be among the first hepatics found by a beginner who examines this type of habitat. The slender stems are commonly 2–4 cm. long and the leaves are very obliquely inserted.

Both the regular 2-ranked arrangement of the leaves and the 2-pronged (bidentate) character of the individual leaf are readily seen with the naked eye. Exactly this type of leaf (drawn out into two fine points) occurs nowhere else among common liverworts except in the closely related species *L. cuspidata*. Thus, in *L. heterophylla* the leaves lack the fine points and the upper ones are quite undivided; and the leaves of *L. bidentata* are three times the size of those of *Cephalozia bicuspidata*. The underleaves are small and deeply divided.

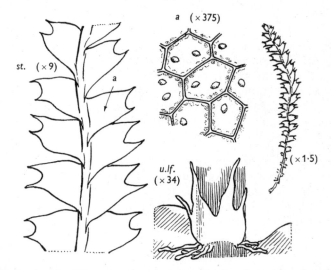

Fig. 167. *Lophocolea bidentata*: *st.* part of shoot in dorsal view showing succubous leaf arrangement; *u.lf.* fragment of shoot in ventral view showing one underleaf and a few rhizoids. One or two oil bodies are seen in each cell at *a*.

Under the microscope the leaf cells appear six- to eight-sided in surface view, and are thin-walled with scarcely any corner thickenings. Average dimensions are $30 \times 35\mu$, but the cells vary much in shape and size, and so afford no safe character for the separation of this species from *Lophocolea cuspidata*.

L. bidentata is dioecious and perianths are rare. In this it contrasts strongly with *L. cuspidata*, which is monoecious and commonly fertile.

Ecology. The important ecological characteristic of *L. bidentata* is that, unlike the other two common species of *Lophocolea*, it grows on soil, not normally on wood or bark. It occurs in grassland and on the ground in woods. Robust and slender varieties are known, apparently correlated respectively with very moist and unusually dry habitats. *L. bidentata* is, perhaps, the commonest hepatic of town lawns.

14. LOPHOCOLEA CUSPIDATA (Nees) Limpr.

L. cuspidata in many respects comes very near *L. bidentata*. Generally, however, it grows in rather compact patches, with the stems only 1–3 cm. long. In its prostrate habit and pale green colour it is like *L. bidentata*. It grows most commonly on decaying logs and stumps in woods, and in this it resembles *L. heterophylla*.

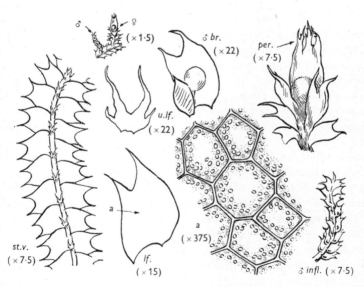

Fig. 168. *Lophocolea cuspidata*: *st.v.* part of leafy shoot in ventral view, *lf.* leaf, *u.lf.* underleaf, *per.* perianth, ♂ *infl.* male 'inflorescence', ♂ *br.* male bract with antheridium. The shoot fragment at the top left shows the position of male (♂) and female (♀) 'inflorescences'.

Known at once as a *Lophocolea* by the two ranks of very obliquely set, bidentate leaves and small, deeply divided underleaves, *L. cuspidata* differs usually from *L. bidentata* in the slightly greater length of the leaf 'prongs' and a leaf shape that is more nearly symmetrical.

The microscope shows the same type of large, thin-walled cells as in the last species, and the greater cell size that is sometimes quoted is not in itself a reliable character, since forms of *L. cuspidata* occur with average leaf-cell size smaller than that of some plants of *L. bidentata*.

L. cuspidata is monoecious and, unlike *L. bidentata*, usually produces perianths and capsules freely. The perianth is acutely 3-angled, and at its mouth is irregularly torn into long fine teeth. In this it differs from *L. heterophylla*, where the mouth of the perianth is fringed with short teeth only. Antheridia occur singly in the axils of curved, asymmetrical bracts. These bracts are arranged in closely overlapping pairs, forming spike-like 'inflorescences' at the ends of special branches.

Ecology. The rotten wood of decaying logs, tree stumps and fallen branches constitutes the chief habitat of this species, but it also grows on the bases of living trees. Less commonly it occurs on rocks, grassy banks and overgrown wall-tops.

15. **LOPHOCOLEA HETEROPHYLLA** (Schrad.) Dum.

This species will be among the first liverworts to be noticed by the beginner who searches the decaying logs and stumps of moist woodland. It is seen at its best in spring, with the prostrate leafy shoot system offset by abundant upright perianths, and capsules in various stages of development. The stems are 1–2 cm. long, and variously branched. They are as a rule more closely attached to the bark than in *L. cuspidata*. The texture of the plant is slightly thicker and more opaque than in that species, and the colour is usually comparatively dark (not whitish) green.

From other common species of *Lophocolea*, *L. heterophylla* may be known by the fact that some of the leaves are nearly entire, not 2-pronged as are the other leaves of this plant and all those of *L. bidentata* and *L. cuspidata*. In their oblique insertion and widely spreading character the leaves resemble those of the other species of *Lophocolea*. *L. heterophylla* also agrees with the allied species in having small, deeply divided underleaves.

The leaf cells agree with the related species in dimensions (about 30μ across) and in their generally thin walls. Small, but perceptible corner thickenings, however, are usually present.

In *L. heterophylla* the male bracts occur in several pairs immediately below the perianth (the 'paroecious' condition), not on separate branches as in *L. cuspidata*; and this species is often very fertile. The shortness of the teeth which fringe the mouth of the perianth will distinguish it from that of *L. cuspidata*. As in other leafy liverworts the seta elongates rapidly when the ovoid, dark brown capsule is ripe, so that on old 'fruiting' material these long white setae and the pale brown 'crosses' formed from the ruptured capsule walls are conspicuous.

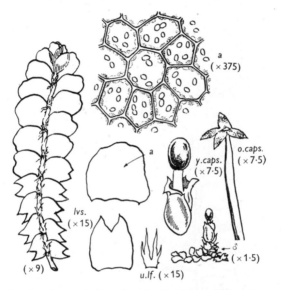

Fig. 169. *Lophocolea heterophylla*: *lvs.* leaves from different parts of shoot, *u.lf.* underleaf, *y.caps.* capsule at stage when seta is just beginning to elongate, *o.caps.* old dehisced capsule. The shoot on the left is seen in ventral view; position of male 'inflorescence' marked ♂; oil bodies are seen in the cells at *a*.

Ecology. As in *L. cuspidata*, decaying logs and tree stumps in moist woodland are the principal habitat, where associated bryophytes are commonly such mosses as *Brachythecium rutabulum*, *Hypnum cupressiforme* and *Bryum capillare*. It also grows on sandy or gravelly hedge banks.

ADDITIONAL SPECIES

Lophocolea alata Mitt. is closely allied to *L. cuspidata*, from which it differs in darker green colour, more asymmetrical leaf outline and very large cells (40–50 μ). It is an uncommon hepatic of moist banks in the southern counties.

16. CHILOSCYPHUS POLYANTHUS (L.) Corda

At first sight there is nothing very striking about this liverwort, which is quite a common plant of wet ground. It forms loose patches of low growth, pale or dull green to brownish in colour, with stems 2–4 cm.

in length. In the very obliquely set, succubous leaves one is reminded of *Plagiochila*, and the plant also bears some resemblance to *Saccogyna viticulosa*. When the leafy shoot is viewed from above, the lower (antical) margin of each leaf is seen to run down the stem a little at its insertion, but not so markedly as in *Plagiochila* spp. Moreover, underleaves are constantly present here, whereas they are normally absent in species of *Plagiochila*.

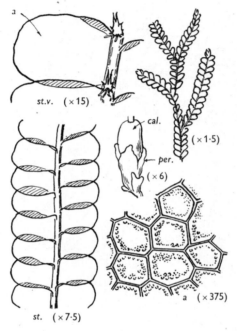

Fig. 170. *Chiloscyphus polyanthus*: *cal.* calyptra, *per.* perianth, *st.* part of leafy shoot in dorsal view, *st.v.* small part of the same shoot in ventral view.

The leaves of *Chiloscyphus polyanthus* are up to 1·5 mm. long and almost equally broad, very shortly oblong with rounded apices. Thus they differ in shape from the ovate-oblong leaves of *Saccogyna viticulosa*, whilst the always entire leaf margin here contrasts with the usually toothed margin of the common species of *Plagiochila*. The underleaves are narrow, 2-pronged structures, quite different from the broad, irregularly toothed underleaves of *Saccogyna*.

The leaf cells are uniformly thin-walled, with no trace of corner thickenings. They measure about 25–35μ across. Plants of all the closely related British genera show some trace of corner thickenings in their leaf cells.

Chiloscyphus polyanthus is monoecious. Antheridia are formed singly in the axils of bracts which occur in several pairs, in the middle or at the end of the stems. Perianths arise on short lateral branches. They are deeply 3-lobed, the edge of each lobe being without obvious teeth. The calyptra extends well beyond the mouth of the perianth. The mature capsule is ovoid.

Ecology. Typical forms of this species occur on wet soil or rocks by streams, also in mountain springs and on wet clayey soil in woods. Some forms are found characteristically on stones submerged in shallow streams. This is especially true of the var. *rivularis* (Schrad.) Nees, a richly branched, dark green plant with rather smaller leaf cells than the typical form.

ADDITIONAL SPECIES

C. pallescens (Ehrh.) Dum. is a plant of ditches, moist banks and decaying logs. It is most readily distinguished from *C. polyanthus* by its pale green colour, very large leaf cells (35×45–60μ), and sharply toothed perianth lobes. The var. *fragilis* (Roth) K. Müll., which occurs in mountain springs and among grass in various wet habitats, tends to form erect tufts and has leaf cells (30–40μ) smaller than in the typical form but larger than in *C. polyanthus*.

17. **MYLIA TAYLORI** (Hook.) Gray (*Leptoscyphus Taylori* (Hook.) Mitt.)

Mylia taylori is a conspicuous liverwort of many moorland areas, especially in north and west Britain. It grows in large, swollen, erect tufts, the yellow-green of the shoots being strikingly variegated with golden brown, chestnut or purple-red. The stems attain 3–8 cm. in length; the leaves, distant below, are close-set and overlapping towards the shoot tips. Abundant rhizoids along the under surface of the stem conceal the small, narrow underleaves. On account of its size and colour *M. taylori* is usually a noticeable liverwort wherever it grows in abundance, but aberrant, small-leaved forms occur at times.

The obliquely inserted, succubous leaves are concave at their extreme bases, but are flattened above, with entire, somewhat wavy margins. The leaf outline varies from broadly oblong to nearly round; sometimes in the shape and arrangement of the upper leaves one is reminded of *Nardia* spp., but the leaves here are not so wide in proportion to their length as in that genus and the whole plant is stouter and more upright in growth.

Under the microscope the leaf of this species is abundantly distinct in its large cells (45–60μ), with very marked corner thickenings, and rough cuticle. This last character is seen very clearly when the edge of a leaf is viewed under high power. The outlines of the cell cavities are often almost circular. A similar type of cell structure is found in the closely related *Mylia anomala*, but that is a plant of very different general appearance, narrower leaves and quite smooth cuticle. Gemmae sometimes occur on the margins of the upper leaves, but are much commoner in *M. anomala*.

M. taylori is dioecious and fertile plants are rare. The oblong perianth extends well beyond the involucral bracts, and is laterally compressed above. The capsule is ovoid-globose, and dark brown in colour.

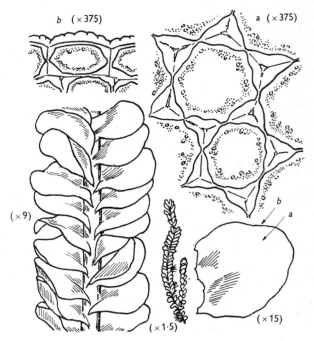

Fig. 171. *Mylia taylori.*

Ecology. It grows on wet rock ledges, but perhaps even more characteristically on wet peaty banks. It is a plant of north-western distribution, being abundant in parts of Wales, north England and Scotland. It is sometimes a prominent bryophyte in clefts of stabilized 'block scree'.

ADDITIONAL SPECIES

M. anomala (Hook.) Gray (*Leptoscyphus anomalus* (Hook.) Mitt.) forms loose tufts in bogs, commonly creeping amongst *Sphagnum* and often associated with *Odontoschisma sphagni*. It resembles *Mylia taylori* in general leaf shape and cell structure, but differs in its yellow-green to brownish colour, quite smooth cuticle and narrow upper leaves which always bear gemmae in abundance. *Odontoschisma sphagni* may be known at once by the small size of its leaf cells ($25\,\mu$ against $50\,\mu$ in *Mylia anomala*).

18. SACCOGYNA VITICULOSA (Mich.) Dum.

This plant, which grows on moist banks and rock ledges in the west of Britain, bears some resemblance to *Chiloscyphus polyanthus*. Like that species, it grows in wide, irregular pale green to brownish patches, with individual stems 2–5 cm. long; but it tends to be a brighter colour, with the older parts of the shoots reddish brown. Like *Chiloscyphus*, it has two regular ranks of oblong to ovate leaves very obliquely inserted and widely spreading. *Saccogyna viticulosa* is, however, of firmer texture than species of *Chiloscyphus*, and the 'pattern' formed by the leafy shoot is quite distinct on account of the leaves being nearly opposite.

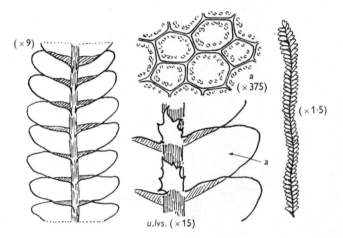

Fig. 172. *Saccogyna viticulosa*: *u.lvs.* part of shoot in ventral view showing two underleaves; in the other drawings the shoot is seen in dorsal view.

The antical margins of the leaves are strongly decurrent. The underleaves, which have their bases joined to those of the lateral leaves, differ from the underleaves of *Chiloscyphus* in being broadly ovate, with numerous teeth.

The leaf cells are about 25μ wide, with walls thin except for minute, but quite distinguishable, corner thickenings. Under a high power the cuticle will be found to be studded, rather irregularly, with minute projections (in *Chiloscyphus polyanthus* the cuticle is smooth).

Saccogyna viticulosa is dioecious, and reproductive organs are very rare. An interesting feature of the fertile condition here is the sac-like outgrowth that forms beneath the developing sporophyte (whence the name *Saccogyna*) and assists in its nutrition.

Ecology. The characteristic habitat is that of moist rock ledges, in deep sheltered valleys and ravines, especially near the sea, in west Britain. It is a member of the non-calcareous rock-ledge community which commonly includes the mosses *Fissidens adianthoides*, *Dicranum majus* and many others.

19. **LOPHOZIA VENTRICOSA** (Dicks.) Dum.

L. ventricosa is a very common liverwort of acid banks and ledges. It forms extensive patches of a bright or yellowish green, the densely massed stems attaining 1–3 cm. in length. Often it is intermixed with other hepatics such as *Diplophyllum albicans* and low-growing mosses such as *Dicranella heteromalla*, but it is usually an easy plant to recognize by the pale green, powder-like clusters of gemmae which almost invariably occur in abundance at the tips of the apical leaves.

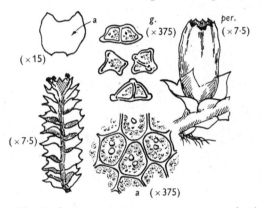

Fig. 173. *Lophozia ventricosa*: *g*. gemmae, *per*. perianth; four of the cells at *a* show the oil bodies that can be seen only in fresh specimens.

This species has shallowly bilobed leaves which approach in outline those of several other species of *Lophozia*. They are less narrowly contracted at the base than those of *Marsupella emarginata*, whilst the lobes are more acute than in *Gymnocolea inflata*; moreover, both these plants lack the bright green colour of *Lophozia ventricosa*. As in all species of *Lophozia*, the leaves are slightly oblique in their insertion, and are succubous. There are no underleaves. The bracts differ in being larger, and 2- to 5-lobed.

The leaf cells are about 25–30μ across, with thin walls which are only very slightly thickened at the corners. The gemmae are 2-celled, and angular in shape. Irregularly shaped, glistening oil bodies may usually be seen in the leaf cells of fresh material.

L. ventricosa is dioecious. The ovoid to barrel-shaped perianth extends well beyond the involucral bracts. It is contracted and toothed at its mouth, and is smooth except for a few furrows near the apex. The capsule is ovoid, and dark red-brown. On the male plants the antheridia are borne in the axils of concave, transversely inserted bracts which occur in several pairs at the ends of the shoots.

Ecology. Much commoner in the north and west than in south-east England, *L. ventricosa* will grow on a wide range of substrata, including sandy and peaty banks, rotten wood and rocks covered with humus. It ascends to the alpine zone. It is, however, confined—or nearly so—to places where the soil reaction is acid.

ADDITIONAL SPECIES

Several other species are by no means rare.

L. porphyroleuca (Nees) Schiffn. occurs on decaying wood and peat in mountain districts. It differs from *L. ventricosa* in being commonly tinged with red, in the leaf cells which are strongly thickened at the corners, and in the long teeth on the mouth of the perianth.

L. alpestris (Schleich.) Evans also comes near *L. ventricosa*, but differs in its brownish colour, smaller leaf cells (19–24 μ) and concave, shallowly bilobed leaves. It is chiefly an alpine plant in Britain.

L. excisa (Dicks.) Dum. is a small species (stems up to 10 mm. long), found on banks and old walls. It is like *L. ventricosa* but differs in its rather larger, very thin-walled cells, and in its paroecious 'inflorescence'. It is also rather commonly tinged with red, and bears purple-red gemmae.

L. incisa (Schrad.) Dum. has short, very thick stems 4–10 mm. long, the whole plant forming compact blue-green patches in very wet moorland habitats. It is very distinct in its coarsely toothed, bilobed leaves and fleshy, brittle stems.

20. **LEIOCOLEA TURBINATA** (Raddi) Buch (*Lophozia turbinata* (Raddi) Steph.)

This very small plant grows in delicate pale green patches on calcareous soil, occurring commonly, for instance, in chalk pits. The stems are very slender, and only 0·5–1 cm. in length. They are usually nearly prostrate, so that the patches as a whole, rather than the individual shoots, attract attention. The leaves are very obliquely inserted and, where close enough to overlap, do so succubously. There are no under-leaves.

The leaf is bilobed to about one-third of its length, the lobes being broadly rounded to moderately acute. Thus the present plant differs from most common species of *Cephalozia*, which have the leaf lobes distinctly acute. Also, the 'sinus' between the two leaf lobes is V-shaped, whereas in *Cephalozia* spp. it is generally U-shaped. In other respects *Leiocolea turbinata* somewhat resembles *Cephalozia* and, indeed, is often mistaken by beginners for a member of that genus or *Cladopodiella*. The very oblique leaf insertion heightens this resemblance.

Another feature of *Leiocolea turbinata* which suggests *Cephalozia* is the cell size. The cells here are about 40μ across, very large compared with those of most closely allied plants, and comparable with the large cells seen in some species of *Cephalozia* and *Cladopodiella*. The walls are uniformly thin, with no corner thickenings.

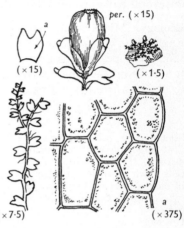

Leiocolea turbinata is dioecious. The antheridia are protected by bracts which occur in 4–8 pairs in the middle of the shoot. On female plants the perianths are conspicuous pear-shaped structures which extend well beyond the slightly enlarged involucral bracts that surround them at their bases. The perianth is smooth except near the apex, and is finely fringed at its contracted mouth.

Fig. 174. *Leiocolea turbinata*: *per.* perianth.

Ecology. It appears to be confined to calcareous rocks and soil, and prefers shade. Chalk pits are a common habitat in southern England, where I have found it associated with the moss *Dicranella varia*. Other habitats are moist rock ledges and grassy slopes in limestone country generally, and wet hollows in calcareous sand dunes.

ADDITIONAL SPECIES

Leiocolea muelleri (Nees) Joerg. (*Lophozia Muelleri* (Nees) Dum.), although rather small, is considerably larger than *Leiocolea turbinata*. It is a dull green to brownish plant of calcareous rock ledges on mountains. The bifid leaves are much more crowded than in *L. turbinata*, and lance-shaped toothed underleaves are present.

21. GYMNOCOLEA INFLATA (Huds.) Dum.

This species forms dark green to nearly black patches on wet moorland and heaths, where its prevailing dark or dingy colour will distinguish it from most other common hepatics. The stems are short (0·5–2 cm.) and vary from nearly prostrate to almost erect. At times they are much embedded in silty material and thus appear even shorter than they are. Some of the small, very dark forms of this plant might be taken for one of the small species of *Marsupella*; but *Cladopodiella fluitans* is the species which most resembles *Gymnocolea inflata*. The two ranks of leaves are rather obliquely inserted, especially near the bases of the shoots, where also they become widely spaced. Normally there are no underleaves.

The leaf is somewhat concave, and is bilobed to one-third of its length. The lobes are notably obtuse. A leaf of similar shape occurs in *Leiocolea turbinata*, but that plant is pale green in colour and occurs chiefly in calcareous places.

Under the microscope it will be seen that the cells are about 25μ across, much smaller than those of *Cladopodiella fluitans*. Their walls are somewhat thickened, but with scarcely any increase of thickening at the corners. Occasionally a few rudimentary, lance-shaped underleaves may be found.

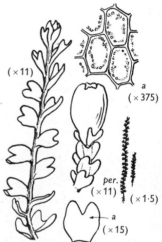

Gymnocolea inflata is dioecious and is rarely fertile. It has, however, the peculiar power of propagating itself by means of the sterile perianths. These are conspicuous, terminal, pear-shaped structures, which break away readily and subsequently become attached by rhizoids to give rise to new plants. The name *Gymnocolea* refers to the exposed position of the perianth, exserted well above the two spreading bracts.

Fig. 175. *Gymnocolea inflata*: *per*. perianth.

Ecology. It grows commonly on moist gravel, silt or sand on heaths and moors, always with an acid soil reaction. To that extent it overlaps in habitat requirements the two liverworts *Plectocolea crenulata* and *Nardia scalaris*; but *Gymnocolea inflata* occurs on peat more often than do either of these, and is common in wetter situations.

22. ORTHOCAULIS FLOERKII (Web. & Mohr) Buch (*Lophozia Floerkii* (Web. & Mohr) Schiffn.)

This is an upland species, occurring on slopes and rock ledges in mountain districts. It grows in tufts or wide patches, the stems commonly 3–5 cm. long; but it is variable, and some forms—especially starved (depauperate) ones—may prove confusing. The colour is sometimes brown or blackish, but is most often some shade of green.

The obliquely inserted, succubous leaves of *Orthocaulis floerkii* are typically 3-lobed, but 2-lobed and 4-lobed leaves may be found at times.

Finely divided, toothed underleaves are present. A leaf, carefully removed from the stem, will show one or more fine projections on its ventrally facing (postical) margin near the base. These are the 'cilia', which are lacking in *O. attenuatus* and *Barbilophozia barbata*; sometimes they are much reduced here. Gemmae are very rare indeed.

Under the microscope the leaf cells are seen to be rounded, about 25μ across, thin-walled, but with evident corner thickenings; the extent of these, however, is variable. An important character lies in the shape of the cells which form the cilia. These cells in *Orthocaulis floerkii* look nearly square in surface view. In certain rarer species they are long and narrow.

O. floerkii is dioecious and fertile material is very rare. Female plants are seen more commonly than male, and sterile female 'inflorescences' occur.

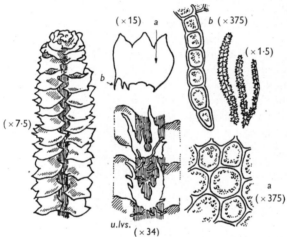

Fig. 176. *Orthocaulis floerkii*: *u.lvs.* small portion of stem in ventral view showing two underleaves.

Ecology. Absent from the south-east, this species is common in the north and west, where it occurs in many of the same habitats as *Lophozia ventricosa*. Moist boulders, crevices and rock ledges at high altitudes are favoured, especially where the soil reaction is at least mildly acid.

23. ORTHOCAULIS ATTENUATUS (Mart.) Evans (*Lophozia attenuata* (Mart.) Dum.)

O. attenuatus grows in loose green to brownish tufts on peaty banks and walls, chiefly in mountain districts. The slender stems attain 1·5–4 cm. in length, and a highly characteristic feature is provided by the still more slender branches which arise at the tips of the stems and bear closely appressed leaves. The obliquely inserted, succubous leaves of the ordinary leafy shoot are typical of the genus. Being mostly 3-lobed, they bear a close superficial resemblance to those of *O. floerkii*. In *O. attenuatus* rhizoids are usually numerous, but there are no under-leaves.

The leaf is lobed to about $\frac{1}{4}-\frac{1}{3}$ of the leaf length, much as in *O. floerkii*. The 2-lobed and 4-lobed leaves are most commonly met with on the specialized slender shoots. The leaves on these shoots are also frequently slightly malformed and bear gemmae at their tips.

The microscope will show the really diagnostic characters for the separation of this species from *O. floerkii*. These are (i) the absence of 'cilia' at the leaf base, and (ii) the relatively small cell size, the cells

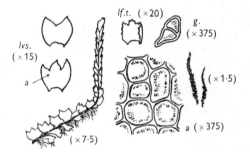

Fig. 177. *Orthocaulis attenuatus*: *lvs.* leaves from lower (normal) part of shoot, *lf.t.* leaf from tip of shoot, *g.* gemma.

here being only about $18-20\mu$ across. The cell walls are slightly thickened throughout, with corner thickenings developed to a variable extent. Minute underleaves occur occasionally. The gemmae are 2-celled, not unlike those of *Lophozia ventricosa*.

Orthocaulis attenuatus is dioecious. The perianth is nearly cylindrical, slightly plicate at the apex, with bristle-like processes ('cilia') at its mouth. The 3- to 5-lobed involucral bracts are larger than the leaves.

Ecology. Its habitats include rotten wood and steep sandy and rocky banks. It favours acid soils and demands some shade. It is a plant of low to moderately high altitudes.

RELATED SPECIES

Several species belonging to a number of allied genera are best mentioned at this point.

Isopaches bicrenatus (Schmid.) Buch (*Lophozia bicrenata* (Schmid.) Dum.) is a minute reddish yellow or red-brown plant, with stems not above 6 mm. long. It occurs in rather dry situations on banks and may be known by its colour and by the rounded-oval cavities of the cells. The latter character will distinguish it from certain forms of *Lophozia excisa* which it may resemble in the field. Clusters of orange-red gemmae are commonly present and visible (in the mass) with a lens.

Barbilophozia barbata (Schmid.) Loeske (*Lophozia barbata* (Schmid.) Dum.) is a well-marked species which occurs locally on rocky slopes and walls in the north and west. The stems are up to 5 cm. long and grow in loose tufts which are most often tinged with brownish yellow. It may be known at once by the four obtuse lobes of the leaves.

Tritomaria quinquedentata (Huds.) Buch (*Lophozia quinquedentata* (Huds.) Cogn.) is a rather robust mountain species which has the 3-lobed leaves of *Orthocaulis floerkii*, but as the leaves are remarkably asymmetrical, with lobes differing greatly in size, it is usually an easy plant to recognize.

Tritomaria exsectiformis (Breidl.) Schiffn. (*Sphenolobus exsectiformis* (Breidl.) Steph.) is a plant of peaty and sandy banks. The transversely inserted leaves are peculiar in form; each leaf is divided to the middle into two very unequal, pointed lobes, the larger lobe being notched at its apex. Bright red masses of gemmae are borne at the shoot tips and are conspicuous under the lens.

Sphenolobus minutus (Crantz.) Steph. resembles *Marsupella funckii* in its transversely inserted, concave, bifid leaves, but the leaves of *M. funckii* are more deeply lobed and the cells in that plant have conspicuous corner thickenings which are lacking here. *Sphenolobus minutus* grows on peaty banks at moderate altitudes.

Anastrepta orcadensis (Hook.) Schiffn., a plant of mainly western distribution, is frequent in the Scottish Highlands, where it may be found on steep grassy slopes among mountain rocks. In its elongated stems (4–7 cm.) and very obliquely inserted leaves it looks superficially like *Bazzania tricrenata*, but the leaves are succubous, bifid (not 3-lobed) at the apex, and each leaf has one margin markedly recurved.

24. JUNGERMANNIA CORDIFOLIA Hook. (*Aplozia cordifolia* (Hook.) Dum.)

Jungermannia cordifolia grows in mountain springs and on rocks in clear streams, and is marked by its tall stems (6–10 cm.), which are erect or nearly so, and form swollen spongy tufts of a dark green or purple-black colour. The long stems are weak and the leaves limp, becoming much shrivelled on drying. Rhizoids are scarce, especially towards the tips of the shoots. The rather close-set leaves are nearly transversely inserted, and are almost erect, overlapping one another considerably. As in all species of this genus, there are no underleaves.

The leaf is relatively long (1·5–2 mm.) and almost equally broad in its broadest part. It varies from heart-shaped to widely ovate in outline. At its base it is contracted into a very narrow insertion. In form it is closest to the leaf of *J. tristis*; in fact it is chiefly with large forms of that plant that *J. cordifolia* is likely to be confused. The common species of *Nardia* differ in their broader leaves and more rigid habit.

The leaf cells (about 30μ across) are uniformly thin-walled, with no corner thickenings. The cell walls are a characteristic dark brown.

Jungermannia tristis, a smaller plant, has minute, but recognizable thickenings at the cell corners, and colourless cell walls.

J. cordifolia is dioecious and is often sterile. At times, however, it is abundantly fertile, and then the long, narrow perianth is distinctive. It projects far beyond the bracts and is notable for its fusiform shape and smooth character, with a few indistinct folds only near its apex. The capsule is narrowly ovoid, and purplish black.

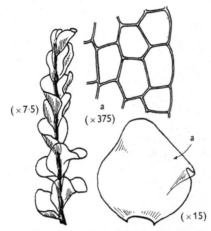

Fig. 178. *Jungermannia cordifolia*: only the terminal part of a shoot is shown.

Ecology. It is sometimes a member of the very characteristic bryophyte flora that develops around springs on mountains—a community that includes the mosses *Philonotis fontana*, *Bryum pseudotriquetrum* and others. Equally typically, it occurs on rocks in mountain streams, where an associated species is *Scapania undulata*.

ADDITIONAL SPECIES

Three further species of *Jungermannia* are by no means rare on wet rock surfaces in mountain districts.

J. sphaerocarpa Hook. (*Aplozia sphaerocarpa* (Hook.) Dum.) resembles *Plectocolea crenulata*, but is somewhat larger; moreover, its leaves are more widely spreading and lack the border of enlarged cells, whilst the cell walls as a whole have distinct corner thickenings.

Jungermannia tristis Nees (*Aplozia riparia* (Tayl.) Dum.) is a variable plant, in its larger states resembling *Jungermannia cordifolia*. It lacks the dark-coloured cell walls of *J. cordifolia*, however, and the whole plant is usually a characteristic dull green. In many districts it is the commoner plant of the two. The club-shaped or pear-shaped perianths are commonly present and afford a good character for identification.

J. pumila With. (*Aplozia pumila* (With.) Dum.) is an uncommon plant, mainly of lowland districts. It is marked by its small size and narrowly oval leaves. The distinctive, pointed perianths are usually present.

25. NARDIA SCALARIS (Schrad.) Gray (*Alicularia scalaris* (Schrad.) Corda)

Typically low-growing and nearly prostrate, the almost unbranched stems of this species are short (1–3 cm.) and would be very inconspicuous but for the fact that they form rather extensive patches of growth. These patches vary from deep or yellowish green to red-brown in colour and are a common feature of gravelly banks and ledges on moors and in the mountains. *Nardia scalaris* is also common on the acid heaths of southern England; indeed, it is our commonest entire-leaved liverwort.

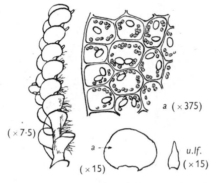

Fig. 179. *Nardia scalaris*: *u.lf.* underleaf; in the cells at *a* are seen large oil bodies and smaller chloroplasts.

A lens will show the characteristic form of the shoot, with its two ranks of somewhat concave leaves that tend to lie pressed together and become larger and more crowded nearer the tip. The very broad and rounded (orbicular) leaf has no trace of the apical indentation of *Marsupella*, its outline being broken only by the narrow, slightly decurrent insertion. Similar leaves are found in other genera (e.g. *Plectocolea, Jungermannia, Odontoschisma*), but a further mark of recognition is provided by the underleaves, which are narrowly lanceolate and project on the under surface of the stem, where long white rhizoids are also numerous. The underleaves are best seen near the tip of the stem.

The margin of the leaf is not marked by enlarged cells, as it is, for instance, in *Plectocolea crenulata*. The cells (25–35μ across) are notably thickened at the corners. Two to four glistening oval oil bodies are almost always present, at least in some of the cells.

Nardia scalaris is dioecious but the fertile state is fairly common. The perianth is pear-shaped, and is free from the bracts only at its apex. The capsule is broadly ovoid.

Ecology. This is a pronouncedly calcifuge species and is thus rare or absent in limestone districts. The substrata most favoured are sand and gravel. It is common alike on heath, moor and at high altitudes on mountains, where it always tends to grow on gritty detritus rather than on peat surfaces. In the south it sometimes grows on sandy woodland rides.

ADDITIONAL SPECIES

N. compressa (Hook.) Gray (*Alicularia compressa* (Hook.) Nees) is a robust species found in mountain streams. In habit, with its large tufts often strongly tinged with purple, it resembles *Jungermannia cordifolia* or a species of *Scapania*, but the leaf shape is near that of *Nardia scalaris* and the leaves of opposite sides tend to lie pressed together.

N. geoscyphus (De Not.) Lindb. (*Alicularia Geoscyphus* De Not.) is an uncommon plant of peaty banks. It is of interest in being one of the few British liverworts to show the 'marsupium'—a bulbous development at the base of the involucre which bears numerous rhizoids and assists in the nutrition of the developing sporophyte. *Nardia geoscyphus* has some of the leaves notched or bilobed, and the oil bodies are never smooth or shining as in *N. scalaris*.

26. PLECTOCOLEA CRENULATA (Sm.) Evans (*Aplozia crenulata* (Sm.) Dum.)

This species resembles *Nardia scalaris* and others in forming low, but typically rather wide, patches of growth on moist banks and bare soil. Red-brown to pale green in colour, the plant generally takes the form of a great number of short stems, each only a few millimetres to 1 cm. long, growing prostrate with ascending tips. With a lens the imbricated, fairly close-set leaves will be seen, those near the shoot tips larger than the rest. Their insertion is slightly oblique.

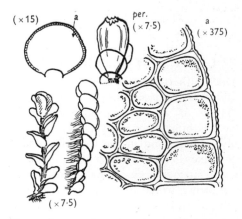

Fig. 180. *Plectocolea crenulata*: two views of the shoot are shown; above them is a single detached leaf; *per.* perianth.

The leaf is about 1 mm. long and almost equally broad. Its nearly circular outline is broken only by the narrow insertion. A leaf of similar shape occurs in *N. scalaris*, but that species may be known by the presence of underleaves.

The marginal row of leaf cells provides the best microscopic character. These (up to 30μ across) are, for the most part, fully twice as wide as the cells of the next row.* Since much of the rest of the leaf is composed of smaller cells, the enlarged marginal row is very conspicuous and gives the effect of a distinct border. The ordinary leaf cells are somewhat thick-walled, with only very slight additional corner thickenings; the walls of the marginal cells are considerably thickened.

Plectocolea crenulata is dioecious, and male plants are more slender than the female. In the latter the bracts are larger than the vegetative leaves and are in part united with the ovoid to pear-shaped perianth. This projects beyond them, however, and is marked by its reddish colour and 4 or 5 prominent longitudinal folds. The rounded teeth on the contracted mouth of the perianth have given the plant its specific name.

Ecology. In favouring sandy or gravelly soil this species resembles *Nardia scalaris* but is less markedly calcifuge. Like that species, it is a plant of footpaths, moist banks, and mountain detritus. Both these liverworts occur commonly with the moss *Oligotrichum hercynicum* as colonists of moist scree.

ADDITIONAL SPECIES

Plectocolea obovata (Nees) Mitt. (*Eucalyx obovatus* (Nees) Breidl.), which occurs on wet mountain rocks, resembles *Jungermannia sphaerocarpa* but may usually be known by its violet-coloured rhizoids.

Plectocolea hyalina (Lyell) Mitt. (*Eucalyx hyalinus* (Lyell) Breidl.) is mainly a plant of lowland banks. Its numerous rhizoids are reddish, and the whole plant, while bearing a general resemblance to *Jungermannia* spp., has a peculiar pale, glistening character of its own.

27. MARSUPELLA EMARGINATA (Ehrh.) Dum.

Two plants which hitherto have usually been regarded as distinct are here included under *M. emarginata*. These are the plant of that name and the much larger *M. aquatica* (Lindenb.) Schiffn., which is now regarded as a mere habitat form of *M. emarginata*.

On moderately high ground in the wet hilly districts of north and west Britain this species is often very plentiful and is likely to be among the first hepatics to attract notice. The typical form grows on moist ledges in greenish brown to red-brown patches, the stems almost erect, 1–4 cm. tall, unbranched or nearly so. The transversely inserted leaves

---

* Whilst this is true of typical forms it should be pointed out that small states occur in which the marginal cells of the leaf are scarcely at all enlarged.

are much larger than those of *Gymnomitrium crenulatum*, and being widely spreading, not appressed, they give the shoots an entirely different appearance. Aquatic forms (*Marsupella aquatica* in MacVicar's *Handbook*) have stems 3–10 cm. long and leaves more widely spreading. Species of *Lophozia* which resemble *Marsupella emarginata* in leaf shape are distinct in the oblique insertion of their leaves.

The leaf is nearly round in outline except for the apex which is indented, commonly to about one-fifth of the leaf length, the lobes thus formed being bluntly pointed. The antical margin of the leaf is recurved. There are no underleaves, and when a fragment of the shoot is examined intact the leaves appear notably trough-like in form. This is particularly true of aquatic states, in which moreover the leaves also tend to be more shallowly indented at the apex.

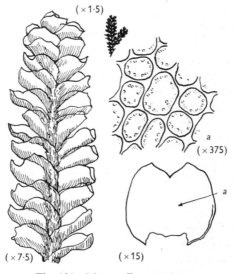

Fig. 181. *Marsupella emarginata.*

The leaf cells are about 20μ in diameter, and have cavities that appear nearly round in outline owing to the marked thickenings at the corners of the cells. These corner thickenings are, however, less pronounced in aquatic forms. In those species of *Lophozia* which approach *Marsupella emarginata* in leaf shape the cells are wider and less thickened at the corners.

This species is dioecious. The leaves are larger at the tips of fertile shoots, the largest and uppermost (bracts) being united with the perianth for half their length. The form of the capsule ranges from broadly ovoid to oblong-cylindrical.

Ecology. The typical form grows on siliceous rocks, gravelly banks and peaty ledges, both in lowland districts and on mountains. It is most abundant as a mountain plant, and according to MacVicar ascends to 3600 ft., but at the highest altitudes is confined to rock habitats. Aquatic forms may be found in and beside mountain streams and lakes, at times submerged to a considerable depth. *M. emarginata* is absent from much of southern England.

ADDITIONAL SPECIES

M. funckii (Web. & Mohr) Dum. is not uncommon on stony soil in hill districts, but it seldom occurs above 1000 ft. It resembles small states of *M. emarginata*, but the leaves are more deeply and acutely lobed, the antical margin is not reflexed and the leaf cells are only 14–18 μ in diameter.

28. GYMNOMITRIUM CRENULATUM Gottsche

G. crenulatum is perhaps the commonest of several British species of *Gymnomitrium* which form small, dense cushions on mountain rocks. It is, however, confined to west Britain. Red-brown or (more rarely) dull green in colour, the shoots of this species are very slender and look like thin wire to the naked eye. Each is commonly under 1 cm. long, although the cushions may be several centimetres across. Numerous much-branched stems compose each cushion. Even with a lens leaves are hard to distinguish, for they are very closely appressed to the stem, so that the shoots (when magnified) look like minute catkins. The transversely inserted leaves overlap one another in two close-set ranks. This plant is darker in colour than other species of *Gymnomitrium* which resemble it structurally. At times it is so black that at a casual glance it might be passed over for the moss *Andreaea*.

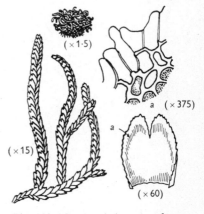

Fig. 182. *Gymnomitrium crenulatum*.

To examine such a plant it is best to scrape some leaves off the stems into a drop of water on a slide and examine the scrapings, or dissect off leaves with needles under a dissecting microscope. The leaf will then be seen to have an indented apex and a margin that is made rough (crenulate) by the projecting tips of cells. These marginal cells are elongated, conical and translucent, whereas the rest of the leaf is composed of shorter, rounded cells, which are thickened at the corners and

Plate XIII. *Ctenidium molluscum* (natural size)

Plate XIV. *Plagiochila asplenioides* (natural size)

have coloured contents. The leaf is only 0·3 mm. long and very concave. There are no underleaves.

Gymnomitrium crenulatum is dioecious. The capsule is spherical, and is raised on a short seta. The perianth, which surrounds the developing capsule in most leafy liverworts, is wanting here; its absence is a generic character. The nearly globose antheridia occur singly, on short stalks, in the axils of bracts which are broader and more concave than the vegetative leaves.

Ecology. It is found mainly on hard siliceous rocks. It is confined— at least in England and Wales—to altitudes above 500 ft. MacVicar states that it does not occur above 2500 ft. This species is frequent to common in some mountainous districts of the west but is absent elsewhere.

ADDITIONAL SPECIES

Four of the remaining six British species of *Gymnomitrium* are moderately common at high altitudes in some parts of the Scottish Highlands, but occur only very locally elsewhere. All are small plants, and all except *G. varians* grow chiefly on rock, rather than soil.

G. concinnatum (Lightf.) Corda has whitish and swollen shoots; the closely imbricated leaves are acutely bilobed, and the cells very thick-walled.

G. obtusum (Lindb.) Pears., another whitish green plant, resembles *G. crenulatum* in its densely imbricated leaves, but differs in its very obtuse leaf lobes and cell walls that are thin except at the corners. It is plentiful in parts of Wales, northern England and the Scottish Highlands, and in some districts is commoner than *G. crenulatum*.

In the two remaining species the leaves are only loosely imbricated so that the shoots do not have a 'plaited' appearance.

G. adustum Nees is a minute plant with concave leaves 0·3 mm. long except near the tips of the fertile shoots (where they are much larger); the 'sinus' between the leaf lobes is minute and rounded.

G. varians (Lindb.) Schiffn. is a slightly larger plant, with longer leaves (up to 0·5 mm.). The deeper, acute 'sinus' gives the leaf outline a resemblance to that of *Marsupella emarginata* in miniature. It grows on moist soil from 2800 ft. upwards and is confined to the Highlands of Scotland.

29. PLAGIOCHILA ASPLENIOIDES (L.) Dum. (Plate XIV)

P. asplenioides is one of the most widespread liverworts of moist shady places. Typical forms are notable for their robust habit (leafy shoot 3–7 mm. broad) and dull green colour. The stems are 3–10 cm. long, erect or ascending, with few branches. Below they are connected with a prostrate creeping rhizome-like stem. The large, obliquely inserted leaves overlap one another in the succubous manner. Underleaves are often absent, but may be represented by very minute, acutely pointed structures, especially near the tip of the stem.

The leaf is broadly rounded-ovate in shape, with no trace of the division into two lobes so common among liverworts. It has a nearly straight antical margin, which is distinctly decurrent, and a rounded postical margin. The latter usually bears numerous small but distinct teeth; but it should be noted that *P. asplenioides* is very variable in leaf margin, as in size and general habit. Forms occur in which only a few poorly developed teeth can be found on the margins, at least of the lower leaves.

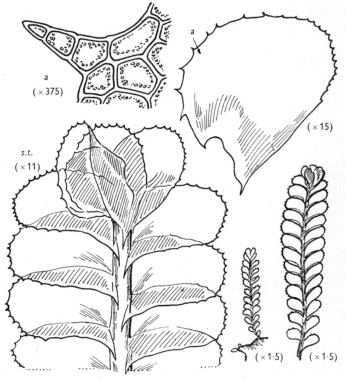

Fig. 183. *Plagiochila asplenioides*: on the right are shoots of two states of the plant, a rather small form and the var. *major*; *s.t.* tip of the var. *major* shoot.

The microscope will show each marginal tooth to be composed of 1–3 cells. The leaf cells are rounded-hexagonal, about 30μ across, with rather thin walls. Recognizable corner thickenings can usually be seen.

P. asplenioides is dioecious. Antheridia are protected by a number of pairs of bracts borne normally at the ends of branches; but subsequent proliferation of the shoot may result in a succession of antheridial regions separated by intervening lengths of vegetative shoot. The perianth is a conspicuously exserted structure, toothed at the mouth. The capsule is ovoid-cylindrical, purple brown in colour.

Ecology. The habitats favoured include loamy banks in woods, soil-covered boulders and walls. It probably prefers a calcareous or nearly neutral soil and is particularly abundant on shaded walls in limestone country. The var. *major* Nees perhaps represents a distinct ecotype; it is a very large pale green form (leafy shoot 5–8 mm. broad) which occurs on the ground in wooded valley bottoms and on slopes in relatively open mountain woods. A very small form grows in moist places in sand dunes and elsewhere.

30. **PLAGIOCHILA SPINULOSA** (Dicks.) Dum.

This is a plant of west Britain, and is rare elsewhere. Growing on banks or rock ledges in wet hill districts, it differs, in its typical state, from *P. asplenioides* in its elongated, scarcely branched stems (3–12 cm.) and rather smaller leaves which do not form so wide an angle with the stem as in the common plant. Its shoots often occur scattered amongst mosses, but conspicuous pure tufts are formed at times. In their succubous, rather crowded leaf arrangement this species and the last agree fairly closely. Variable underleaves are sometimes present. *P. spinulosa* is usually of a light yellowish green colour, very different from the dull mid-green of typical *P. asplenioides*.

The leaf in *P. spinulosa* is usually narrower in outline than in *P. asplenioides*, and the postical margin, instead of being evenly beset with small teeth, bears comparatively few, large and irregular, finely pointed teeth.

Each leaf tooth is composed of a number of cells at its wide base, whilst at its tip the cells are long and narrow. The cells in mid-leaf

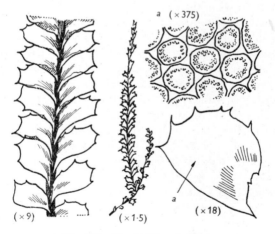

Fig. 184. *Plagiochila spinulosa.*

differ from those of *P. asplenioides* in their much more prominent corner thickenings. Also, the cell dimensions tend to be rather smaller (about 25 μ across cell). This species, however, like *P. asplenioides*, is notably variable in many of its characters, especially in size and habit. It is sometimes difficult to distinguish it from another western species, *P. punctata*, but that plant may be known usually by its more nearly oval leaves and the still larger corner thickenings of the cells.

P. spinulosa is dioecious. The perianth is pitcher-shaped, its wide mouth coarsely and irregularly toothed.

Ecology. Perhaps most typically a plant of moist, soil-capped rock ledges and steep rocky banks in sheltered and shaded situations, it will also grow on the bases of trees. It is almost confined to the west, where it is found from sea-level up to moderate altitudes. *P. spinulosa* is a calcifuge, *Dicranum majus* and *Rhytidiadelphus loreus* being two of the mosses that grow with it both in mountain woods and on stabilized 'block scree'.

ADDITIONAL SPECIES

Plagiochila punctata Tayl. is rather common in some of the western counties of Scotland and Ireland, but is rare elsewhere. It is closely related to *P. spinulosa*, but differs normally in being smaller and more freely branched, whilst the leaves are more nearly oval in shape, have longer marginal teeth and cells with much larger corner thickenings. Also the plant is less yellowish in colour.

31. DIPLOPHYLLUM ALBICANS (L.) Dum.

This is in general the commonest British liverwort of non-calcareous soils and is certain to be among the first species met with by the beginner on heathy banks and rocky ledges. It is also one of the most easily recognized. It grows in tufts or wide patches, with numerous ascending or upright leafy shoots arising from a creeping primary stem. These shoots vary in length from 1 to 5 cm., and their colour ranges from light green to reddish brown; sometimes they have a whitish appearance. *D. albicans*, however, may always be known by its very regularly 2-ranked leaves of peculiar form and cell structure.

Each of the close-set leaves is divided to near its base into two lobes, with the smaller (antical) lobe bent back so as to lie in part upon the larger (postical) one. This is apt to give the impression, at first sight, that there are four rows of leaves. Each leaf lobe is roughly oblong in outline, narrowing slightly at the irregularly toothed tip. There are no underleaves.

The best diagnostic character for *D. albicans* lies in the long, narrow, clear cells which form a band down the middle of each leaf lobe. This narrow 'vitta' may be easily seen with a lens and will identify this species in the field. The remaining leaf cells are short and rounded.

Unicellular gemmae quite commonly form yellowish green clusters on the margins of some of the upper leaves.

D. albicans is dioecious. Antheridia are formed in pairs in the axils of slightly swollen male bracts which occur in 4–6 pairs on specialized shoots. They may be terminal or in the middle of the shoot. The broadly obovoid perianth extends far beyond the involucral bracts. It is furrowed towards its apex, and has a narrow, toothed mouth. Perianths and capsules are quite common, the latter ripening in late spring.

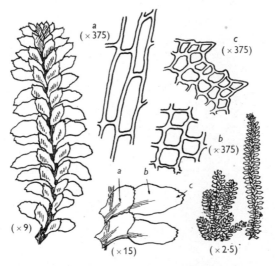

Fig. 185. *Diplophyllum albicans*: bottom centre, two leaves and a fragment of stem detached from shoot on left; the smaller antical and larger postical lobes are seen, each with a vitta.

Ecology. *D. albicans* is strictly calcifuge, but will grow on very various substrata, including many types of soil, peat, sand and rock. MacVicar states that it occurs on trees, but it is certainly rare on wood, dead or living. It ascends to the highest altitudes, and indeed is as typical a member of the mountain-ledge community as it is of the lowland acid bank flora.

32. SCAPANIA NEMOROSA (L.) Dum.

S. nemorosa is a rather common plant of shaded banks, where it grows in loose tufts of a green or brownish colour. The numerous ascending stems, which spring from a creeping primary stem, are 1–6 cm. long, and the habit of the plant is more slender than in most forms of the very common related species *S. undulata*.

The form of the two-lobed leaf is a constant feature of the genus *Scapania*, the smaller (antical) lobe being folded back over the larger (postical) one, much as in *Diplophyllum*; but each lobe is proportionately broader than in that genus. Careful attention to detail is necessary to distinguish the species. In *Scapania nemorosa* the diagnostic characters are (i) the smaller lobe less than half the size of the larger one and pointed, and (ii) the margins of both lobes, but especially the

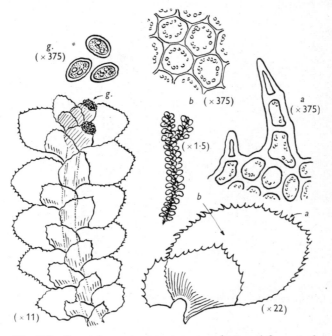

Fig. 186. *Scapania nemorosa*: *g*. gemmae; bottom left, part of gemmiferous leafy shoot in antical view.

larger one, bearing regular projections termed 'cilia'. The figure shows that these, which are 2–3 cells long, are quite different from the short teeth that are borne on the leaf margins in some forms of *S. undulata*, and in *S. gracilis*.

The corner thickenings of the small, rounded-polygonal leaf cells vary from poorly defined to quite distinct, but they are usually less pronounced than in *S. gracilis*.

Gemmae occur at times in clusters on the tips of the stem and upper leaves. They are always unicellular, whilst in *S. gracilis* they are 1- or 2-celled. The clusters as a whole stand out dark brown against the green of the leaves.

S. nemorosa is dioecious. The tubular perianth is fringed at its mouth with thread-like 'cilia'.

Ecology. It grows on humus, on the ground in woods, and less commonly on decaying logs. It demands shade and moisture, and favours acid soils.

33. SCAPANIA UNDULATA (L.) Dum.

Two plants, which in their extreme forms look very different, but are in fact bridged by intermediates, are now included under *S. undulata*. Hitherto these have generally been known as *S. dentata* Dum. and

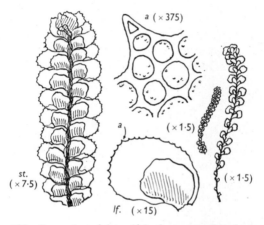

Fig. 187. *Scapania undulata*: *lf.* leaf, *st.* part of leafy shoot; the shoots on the right illustrate something of the range of variation within the species.

S. undulata (L.) Dum. respectively, the former reddish purple, with toothed leaf lobes, the latter green, with entire leaf lobes. All forms, however, of *S. undulata* in the broad sense are alike in being robust plants of swiftly flowing mountain streams, boggy ground and wet rock ledges. Indeed, this is one of the commonest hepatics in such places, where it makes splashes of vivid green or purplish crimson that catch the eye. The leafy shoots are erect or ascending, scarcely branched. On ledges they are commonly only 2–3 cm. long, but in streams and springs may reach 12 cm.

The bilobed, folded leaf has the smaller lobe rather more than half the size of the larger one. Both are broadly rounded in outline, the margins entire or toothed. Each tooth is composed of only one or two cells. The hind margin of the larger (postical) lobe is narrowly recurved.

The leaf cells lack obvious corner thickenings, but vary much in general wall thickness; many of the larger green forms have relatively thin-walled cells. The cell width varies between 15μ and 30μ. Pale green 1-celled or 2-celled gemmae may be found on the tips or margins of the upper leaves.

S. undulata is dioecious. The perianth is broadly tubular, compressed, with a wide, toothed mouth. The male plants bear the globose to oval antheridia in the axils of bilobed bracts. These bracts occur in 3–5 pairs at the tips of stems or branches and are marked by swollen bases and nearly equal lobes.

Ecology. It is a plant of several distinct habitats. Most typically and in its most robust states it grows in fast-flowing streams, and it is very widespread in this habitat in the north and west. It is common in and about mountain springs, where associated species in the spongy bryophyte mat are the mosses *Philonotis fontana* and *Bryum pseudotriquetrum*. It also occurs on wet mountain ledges. *Scapania undulata* is found more sparingly in the south-east of Britain, in and by small streams in the shade of woods.

34. **SCAPANIA GRACILIS** (Lindb.) Kaal.

S. gracilis is a plant of drier, more exposed habitats than either *S. nemorosa* or *S. undulata*. It forms compact, moderately robust tufts, with the upright leafy stems 3–12 cm. in length. Mature plants are of a characteristic dull yellowish brown or brownish green colour. A large *Scapania* of this colour, on an exposed rocky bank or wall-top, is likely to be this species. Smaller forms, occasionally met with in calcareous districts, may be difficult to separate from the rarer *S. aspera*.

In the bilobed leaf the smaller (antical) lobe is two-thirds of the size of the larger one. A good character lies in the coarse teeth which occur almost always near the base of the antical lobe. This lobe is also distinct in two further respects: (i) it extends across the full width of the stem; and (ii) it stands out horizontally with its outer part reflexed, thus differing from most other British species of *Scapania* in

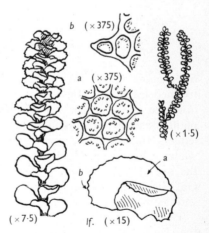

Fig. 188. *Scapania gracilis*: *lf.* single detached leaf showing smaller antical and larger postical lobe.

which this lobe lies flat upon the larger (postical) one. The postical lobe has the obtusely rounded outline, reflexed hind margin, and shallow distant teeth of some forms of *S. undulata*.

A good microscopic character is furnished by the well-defined corner thickenings of the leaf cells. This contrasts with *S. undulata* where such thickenings are practically wanting. Cells at the leaf margin will show the nearly smooth cuticle, whereas in *S. aspera* the cuticle is rough with coarse papillae. Gemmae occur in clusters at the tips of the upper leaves. They are 1- to 2-celled, and pale green in colour.

S. gracilis is dioecious, but perianths are quite common. The bracts are slightly larger than the vegetative leaves; the perianth is sharply toothed at its mouth.

Ecology. Principally a calcifuge species, it grows on rock and peat surfaces, and on the bases of trees. Perhaps its most typical habitats are overgrown stone walls and rocky banks, in open situations. Often common in the open type of 'highland' wood, it is abundant near the west coast, in hill country, but is rare or absent in most of southern, midland and eastern England.

ADDITIONAL SPECIES

Of the numerous remaining species of this genus, the following six are by no means rare.

S. compacta (Roth) Dum. is a short, rigid plant with nearly equal leaf lobes. It grows on rather dry banks and boulders and may be distinguished from *S. subalpina* by its paroecious 'inflorescence' and usually brownish green colour.

S. subalpina (Nees) Dum., in which the leaf lobes are again nearly equal, is usually a taller plant (stems 3 cm. or more, against 1–2 cm. in *S. compacta*), and is dioecious. Its habitat is quite different, for it grows chiefly on moist detritus by mountain streams.

S. aspera Bernet resembles *S. gracilis* in size and toothed leaf margin, but lacks the recurved antical lobe of that species. It is also distinct in its cuticle which is coarsely papillose; it is found on calcareous rock ledges, chalk and limestone grassland, and sand dunes.

S. aequiloba (Schwaegr.) Dum. resembles *S. aspera* in form and in its rough cuticle; moreover it is also calcicolous and grows in similar habitats. However, it is usually a more slender plant (stems 3–4 cm., against 3–10 cm. in *S. aspera*) and it has the antical leaf lobe more pointed and usually reflexed.

S. irrigua (Nees) Dum., a plant of moist woodland rides and marshy ground, may be known from green entire-leaved forms of *S. undulata* by the presence of distinct corner thickenings to the cells and the fact that rhizoids extend along the whole length of the stem except the young apical region; also by the apiculate leaves and smaller size of the plant.

S. curta (Mart.) Dum. resembles small forms of *S. irrigua* (stems 1–2 cm., against 2–5 cm. in *S. irrigua*) but has the leaf lobes ovate rather than heart-shaped. It grows on moist banks (on loam, peat or sand) and in woodland rides.

35. PORELLA PLATYPHYLLA (L.) Lindb. (*Madotheca platyphylla* (L.) Dum.)

This is one of the most conspicuous and abundant hepatics on shaded banks and about tree bases in calcareous districts. A robust plant, it grows in fairly compact, flat, dark or (more rarely) yellowish green patches. The stems are twice or thrice pinnately branched and 3–8 cm. in length. The closely overlapping (imbricate) leaves are arranged incubously. The underside of the shoot gives the impression of five

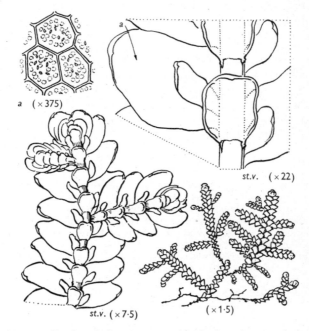

a (×375)

st.v. (×22)

st.v. (×7·5)   (×1·5)

Fig. 189. *Porella platyphylla*: *st.v.* part of leafy shoot in ventral view; the plant at bottom right is seen in dorsal view and in dry condition.

ranks of leaves, but this is because each leaf in the two lateral ranks is deeply divided into two very different looking segments, whilst there is an additional rank of large and conspicuous underleaves. The only other robust liverworts to show this apparently 5-ranked leaf arrangement are species of *Frullania*, which are, however, distinct in their brown colour, and in the details of their leaf structure.

The larger (antical) lobe of the leaf is broadly ovate, with rounded apex; the much smaller (postical) lobe (which looks like a separate leaf) is narrower in form and more acute at the apex. The underleaves are broadly rounded, with margins narrowly recurved, and entire except for an occasional tooth at the base. In this *Porella platyphylla* differs from *P. laevigata*, in which the postical leaf lobes and underleaves have toothed margins; *P. cordeana* is distinct in that both postical lobes and underleaves are strongly decurrent and form irregularly toothed wings extending down the stem. The cells, with minute corner thickenings, are without special diagnostic value.

P. platyphylla is dioecious. The perianth, which arises at the tip of a short lateral branch, is toothed at its mouth. The antheridia occur singly in the axils of bracts. In these paired, closely overlapping bracts the lobes are more nearly equal in size, and the male branches (like short broad catkins in miniature) are easily recognized.

Ecology. It is mainly a plant of chalk and limestone districts, where it will grow on rock, tree roots and soil. It demands some shade and is sometimes the chief hepatic on the ground in beech woods on chalk. On banks and the bases of stone walls a commonly associated species is the moss *Anomodon viticulosus*.

ADDITIONAL SPECIES

Porella laevigata (Schrad.) Lindb. (*Madotheca laevigata* (Schrad.) Dum.), which occurs locally in the same habitats as *Porella platyphylla*, is known at once by its peppery taste and, structurally, by its toothed underleaves and cells with large corner thickenings. It is perhaps more a mountain plant than *P. platyphylla*.

P. cordeana (Hüb.) Evans (*Madotheca Cordeana* (Hüb.) Dum.), which usually grows on trees and stones by water, has a small, acute and frequently twisted postical leaf lobe and widely spaced underleaves. As mentioned under *Porella platyphylla*, *P. cordeana* also differs from that species in that the underleaves and the postical lobes of the lateral leaves have toothed decurrent wings. It is not a common plant.

36. RADULA COMPLANATA (L.) Dum.

This is the only common British species of a genus remarkable for the fact that the rhizoids are borne on the smaller (postical) lobes of the leaves, and not directly upon the stem. *R. complanata* occurs on tree trunks and less frequently on rocks, forming low patches of a pale yellowish green colour. The stems are 1–5 cm. in length and irregularly branched. The close-set leaves differ from those of *Scapania* in their incubous arrangement and in the fact that the smaller, infolded lobe is here the postical one, i.e. it is on the side away from the observer when the prostrate leafy shoot is viewed from above. This is the reverse of the condition in *Scapania* and *Diplophyllum*, but agrees with that of *Porella*, *Lejeunea* and *Frullania*. *Radula*, however, differs from all these genera in its lack of underleaves.

The smaller, postical leaf lobe in *R. complanata* is less than half the size of the broadly rounded antical lobe. The leaf cells are rounded-hexagonal, about 20μ wide, thin-walled, with minute corner thickenings.

The cell contents are dense and commonly include 1–3 oil bodies. Disc-shaped, multicellular gemmae sometimes occur on the margins of the leaves.

A feature which distinguishes *R. complanata* from the other British species of *Radula* (all of which are rare or local) is the paroecious 'inflorescence', the antheridia occurring in the axils of swollen bracts on special lateral shoots formed just below the perianths. This species is very commonly fertile and the perianths are then conspicuous, projecting far beyond the involucral bracts, and notably compressed and wide-mouthed.

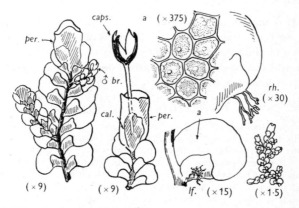

Fig. 190. *Radula complanata*: portions of two shoots are shown, both in postical view (smaller postical lobe of leaf uppermost); *cal.* calyptra, *caps.* capsule (old), *per.* perianth, ♂ *br.* male branch, *lf.* single leaf with rhizoids attached to postical lobe, *rh.* part of the same leaf enlarged to show rhizoids.

Ecology. Tree-trunks are its chief habitat. On rock surfaces, where it occurs less frequently, it demands some shade. It is most plentiful in districts of high rainfall.

ADDITIONAL SPECIES

Two further species of *Radula* are mainly western in distribution.

R. lindbergiana Gottsche forms green patches, mainly on wet rocks, and differs from *R. complanata* in being dioecious and thus lacking the characteristically swollen subinvolucral bracts of that species.

R. aquilegia Tayl., which occurs on rocks or trees, is quite distinct in its dark reddish brown or yellowish brown colour and in its leaf structure. Each leaf has the postical lobe strongly inflated at its base, the upper margin lying flat upon the larger antical lobe. It is dioecious.

37. PLEUROZIA PURPUREA (Lightf.) Lindb.

This is one of the most easily recognized British hepatics, and is a very distinctive plant with its robust, more or less erect, deep purplish red or reddish yellow leafy shoots. On parts of the wet moorland of the West Highlands of Scotland, where it abounds, its highly coloured tufts

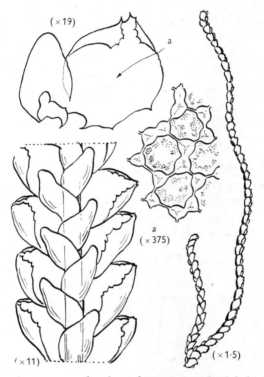

Fig. 191. *Pleurozia purpurea*: the shoot fragment on the left is seen in ventral (postical) view; the single detached leaf above shows the larger antical and the smaller, hood-like, postical lobe.

are indeed a striking feature of the vegetation. The leafy shoots, which arise from a creeping primary stem system, are 4–14 cm. in length, and little branched. Each stem bears two ranks of closely overlapping leaves of unusual and characteristic structure. The colour and size of this plant may recall some forms of *Scapania undulata*, but that plant lacks the rigid habit of *Pleurozia*, and in leaf structure is totally different.

Each leaf is made up of two parts, a large, somewhat concave antical lobe, which is toothed at its shallowly cleft apex, and a much smaller, deeply concave and hooded postical lobe, or water-holding sac. This sac possibly also serves for trapping small aquatic animals and has a single small opening at its base which allows the entry of water but which, through a valve mechanism, prevents its escape.

The cell structure of *P. purpurea* is remarkable for the great size of the corner thickenings, which frequently run together to make areas of thickening almost as large as the cavities of the cells.

This species is monoecious, with short male branches which arise near the female. Only incompletely developed perianths have been found. The sporophyte is unknown.

Ecology. Its chief habitat is wet peaty moorland in the west of Scotland and Ireland. There it often occurs in pure tufts, or as scattered stems among heather. It also grows on moist rock ledges, ascending to the summits of the hills. Then it is commonly associated with a number of other liverworts, including *Mylia taylori*, *Diplophyllum albicans* and the rarer *Bazzania tricrenata* and *Anastrepta orcadensis*.

38. FRULLANIA TAMARISCI (L.) Dum.

This is usually a robust and conspicuous hepatic and is likely to be among the first species to attract the attention of the beginner. It is especially plentiful in the west and north, where it forms extensive mats on boulders and wall-tops, or creeps over the surface of the stems

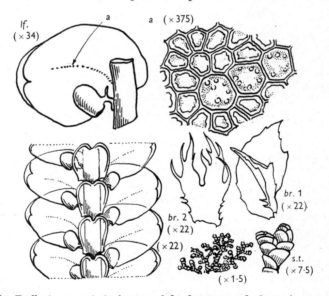

Fig. 192. *Frullania tamarisci*: bottom left, fragment of shoot in ventral view showing underleaves; *s.t.* shoot tip, *lf.* single detached leaf showing antical and (much smaller) postical lobe; *br.*1, bract; *br.*2, bracteole from female 'inflorescence'.

of moorland shrubs such as gorse and heather. In the southern counties it is much rarer. *F. tamarisci* is notable for its mat of richly (bipinnately) branched leafy stems and a colour that ranges from reddish olive to deep purple-brown. The leafy shoots are often glossy with a sheen like burnished copper. *F. tamarisci* can be readily distinguished, even in the field, from all other hepatics except certain relatively rare species of the same genus.

The two lateral rows of leaves are close-set on the stems and each leaf is completely bipartite to form a large, expanded antical lobe and a small helmet-shaped postical lobe. The pointed tip of the wide antical lobe is usually decurved; the pitcher-like postical lobe bears a short projection (the stylus) near its attachment. A row of close-set underleaves is present, each underleaf being shallowly bilobed and having a narrowly recurved margin.

A good microscopic character is the line of enlarged cells that traverses the antical leaf lobe from its base to near the apex. In size and content they are distinct from the surrounding cells and form a conspicuous 'stripe'. Sometimes these enlarged cells form a broken line.

F. tamarisci is dioecious. Bracts and bracteole are irregularly and sharply toothed, the bracteole with longer, finer teeth than in the much rarer related species. The perianth is 3-angled and free from tubercles; the capsule is globose. The male plant is more slender, and produces antheridia in the axils of bracts which occur in 3–4 pairs on short lateral branches.

Ecology. As compared with *F. dilatata*, it grows more often on rocks or on the ground, and less frequently on trees, though in many western districts it is abundant in all these habitats. It will often colonize relatively exposed boulders and wall-tops, where lichens predominate, and associated bryophytes include the mosses *Rhacomitrium fasciculare* and *R. lanuginosum*, siliceous rock being favoured.

39. FRULLANIA DILATATA (L.) Dum.

This species is smaller than *F. tamarisci*, and, unlike that plant, is commoner on trees than on rocks. It lacks the glossiness of *F. tamarisci* and forms flat patches closely appressed to the bark, that vary in colour from dull green to reddish brown. It is common in fairly sheltered places throughout Britain. The stems are pinnately or bipinnately branched.

A lens will show the characteristic *Frullania* organization of the shoots—the two lateral ranks of leaves and the underleaves, each lateral leaf being divided into two distinct parts—the expanded antical lobe of rounded outline and the postical lobe modified to form a helmet-shaped structure. The helmets here are nearly as large as, or even larger than, the underleaves, and about half the breadth of the expanded antical lobes. In no other British species of *Frullania* are they proportionately so large. Each underleaf is divided for one-third of its length into two acute lobes. Each lobe usually bears one small tooth on its margin, but the margin is not recurved as it is in *F. tamarisci*.

The best microscopic character is the absence of the enlarged cells which, in *F. tamarisci*, form a 'stripe' on the expanded antical leaf lobe, and in the rarer *F. fragilifolia* occur scattered over the surface of the

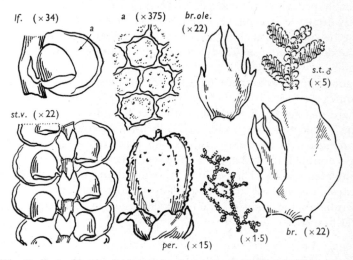

Fig. 193. *Frullania dilatata*: *br.* bract; *br.ole.* bracteole; *lf.* single leaf showing larger antical and smaller postical lobe; *per.* perianth; *st.v.* part of shoot in ventral view; *s.t.* ♂, shoot tip of male plant showing three antheridial branches.

lobe. Rounded multicellular gemmae occur at times on the surfaces and margins of the leaves.

F. dilatata is dioecious, but perianths are common. Each is a pear-shaped structure and bears numerous prominent tubercles on its surface. In all the other British species of *Frullania* the perianth is smooth, so that even *F. germana*, which resembles this species in the absence of enlarged cells in the leaf, is distinguished at once on perianth characters. The bracts and bracteole are irregularly lobed, but not toothed as they are in *F. tamarisci*. The male plants are less elaborately branched, and antheridia occur in the axils of bracts which are borne in 6–20 pairs on short lateral branches.

Ecology. It will tolerate much drier conditions than *F. tamarisci* and is much commoner than that species in the south and east. It grows chiefly on trees, elm, ash and elder all being favoured. Often it is the principal liverwort of a community in which the mosses *Zygodon viridissimus* and *Cryphaea heteromalla* occur. It is not rare on rocks, but never occurs on the ground (as *Frullania tamarisci* sometimes does).

ADDITIONAL SPECIES

F. fragilifolia Tayl. is a small reddish plant of western distribution. The leaves are attached by a narrow insertion and so break off readily when the plant is handled. The species is further known by its wedge-shaped underleaves and by the irregular distribution of enlarged cells in the rounded antical leaf lobe.

F. germana Tayl. resembles *F. tamarisci*, but differs in its usually duller colour and in the absence of enlarged cells in the leaf. The segments of the bracteole have entire margins, and this affords the surest distinction. *F. germana* is rare except in west Ireland, the West Highlands and the Hebrides.

RELATED SPECIES

Jubula hutchinsiae (Hook.) Dum. is a rather rare plant of wet, shaded rocks in the west. It is of robust habit and a characteristic dark green colour. It resembles *Frullania* spp. in its pinnate branching and helmet-like postical leaf lobes; but each helmet is prolonged into an evident beak and the larger, antical leaf lobe is coarsely toothed. Thus, under the microscope, it can be confused with no other British liverwort.

40. LEJEUNEA CAVIFOLIA (Ehrh.) Lindb.

L. cavifolia is perhaps the most frequent British species of the genus. It forms fairly extensive pale or rather vivid green patches in moist, shaded places. Although the irregularly branched stems are usually less

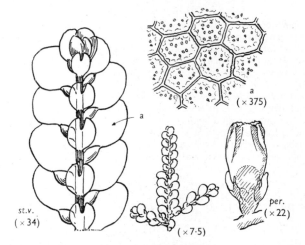

Fig. 194. *Lejeunea cavifolia*: *st.v.* part of leafy shoot in ventral view, *per.* perianth with bracts etc.; numerous small oil bodies are visible in each cell at *a*.

than 2 cm. long, the low mats which they form are often fairly conspicuous owing to their light green colour. *L. cavifolia* can be confused only with *L. lamacerina* and certain rarer species of the genus.

The leaves are very close-set, and incubous. Each leaf consists of a large, ovate-rounded antical lobe and a minute postical lobe of swollen sac-like form. There is a series of rounded underleaves, each cleft to about the middle. It will be noted that each underleaf is much larger than the sac-like postical lobe of a lateral leaf, a mark of distinction from *L. patens*, where the sac-like lobes equal or exceed the underleaves in size.

The leaf cells are rounded-hexagonal, thin-walled with minute corner thickenings, and abundantly filled with chloroplasts. They measure 20–30μ across, being somewhat larger than in *L. patens* where they are usually only 15–20μ. Each cell commonly contains, when fresh, numerous simple, glistening oil bodies.

This species is monoecious. The perianth is conspicuous, extending for half its length beyond the bracts. It is ovoid, becoming sharply 5-angled near its apex. The capsule is globose.

Ecology. It grows on moist rocks and stones; also on soil-capped ledges and, more rarely, on earthy banks and on trees. On the whole calcicole, it appears to tolerate some range of soil reaction. *L. cavifolia* is rare or absent in many south-eastern counties, and even in the west is not everywhere the commonest species of the genus.

ADDITIONAL SPECIES

L. lamacerina Steph. (*L. planiuscula* (Lindb.) Buch) is as common as *L. cavifolia* in the west (often commoner), growing chiefly on rocks. It is often lighter in colour and the antical lobe of the leaf is rounder than in *L. cavifolia*. The underleaves are relatively small (not more than one and a half times as large as the postical lobes of the lateral leaves), and are more distantly placed on the stems than are those of *L. cavifolia*.

L. patens Lindb., which grows chiefly on wet rocks, is most plentiful in the west and north of Britain. It is distinct from *L. lamacerina* in the much larger postical lobe of the leaf (about half as big as the antical lobe and twice as big as the underleaves). The colour is very light, often whitish. It is doubtful, however, whether colour in itself is a very valuable character in this difficult genus. Leaf proportions are more reliable, but it is essential that these should be examined on mature main stems.

41. MICROLEJEUNEA ULICINA (Tayl.) Evans

Although locally common, at least in the south and west, this plant is so small as to be readily overlooked. It forms irregular dull or yellowish green patches on the bark of trees, the sparingly branched stems being only 4–8 mm. in length and the leaves almost too small to be distinguished with the naked eye. The leaf length here is only one-third of that of *Lejeunea cavifolia*. Without a lens the patches of growth look like exceedingly fine wefts of green cottony material, the leafy stems appearing so thread-like that they might almost be mistaken for algal filaments.

The leaves are somewhat distantly spaced, and each leaf consists of a large rounded antical lobe and a slightly smaller, inflated postical lobe. Near the tip of this inflated lobe will be seen a small tooth composed of a single cell, and, beside it, an even smaller papilla. The presence of a series of widely spaced, deeply 2-lobed underleaves at once distinguishes this plant from the rarer species of the genus *Cololejeunea*, which it resembles in size and in some other respects.

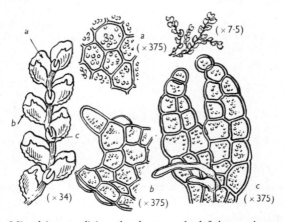

Fig. 195. *Microlejeunea ulicina*: the shoot on the left is seen in ventral view; *c* shows an underleaf with two rhizoids emerging at the point where it is attached to the stem.

The leaf cells are rounded-hexagonal, 15–20μ across, with thin walls and minute corner thickenings. Each cell normally appears much swollen, so that the keel of the folded leaf presents a series of convexities.

Microlejeunea ulicina is dioecious, and the 5-angled pear-shaped perianths are rather rare. The involucral bracts are much larger than the vegetative leaves. In the bracts of the male branches the two lobes are nearly equal in size.

Ecology. Its normal habitat is the bark of trees and shrubs. The specific name *ulicina* suggests a preference for growing on gorse (*Ulex* spp.), but *Microlejeunea ulicina* may be found in fact on a fairly wide range of (chiefly smooth-barked) trees. Only very rarely does it occur on rock. It demands a moist, sheltered situation.

RELATED SPECIES

Cololejeunea calcarea (Lib.) Schiffn., although rare, may be mentioned as an example of a genus of minute plants of which four species occur in Britain. The genus is marked by the absence of underleaves and by the inflated postical lobes of the lateral leaves. *C. calcarea* forms very small green patches on moist, shaded calcareous rock surfaces. The margin of the relatively small postical lobe is entire.

Colura calyptrifolia (Hook.) Dum. is a rare plant of western distribution in Britain. It has stems only 2–4 mm. long and forms minute pale green patches, which stand out against the dark surfaces of the moist rocks or *Frullania* plants on which they grow. In structure it is remarkable, both in its lateral leaves with their beaked and hooded antical lobes and small incurved postical lobes, and in having two bifid underleaves (instead of only one) to each pair of lateral leaves.

Marchesinia mackaii (Hook.) Gray is a dark green plant found on shaded rocks (especially limestone) in west Britain. In size and structure it somewhat resembles *Radula complanata* or a species of *Frullania*; but it is distinguished from the former by the presence of underleaves and is often almost black in colour. The underleaves are almost circular in outline, not bilobed as in *Frullania*, and the small postical leaf lobes are not helmet-shaped as in that genus.

METZGERINEAE

(JUNGERMANNIALES ANACROGYNAE)

42. FOSSOMBRONIA PUSILLA (L.) Dum.

Fossombronia is the only common genus of the Jungermanniales Anacrogynae in which well-defined leaves occur. Despite the foliose character of the plant, however, *F. pusilla* is not very likely to be confused with any of the 'leafy liverworts' which compose the great group of the Jungermanniales Acrogynae. It is a small plant, with creeping stems only about 1 cm. long. It forms characteristic pale yellowish green patches on banks or along woodland rides, and the form of the leaf is scarcely like that of any other genus. It is a less clearly defined structure than that of most leafy liverworts, the leaves on a single stem varying considerably in shape; they are very obliquely inserted and somewhat crowded, especially near the stem apex. The breadth of the leaf usually exceeds its length; the antical margin is decurrent, and the whole leaf has an irregularly wavy, lobed outline. The underside of the stem bears numerous purple rhizoids.

The cells of the leaf are large and thin-walled, with no obvious corner thickenings. They measure approximately $60 \times 40\mu$, and are densely filled with chloroplasts. The leaf is two or more cells thick near its base. Most of the above characters are true of the genus *Fossombronia* as a whole, and for the determination of the species spores must be examined. *F. pusilla* is paroecious, the archegonia occurring on the upper surface of the stem, chiefly near the apex, the antheridia being further back, and conspicuous by their bright orange colour. The spore of *F. pusilla* is about 40μ in diameter, and its surface is marked by a number of nearly parallel wings of tissue (lamellae) which join in some cases to form something of a network as seen in surface view, and project on the margin of the spore as 16–26 apparent spines. No other species of *Fossombronia* with violet rhizoids has spores precisely like this. In addition to the delicate calyptra there is an outer bell-shaped pseudoperianth surrounding the base of the short seta.

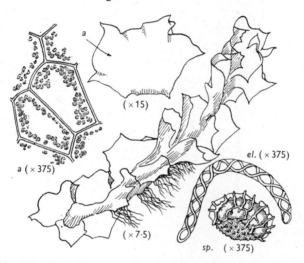

Fig. 196. *Fossombronia pusilla*: *el.* elater, *sp.* spore (lateral view).

Ecology. It occurs in damp situations on clay or loamy soils. Fallow fields, banks and woodland rides are its principal habitats. It is among the numerous bryophytes which often colonize the moist sides of old cart ruts in little-used grassy by-ways, associated species frequently including *Dicranella heteromalla*, *D. varia*, *Pleuridium* spp. and other small mosses.

ADDITIONAL SPECIES

Only two of the remaining British species are moderately common.

Fossombronia wondraczeki (Corda) Dum. may be found in the same habitats as *F. pusilla*. It has spores with more close-set lamellae than those of *F. pusilla* (about 4μ apart, as against 8μ in *F. pusilla*) and they project as 28–36 spines around the margin of each spore.

F. dumortieri (Hub. & Genth.) Lindb. occurs occasionally on moist heaths and low moorland. It is more strongly odorous and larger than most other species and the spores are quite distinct in having lamellae which form a regular network, so dividing the spore surface into a number of polygonal areolae.

43. PELLIA EPIPHYLLA (L.) Corda

This plant, usually taken as a type for study in elementary courses in botany, is one of the commonest and most noticeable hepatics of moist banks and ditches, being abundant in all districts except those which are highly calcareous. It is marked by its relatively robust habit, with thallus up to 1 cm. broad, many centimetres long, dark shining green in the older parts and pale green at the tips, irregularly branched

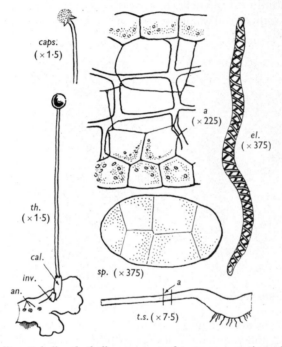

Fig. 197. *Pellia epiphylla*: *th.* thallus, *t.s.* part of transverse section of the thallus, *an.* antheridia, *cal.* calyptra, *inv.* involucre, *caps.* capsule (dehisced), *el.* elater, *sp.* spore (multicellular).

and often completely carpeting the moist loamy or peaty banks where it grows. The upper surface lacks the areolate markings found in *Marchantia* and its allies, and the thallus has a shining translucent quality which is not readily mistaken. From *Riccardia* it differs in its much larger size and the presence of a midrib; but the midrib is ill defined, never clear-cut as in *Metzgeria*. On the underside are numerous long rhizoids but no ventral scales. Erect-growing, tufted forms of the plant are not uncommon in and by mountain rills.

A transverse section of the thallus should be cut to distinguish *Pellia epiphylla* with certainty from *P. fabbroniana*, although the latter occurs exclusively in calcareous habitats whilst *P. epiphylla* avoids lime. The section will almost always show, here and there, strands of thickening which form a kind of network throughout the otherwise fairly uniform tissue of the thallus. In *P. fabbroniana* no such thickening strands occur. *P. neesiana*, the only other British species, is a north-western plant. It is usually tinged with dark red and the thallus has strands of thickening in its tissues.

P. epiphylla has male and female organs on the same branch (the paroecious condition). The antheridia occur singly, immersed in the thallus, forming slightly raised, dark reddish spots over an area a short distance back from the apex. The group of archegonia, protected by a flap-like involucre, are nearer to the apex. This type of involucre contrasts with the complete cylinder which is found in *P. neesiana* and the longer, tubular involucre of *P. fabbroniana*. The glossy greenish black, globose capsule of *P. epiphylla*, raised on a pale seta 5 cm. long, and surrounded at its base by the rough-edged calyptra, is a familiar feature of moist ditch banks in spring. It will be found to contain very large, ovoid, multicellular* spores and spirally thickened elaters.

Ecology. It occurs on the sides of ditches and on moist loamy banks by streams throughout Britain. It is also common on wet peat surfaces in moorland and mountain country. Wet rocks are a less usual habitat. *P. epiphylla* favours an acid soil reaction.

ADDITIONAL SPECIES

The two remaining species of *Pellia* are both dioecious.

P. neesiana (Gottsche) Limpr. is found in wet grassy places, chiefly in north Britain. The thallus is commonly tinged with red, and in the fertile plant the shortly cylindrical involucre is characteristic.

P. fabbroniana Raddi (*P. calycina* Nees) has a narrow thallus (usually about 5 mm. wide) which is often repeatedly forked at the tips in autumn and winter and lacks the internal thickenings found in the other two species. The tubular involucre is 3–4 mm. long. It occurs in wet calcareous places.

SPECIES OF RELATED FAMILIES

Blasia pusilla L. is an uncommon plant of moist banks and wet clayey or gravelly ground; it has something of the appearance of a small pale state of *Pellia epiphylla*. The lateral lobes of the thallus of *Blasia*, however, represent ill-defined leaves and ovate, toothed underleaves are also present. Often the most conspicuous feature of the plant will be the gemma receptacles shaped like long-necked flasks and containing round, multicellular gemmae. Larger, scale-like gemmae of irregular shape are also produced individually on the dorsal surface of the thallus; whilst on the ventral surface are hollow outgrowths which look like dark spots owing to the presence within them of the blue-green alga, *Nostoc.*

* This multicellular condition results of course from precocious germination of the spores within the capsule.

Moerckia flotowiana (Nees) Schiffn. is a species belonging to a different (but closely related) family. It is an uncommon plant of moist sandy ground, usually near the sea. The bright or yellow-green thallus is notable for its greatly crisped and wavy margins, and a midrib which is prominent (and covered with rhizoids) on the underside. A transverse section of the thallus will show the pair of conducting strands. The whole plant has a rather strong smell.

44. METZGERIA FURCATA (L.) Dum.

The flattened, yellowish green or mid-green patches formed by *M. furcata* are common on tree-trunks in woodland. At once distinguished from that of allied genera by the clearly defined narrow midrib, the thallus is irregularly forked, the branches adhering closely to the bark on which the plant commonly grows. Scattered about the under surface will be found a varying number of delicate hair-like rhizoids. Though reaching 2·5 cm. in length, the thallus branches are only 0·5–1 mm. broad. This liverwort, therefore, may sometimes be noticed only when a tree-trunk or branch is examined closely; but at other times the patches formed are so extensive as to attract attention.

Under the microscope it can be seen that whereas the midrib is composed of several layers of cells, the relatively broad 'wings' of the thallus are only one cell thick. The cells of this wing tissue are usually hexagonal in outline (as viewed from above), 30–40μ across, and show slight thickening at the corners. The hair-like (and of course unicellular) rhizoids are often somewhat expanded at their tips. Sparse or rather numerous, they occur singly, not in pairs as they usually do on the thallus margin in the allied *M. conjugata*. Multicellular, discoid gemmae are produced in abundance on some forms of the plant.

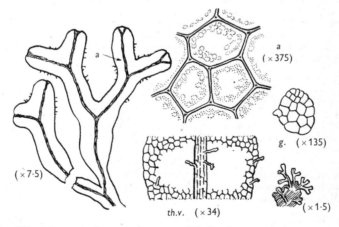

Fig. 198. *Metzgeria furcata*: *th.v.* part of thallus in ventral view, *g.* gemma.

M. furcata is dioecious. Antheridia occur on specialized, inrolled, almost globose branches. The female branch is also specialized in form, serving a protective function. It is hollowed out to form a small, rather densely hairy involucre. In the young fruiting condition hairs are conspicuous on the pear-shaped calyptra that envelops the capsule. The ripe capsule, ovoid-globose and red-brown in colour, is raised on a very short seta 1·5–2·5 mm. in length.

Ecology. It has been noted as occurring chiefly on the comparatively dry habitats afforded by the trunks and branches of living trees. It is much less common on dead wood, but is often plentiful on rock surfaces.

ADDITIONAL SPECIES

The remaining three species of *Metzgeria* occur chiefly on rocks in mountain districts.

M. conjugata Lindb. is distinguished from *M. furcata* by its broader thallus (2 mm.) and by the fact that the hairs on the under side are borne normally in pairs; also the cells of the wing tissue average 50 μ in width (against 35 μ in *M. furcata*). It occurs mainly on rocks and is rare on trees.

M. hamata Lindb., a plant of Atlantic type of distribution, is very distinct in its strongly convex thallus and paired, curved hairs on midrib and margin.

M. pubescens (Schrank) Raddi occurs on calcareous rocks; it differs from all the other species in being pinnately branched and having both surfaces of the thallus almost uniformly covered with short, nearly straight hairs.

45. RICCARDIA PINGUIS (L.) Gray (*Aneura pinguis* (L.) Dum.)

Riccardia pinguis forms irregular, prostrate patches of bright or deep green colour in moist situations. Rather lacking in distinctive characters, it may not attract attention at first; indeed beginners might pass it over as a small and ill-developed *Pellia*. The thallus of *Riccardia*, however, is much narrower, rather brittle, and always thicker in proportion to its width. Also it lacks the fairly well-defined midrib of *Pellia*, branching tends to be pinnate rather than forked, and the tips of the thallus lobes are rounded. It attains only 2–3 cm. in length and 2–6 mm. in breadth. From other species of *Riccardia*, *R. pinguis* may be separated by its somewhat brittle and fleshy texture, greasy appearance and the much broader segments of the thallus. *R. multifida* is quite distinct in its much narrower thallus segments and twice to thrice pinnate branching.

It should be borne in mind that most species of this genus are very variable and it is wise to confirm the determination under the microscope. A transverse section shows the comparatively uniform nature of the tissue, green above and below, nearly colourless in the deep-seated layers, but without the elaborate internal air space system of the Marchantiaceae. The section tapers gradually, from ten to twelve cells thick in the middle, into the narrow wings which are only 1–3 cells thick. The thallus is thus thicker in the middle than is that of any other species of *Riccardia*.

Although *R. pinguis* is dioecious, the fertile condition is not rare. The comparatively small male plants bear groups of antheridia on short lateral branches. The young ovoid capsules, each enclosed in its fleshy calyptra, arise from beneath the margins of the thallus lobes, and are at first deep green. Later the calyptra is broken through and remains

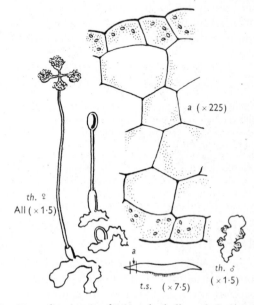

Fig. 199. *Riccardia pinguis*: *th.* ♂, male thallus; *th.* ♀, female thallus showing (right to left) three stages in the development of the sporophyte; *t.s.* section of thallus.

as a collar about the base of the elongated seta. The capsule wall becomes dark brown and opens by four longitudinal splits. When old, the four spoon-shaped segments into which the capsule has split are spread out cross-wise, and each bears a tuft of persistent elaters.

Ecology. Provided the surface on which it is growing is adequately moist, it will tolerate a wide range of habitats; these include wet rock surfaces, peat in fens and bogs, dune slacks and other partially dried-out pools, and the moist clayey surfaces of ditches and pond margins.

46. RICCARDIA MULTIFIDA (L.) Gray (*Aneura multifida* (L.) Dum.)

This is among the most slender in habit of our common thallose liverworts. In length it attains only 2–4 cm. and the ultimate branches of the twice or thrice pinnate thallus are only about 0·5 mm. broad. The absence of a well-defined midrib at once distinguishes it from species of *Metzgeria*. Most species of that genus are also quite different in their forked branching. No other species of *Riccardia* has quite such a finely divided thallus, and the thin transparent margins are especially dis-

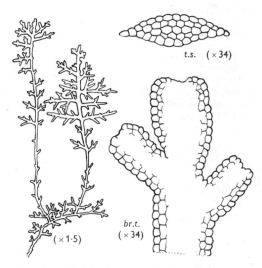

t.s. (×34)

br.t. (×34)

(×1·5)

Fig. 200. *Riccardia multifida*: *br.t.* tip of branch showing clearly defined margins one cell thick, *t.s.* section of thallus.

tinctive. *R. multifida* may be met with in a variety of moist situations, portions of the thallus often appearing here and there amid the denser growth of marshland mosses. It varies in colour from dark green to reddish brown.

With a little practice one will come to recognize this species at sight, so that the section of the thallus will not normally be necessary for determination. It does, however, furnish a useful confirmatory character. Biconvex in shape, the section will show clearly the nature of the margins or 'wings' mentioned above. In them (at least in the ultimate branches) the tissue is only one layer of cells in thickness. In the mid-region the thallus is 6–8 cells thick, the internal cells being very large.

The ends of the branches not uncommonly bear 2-celled gemmae;

internally produced gemmae of a similar type occur in a number of species of *Riccardia*.

R. multifida is monoecious but is not commonly fertile. Antheridia arise on special short oblong branches. The young capsule, enveloped in its calyptra, is at first narrowly club-shaped and green in colour. Later the seta lengthens and the capsule, now brown, opens by the four valves characteristic of this Order of liverworts.

Ecology. It is chiefly an upland species, its principal habitats consisting of wet rocks in or near streams and moist rock ledges on mountains. It occurs more sparingly in lowland marshes and ditches. I do not know its range of tolerance as regards soil reaction.

ADDITIONAL SPECIES

R. sinuata (Dicks.) Trev. (*Aneura sinuata* (Dicks.) Dum.), which is fairly common in very wet places, differs from *Riccardia multifida* chiefly in its paler colour, less freely and less regularly branched thallus, and in the thallus margins which are always more than one cell thick. It will grow on damp soil, on dripping rocks or submerged in pools, and varies much in form according to habitat. Large, freely branched forms (up to 4 cm. long) occur at times.

Plate XV. *Marchantia polymorpha* with gemmiferous cups (×4)

Plate XVI. *Marchantia polymorpha* with female receptacles (× 1·5)
Also seen (top right) is *Porella platyphylla*

MARCHANTIALES

47. MARCHANTIA POLYMORPHA L. (Plates XV and XVI)

M. polymorpha, the only British species of a large genus, is common in many of the same habitats as *Lunularia*, including the soil of flower-pots in greenhouses, which very frequently become infested with its dark green, prostrate, richly branched thallus. But, unlike *Lunularia*, it is also found on damp moorland and by streams. This plant is at once

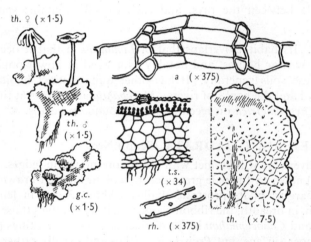

Fig. 201. *Marchantia polymorpha*: *g.c.* gemma cups; *th.* part of thallus to show surface markings; *th.* ♂, part of thallus with male receptacle; *th.* ♀, part of thallus with female receptacle; *rh.* part of tuberculate rhizoid; *t.s.* part of transverse section of thallus; an air pore is shown at *a*.

separated from *Lunularia* by its darker green colour and larger size, as by the goblet-shaped gemma-cups with toothed margins, which are very different from the semilunar ridges of *Lunularia*. The hexagonal areolae which are clearly seen on the upper surface of the thallus resemble those of *Conocephalum conicum*, but the pore in the centre of each is not normally visible to the naked eye. The thallus reaches 2–10 cm. in length and 7–20 mm. in breadth. The midrib is less distinct than in *Conocephalum*. As the raised, symmetrical gemma-cups are nearly always present, they will usually serve to distinguish *Marchantia* without difficulty.

A transverse section of the thallus will show the barrel-shaped pores which are of entirely different structure from the simple pores of

Conocephalum and *Lunularia*, the barrel-shaped channel of the pore being bounded by regular tiers of cells (cf. *Preissia quadrata*). There is a gradual transition from the midrib to the wings of the thallus. On the under surface the ventral scales are in three rows, the marginal ones tongue-shaped. As is usual in the Marchantiales both smooth and tuberculate rhizoids may be seen. The shallowly bilobed, discoid gemmae represent an important method of vegetative propagation.

In the fertile state this dioecious liverwort is known immediately by the peculiar 'umbrella-like' form of the male and female receptacles. The figures will show their precise appearance, disc-shaped in the male and 9-rayed in the female. The numerous sporogonia develop on the underside between the rays, and each capsule contains spores and elaters.

Ecology. Its habitat overlaps that of *Lunularia* in that it occurs very commonly near houses, and notably as a weed in flower-pots and on greenhouse soils generally. It has other quite distinct habitats—wet moorland and the banks of rivers. *Marchantia* also grows at times with *Funaria hygrometrica* on recently burnt ground on heaths.

48. PREISSIA QUADRATA (Scop.) Nees

This liverwort will be met with chiefly on mountain ledges, where it may be known by the pale green general colour and red-brown or deep violet margins of its prostrate thallus. About 3 cm. in length and 0·5–1 cm. in breadth, the ribbon-shaped thallus resembles those of *Marchantia* and *Conocephalum* in the distinct hexagonal markings (areolae) of its upper surface, but *Preissia* is smaller and usually rather paler in colour than either of these. The under surface, like the margin, commonly becomes purplish brown in older parts of the thallus. Gemmae never occur.

The pores, which are clearly visible on the upper surface with a lens, are seen in transverse section to be barrel-shaped, as in *Marchantia*. The underlying green tissue lacks both the conical end cells of *Conocephalum* and the elaborate air-space system seen in *Reboulia hemisphaerica*. On the under surface there is a row of overlapping (imbricate) ventral scales on either side of the midrib, which itself is rather prominent below.

Fertile material of this normally dioecious hepatic is not uncommon. The 4-lobed female receptacles, raised on stalks up to 5 cm. long, are highly characteristic. The only genus with which confusion might then arise is *Reboulia*, in which the female receptacle is 4- to 7-lobed; but in *Reboulia* hair-like structures (lacking in *Preissia*) will be found at both base and apex of the stalk; and the stalk is traversed by only a single rhizoid furrow, not two as in the present plant. The male heads of *Preissia* are less deeply lobed than the female, and are borne on shorter stalks.

Ecology. *Preissia* is a pronounced calcicole, growing with such mosses as *Tortella tortuosa* and *Neckera crispa*, on moist base-rich mountain ledges. Occasionally it grows in moist hollows in calcareous sand dunes, and on wet calcareous rocks at low altitudes. *Preissia* is essentially a plant of the north and west, and grows nowhere in south-east England.

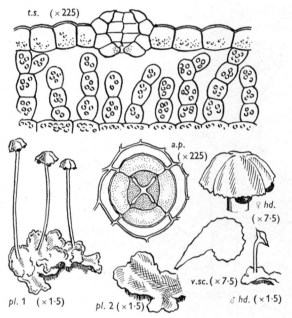

Fig. 202. *Preissia quadrata*: *pl.* 1, normal fertile plant with three female heads; *pl.* 2, part of larger shade form; *a.p.* air pore in surface view; *t.s.* part of transverse section of thallus showing pore in section; *v.sc.* ventral scale; ♂ *hd.* and ♀ *hd.*, male and female heads respectively, the latter with two capsules in view.

Note. The minute appendage has been lost from the ventral scale drawn.

49. CONOCEPHALUM CONICUM (L.) Dum. (Plate XVII)

This plant, which is common on wet rocks in a wide range of habitats, is characteristically robust, and forms extensive, flattened, dark green, shining patches. The irregularly forked branches of the broadly ribbon-shaped thallus attain 10–20 cm. in length and about 1 cm. in breadth. Superficially *C. conicum* is somewhat like *Marchantia polymorpha*, but it lacks the goblet-shaped gemma-cups of that plant and the pores which occur in the upper surface of the thallus are rather more clearly visible to the naked eye. Each pore marks the centre of one of the approximately hexagonal areolae which give the upper surface of the thallus so distinctive a pattern. In *Marchantia* the areolae are present too, but are smaller. There is no dark line along the middle of the dorsal surface of the thallus, but the midrib here is more clearly defined than in *Marchantia*, being prominent on the ventral surface. Other field characters are the violet-edged appendages of the ventral scales, and the pleasant scent of the thallus when bruised.

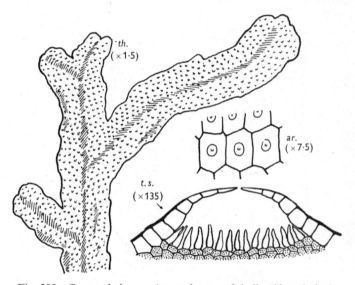

Fig. 203. *Conocephalum conicum*: *th.* part of thallus (dorsal view), *ar.* areolae and pores, *t.s.* section through pore (chlorophyllose cells stippled, somewhat diagrammatic).

A transverse section of the thallus will confirm the determination. It will show the 'simple', not 'barrel-shaped' pore structure, and the floor of each air chamber covered by short chains of chlorophyll-bearing cells, the end cell of each chain being colourless and elongated-conical in shape. These end cells are diagnostic.

Conocephalum conicum is dioecious, but is not rare in fruit. The groups of antheridia form slightly raised, purplish cushions on the upper surface of the thallus, near the apex. Archegonia are borne on the under side of a conspicuous conical 'receptacle' which is itself carried up on a pale stalk 3–8 cm. in length. At maturity the 5–8 capsules, each on a short seta, project downwards from the base of the receptacle. The minute lid is shed and the split walls become recurved on the release of the

Plate XVII. *Conocephalum conicum* (natural size)

Also seen are four stems of *Atrichum undulatum* and a general
background composed largely of *Brachythecium rutabulum*

Plate XVIII. *Lunularia cruciata* with gemmae (×2)

Also seen are a few shoots of the moss *Brachythecium rutabulum*

spores. These, 70–90 μ in diameter, are already multicellular, the result of precocious germination.

Ecology. It will grow in a considerable variety of rock habitats, provided these are moist and to some extent shaded. Thus it occurs on walls, rocks by streams, weirs, the entrances of disused mines, and moist rock ledges on mountains. Calcareous rocks are favoured, but *C. conicum* is possibly not strictly calcicole. It has been found to be more commonly fertile in south than in north Britain.

50. LUNULARIA CRUCIATA (L.) Dum. (Plate XVIII)

On flower-pots and moist brickwork in gardens this is a very common liverwort; indeed it is often a troublesome weed. The prostrate, irregularly branched thallus is thinner and more slender than that of either *Conocephalum* or *Marchantia*. It is also of a lighter green colour. It is commonly only 1·5–2·5 cm. long, and 5–10 mm. wide. It is usually possible to identify *Lunularia cruciata* at sight by the groups of gemmae which are almost always present on the surface of the thallus. These gemmae, instead of being borne in raised cups, as in *Marchantia*, are merely protected on one side by a crescent-shaped ridge of tissue (which explains the generic name of the plant). Older parts of the thallus become a characteristic brownish yellow in colour.

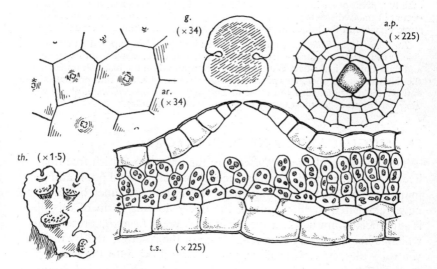

Fig. 204. *Lunularia cruciata*: *th.* thallus (with gemmae), *ar.* areolae on dorsal surface of thallus, *g.* gemma, *a.p.* air pore, *t.s.* section through thallus near the margin showing pore, underlying chamber and chlorophyllose cells.

Microscopic examination will be necessary only to confirm the determination. A transverse section of the thallus will show the pores (which in surface view can be seen in the centre of the rather indistinct areolae) to be of simple structure, not barrel-shaped as in *Marchantia*. The underlying green tissue lacks the conical end cells of *Conocephalum*. The midrib passes over very gradually into the expanded wings of the thallus. The ventral surface shows a single row of delicate scales, half-moon-shaped, with rounded appendages.

Lunularia cruciata, the only species of the genus, is dioecious, and the fertile state is very rare. The male receptacle is borne on the end of a short branch; the female receptacle is remarkable for its cross-shaped (cruciate) form, the four bottle-shaped involucres drooping down from the summit of a short stalk.

Ecology. It will grow on flower-pots, brickwork, rockery stones and garden paths. From gardens it will spread to moist rocks in the vicinity, but away from habitations it seems to be rare. It is perhaps not a true native of Britain.

SPECIES OF RELATED FAMILIES

The genera *Reboulia* and *Targionia* have one British species each.

Reboulia hemisphaerica (L.) Raddi looks very like *Preissia quadrata* but it may be known at once by a section of the thallus, which is honeycombed with air spaces. The differences between the two when fertile are mentioned under *Preissia*. *Reboulia* is mainly a plant of sheltered hedge banks and soil among rocks at low altitudes, and occurs in many districts where *Preissia* is unknown. It is notably plentiful in limestone crevices in the hill country of west Yorkshire, but is not a strict calcicole.

Targionia hypophylla L. is an uncommon plant found in exposed sunny places, chiefly in limestone districts. Its most distinctive feature, apart from its small size and dark colour, is the dark purple involucre produced on the blackish underside of the thallus near the apex. This protects the archegonia, which are thus produced in an entirely different manner from those of *Lunularia* and *Conocephalum*, plants which *Targionia* approaches most nearly in general thallus structure.

51. RICCIOCARPUS NATANS (L.) Corda

This species and *Riccia fluitans* are the only two British free-floating liverworts. *R. natans* is a rather uncommon plant of still pools and pond margins. It is markedly distinct from *R. fluitans* in its much broader and never elongated thallus; indeed, the freely floating form of this plant is superficially not unlike a species of duckweed (*Lemna*), a resemblance which is heightened by the breaking-up of the rosettes, and by the long ventral scales, of violet colour, which hang down into the water. The thallus branches of the terrestrial state tend to form crowded, irregular rosettes; and on them, whilst rhizoids are numerous, ventral scales are greatly reduced.

The thallus attains 5–10 mm. in length, and expands to a width of about 5 mm. It is commonly twice or thrice forked, each lobe with an evident median furrow. The toothed, ribbon-shaped ventral scales, which project beyond the margins of the thallus in the aquatic form, are diagnostic.

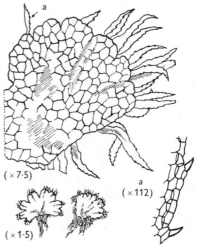

The transverse section of the thallus is nearly oblong in shape, becoming slightly convex only near the median furrow and along the ventral surface; the margins are rounded. Almost the entire tissue is composed of large air chambers, separated by walls that are only one cell thick. These air chambers give the thallus an 'areolate' appearance when viewed from above. The centres of the areolae are marked by minute pores surrounded by 6–8 cells with slightly thickened walls.

Fig. 205. *Ricciocarpus natans.*
Note. The markings on the uppermost sketch indicate areolae, not cells.

Antheridia occur in a toothed ridge in the median furrow of the thallus. The rudimentary involucre that surrounds the archegonium is a generic character. The immersed capsule contains spores which are about 50μ in diameter. This dioecious liverwort, however, is very rarely fertile in Britain.

Ecology. Its habitat agrees closely with that of *Riccia fluitans*, consisting of eutrophic waters such as lowland ponds, lakes and canals; it may be found floating on the surface, or (rarely) on the mud at or near the margins. A plant of almost cosmopolitan distribution, *R. natans* is so abundant in some countries (e.g. Kenya) as to form dense scums that profoundly disturb the fresh-water habitat from the fishery point of view.

52. RICCIA FLUITANS L.

This species occurs in two forms,* one free-floating, with delicate narrow thallus, the other terrestrial, with thicker, broader thallus. The former, which is much the commoner, grows on the surface of the water of ponds; whilst the terrestrial form occurs along the muddy margins, or submerged on the mud at the bottom of a pond.

* These are the extreme states of a plant which closer study may show to be polymorphic.

The thallus is 1–5 cm. long but only 0·5–1 mm. broad in the slender floating form. The terrestrial plant has not only a much broader (up to 2 mm.) and thicker thallus, but also rhizoids and rudimentary ventral scales, both of which are normally lacking in the free-floating state. Furthermore, the thallus segments are broadly but shallowly channelled.

The transverse section of a thallus segment is broadest in the mid-region, narrowing somewhat at the obtusely rounded margins. Large air chambers occur throughout the spongy thallus section, but the

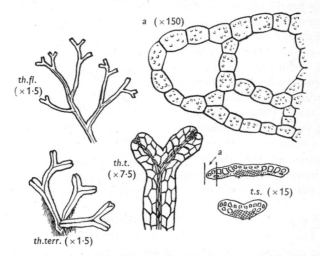

Fig. 206. *Riccia fluitans*: *th.fl.* and *th.terr.*, floating and terrestrial thalli respectively; *th.t.* tip of terrestrial thallus; *t.s.* transverse sections of two different terrestrial thalli. The pattern on *th.t.* is caused by the distribution within the thallus of air chambers of different sizes, and does not imply a true midrib.

detailed structure varies according to habitat. The air chambers, however, give the thallus section an appearance entirely different from that seen in all the common land species of *Riccia*.

R. fluitans is monoecious. In the fertile state (which is very rare) the capsule forms a spherical swelling on the under surface of the thallus. The spores are 75–90 μ in diameter. The surface of each spore is marked by large 'areolae', and the margin bears a distinct wing-like expansion.

Ecology. *R. fluitans* is a plant of ponds and ditches rich in mineral nutrients (eutrophic). It is commonly associated with the duckweeds (*Lemna* spp.). It is widely distributed but not generally common.

53. RICCIA GLAUCA L.

This is perhaps the commonest British species of a genus of small liverworts, most of which form characteristic neat rosettes on moist soil. *R. glauca* should be looked for from early autumn to late spring on clayey soil in fallow fields or gardens. The frond-like branches of the thallus overlap one another, and tend to form a rosette of almost circular outline, a little over 1 cm. in diameter, of a light glaucous green colour.

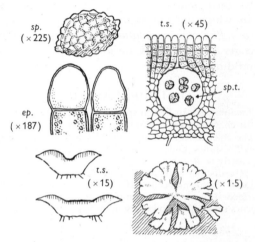

Fig. 207. *Riccia glauca*: *ep.* epidermal cells in section,
t.s. transverse sections of thallus, *sp.* spore,
sp.t. spore tetrads in old sporogonium.

The thallus is regularly forked, and in each segment a shallow and indistinct channel may be seen—very much shallower and less distinct than the deep furrow of *R. sorocarpa*. Thus, a section through one of the terminal segments of the thallus will show tapering margins and a broad, depressed central region. It contrasts sharply with the steep-sided section of *R. sorocarpa*, where the upper surface shows an evident 'V'. There is little differentiation of tissues and no air chambers such as are found in the Marchantiaceae; but the upper part of the thallus is composed of regular columns of cells separated by narrow, slit-like air spaces. The greenish ventral scales are readily lost.

The structure of the upper epidermis is a further useful microscopic character. In *R. glauca* it is composed of a single layer of thin-walled, rounded pear-shaped cells. In *R. sorocarpa*, immediately beneath the thin-walled, almost conical cells of this outermost layer, there is a layer of thick-walled cubical cells which are lacking in *R. glauca*.

As in all species of *Riccia*, the rounded antheridia and flask-shaped archegonia are sunk within the tissues of the thallus. The capsule, which remains embedded in the thallus, contains no sterile cells, only spores, which are large and comparatively few in number. In *R. glauca* the spores measure 75–90μ across, larger on the average than in *R. sorocarpa*, where the corresponding figures are 65–80μ.

Ecology. Its chief habitat is cultivated land, whether arable fields or garden plots, provided the soil is of a type to retain moisture. This species and *R. sorocarpa* are probably the most characteristic liverworts of a community which includes many of the 'ephemeral' mosses, such as *Pottia truncata*, *Phascum cuspidatum* and *Ephemerum serratum*. *Riccia glauca* also occurs on clay banks, on moist woodland rides and in wet hollows in sand dunes.

ADDITIONAL SPECIES

Several of the remaining eleven British species of *Riccia* are rare.

R. sorocarpa Bisch., however, is almost as common as *R. glauca* and occurs in similar places. The chief differences are pointed out under that species. The upper surface of the thallus shows, under the lens, something of the glistening appearance seen even more clearly in *R. crystallina*.

R. crystallina L. is distinct in its close, spongy rosettes 1–2 cm. in diameter, with a thallus section that reveals numerous large air cavities, some of which are exposed on the upper surface. It is an uncommon plant of wet mud by ponds and in damp fields; it also occurs in damp hollows in sand dunes.

SPHAEROCARPALES

Sphaerocarpus is the only British genus of this small Order of liverworts. Two species, *S. michelii* Bellardi and *S. texanus* Aust., occur occasionally on cultivated land. Minute plants with delicate thalli of simple structure, they are chiefly remarkable for the large 'involucres' which surround the spherical capsules during their development. The areolate spores always remain united in tetrads, even when ripe. In *S. michelii* the involucre is pear-shaped and the areolae of the spores bear spines; in *S. texanus* the involucre is club-shaped (narrowed at its tip) and the areolae are rather different, so that the tetrad of spores appears to be surrounded by a narrow wing-like margin. Average diameters for the tetrads are 100μ for *S. michelii* and 135μ for *S. texanus*.

ANTHOCEROTALES

54. ANTHOCEROS PUNCTATUS L.

Although widely distributed, this is not a very common plant. It grows on moist soil, forming patches that are pale green when young, but become dark green or blackish when old or dried-out. The thallus tends to form a rosette 0·5–1·5 cm. in width; there is no clearly defined midrib, and the margin is characteristically lobed and wavy, so that the plant has a 'frilly' appearance that is normally lacking in *A. laevis*. *A. punctatus* is an annual, whereas *A. laevis* and *A. husnoti* are perennial.

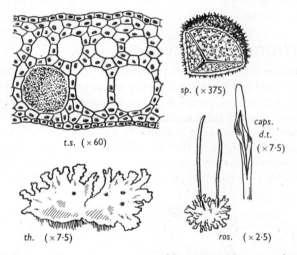

sp. (×375)

t.s. (×60)

caps. d.t. (×7·5)

th. (×7·5)

ros. (×2·5)

Fig. 208. *Anthoceros punctatus*: *ros.* rosette with two capsules, *caps. d.t.* tip of ripe capsule dry, *sp.* spore, *th.* isolated pieces of thallus (dark spots are *Nostoc* colonies), *t.s.* part of transverse section of thallus with *Nostoc* colony in one cavity.

A transverse section of the thallus shows a number of features of interest. The thallus bears numerous long smooth rhizoids, and, as in all species of *Anthoceros*, each chlorophyllous cell contains only a single very large chloroplast. In this last character *Anthoceros* is unique among the genera of British liverworts. The numerous large mucilage cavities, an obvious feature of the section, will distinguish this species with certainty from *A. laevis* (which lacks them). Some of the cavities are occupied by colonies of the blue-green alga, *Nostoc*. The thallus is 8–12 (occasionally up to 20) cells thick in its thickest part.

Anthoceros punctatus is monoecious, and ripe capsules may be found from July to November. Antheridial cavities are seen on the upper surface of the thallus as glistening pimples (owing to mucilage). Each cavity contains 4–14 antheridia, the body of each antheridium being 50–90μ long (smaller than in *A. husnoti*). The archegonia are sunk in the thallus, nearer its apex. The capsules are 1–3 cm. long and less than 0·5 mm. broad, so that when young they remind one of the delicate shoots of grass seedlings. Each is surrounded at its base by a short, tubular involucre. The ripe capsule splits longitudinally along two lines. Each spore is brownish black, about 40μ wide, and covered with spines on its convex face.

Ecology. In some districts it is a member of the bryophyte flora that flourishes on retentive soil in damp fallow fields from August to November or somewhat later. It also occurs on the sides of ditches and on wet clay banks.

55. ANTHOCEROS LAEVIS L.

This species differs from *A. punctatus* in a number of well-defined characters. The thallus is normally smooth (not 'frilly') and is perennial. It forms dark green rosettes, 1–3 cm. in diameter, and favours moist clay soil on banks and the sides of ditches. In many districts it is commoner than *A. punctatus*, but is not everywhere a common plant. While typical forms cannot be mistaken for *A. punctatus*, more care is needed with plants growing under abnormal or crowded conditions, for they may have a deceptively crisped appearance, the lobes of the thallus becoming

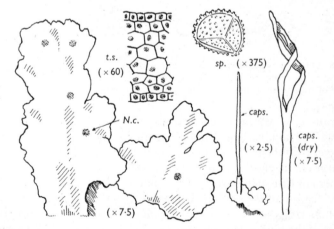

Fig. 209. *Anthoceros laevis*: *caps.* capsule, on left a young specimen, on right showing the characteristic twisting of the two valves when dry; *N.c.* dark spots on the thallus caused by *Nostoc* colonies; *sp.* spore; *t.s.* transverse section of thallus.

channelled, and having wavy margins. It may then be necessary to cut sections of the thallus.

A low-power examination of the surface of the thallus will show the large chloroplasts, one to each cell, that indicate the genus *Anthoceros*; also colonies of *Nostoc* and small tubers may sometimes be seen. A transverse section will show the thallus to be only 6–8 cells thick at its thickest part. This fact, and the absence of the large mucilage cavities that distinguish the thallus of *A. punctatus* and *A. husnoti*, will immediately confirm the determination.

Recent detailed studies have shown that the antheridia and archegonia are borne on separate thalli, although these are commonly crowded together in the same rosette. The male thalli are narrower than the female. The antheridia occur in groups of 2–4 within each antheridial cavity.

The capsules ripen in summer and are chiefly distinct from those of *A. punctatus* in their greater average length (usually 3–4 cm. and occasionally up to 9 cm.) and in the spore colour, which is yellow.

Ecology. It grows commonly on steep clay banks and slopes which either have a constant trickle of water or are at least permanently more or less moist. Sometimes it is found in woodland rides associated with *Fossombronia* spp. and *Pellia epiphylla*; more rarely in fallow fields; but many of the most suitable habitats for it are by roadsides.

ADDITIONAL SPECIES

Anthoceros husnoti Steph. is very near *A. punctatus* but is larger in all its parts. Thus the section of the perennial thallus may be up to 30 (usually 12–22) cells thick in the middle, the antheridia average 120μ long and capsules commonly attain 4–6 cm. and are at times much longer.

HABITAT LISTS

The following habitat lists are intended to give the beginner an idea of the range of species most likely to be encountered in different habitats. Rare species are of course omitted, as they are outside the scope of this book; and neither the range of habitats nor the lists themselves can be claimed to be in any sense exhaustive. Nevertheless, it is thought that the novice who has made a collection in any specified type of country will find that the majority of plants in his collection are represented on the appropriate list.

1. *Arable land, etc.*

(a) On arable and garden soil:
Barbula fallax
B. unguiculata
Brachythecium rutabulum
Bryum bicolor
B. erythrocarpum
Ceratodon purpureus
Eurhynchium praelongum
E. swartzii
Funaria hygrometrica
Phascum cuspidatum
Physcomitrium pyriforme
Pottia davalliana
P. truncata

Anthoceros laevis
A. punctatus
Riccia glauca
R. sorocarpa

(b) In greenhouses:
Ceratodon purpureus
Funaria hygrometrica
Leptobryum pyriforme
Pottia davalliana
P. truncata

Lunularia cruciata
Marchantia polymorpha

2. *On clay banks, etc. (including neutral to mildly acid, clayey soil of woodland rides)*

Amblystegium serpens
Atrichum undulatum
Barbula cylindrica
B. unguiculata
Brachythecium albicans
B. rutabulum
Dicranella varia
Eurhynchium praelongum
Fissidens bryoides
F. taxifolius
Hypnum cupressiforme
Mnium affine

M. longirostrum
Pleuridium acuminatum
Pohlia delicatula
Pseudephemerum nitidum
Rhytidiadelphus squarrosus
R. triquetrus
Thuidium tamariscinum
Weissia controversa

Anthoceros laevis
A. punctatus
Cephalozia bicuspidata

Cephaloziella spp.
Diplophyllum albicans
Fossombronia pusilla
Nardia scalaris

Pellia epiphylla
Riccardia pinguis
Scapania irrigua
S. nemorosa

3. *Colonizing the bare mud of pond margins*

Acrocladium cuspidatum
Brachythecium rivulare
B. rutabulum
Bryum bicolor
Ephemerum serratum
Leptodictyum riparium
Physcomitrella patens
Pleuridium subulatum

Pseudephemerum nitidum

Riccardia multifida
R. pinguis
R. sinuata
Riccia crystallina
R. fluitans

4. *In and by lowland streams (on tree roots, stonework, etc.)*

Amblystegium serpens (*t*)
Brachythecium rivulare
Cinclidotus fontinaloides
C. mucronatus (*t*)
Eurhynchium riparioides
Fissidens crassipes

Fontinalis antipyretica
Leptodictyum riparium (*t*)
Leskea polycarpa (*t*)
Orthotrichum rivulare (*t*)
Tortula latifolia (*t*)

(*t*) Especially on tree-roots by water.

5. *On town walls, stones and brickwork*

Barbula convoluta
Brachythecium rutabulum
B. velutinum
Bryum argenteum
B. bicolor
B. caespiticium
B. capillare
Ceratodon purpureus

Eurhynchium confertum
Grimmia pulvinata
Hypnum cupressiforme
Orthotrichum diaphanum
Tortula muralis

Lunularia cruciata

6. *On mud-capped walls*

Aloina ambigua (*c*)
Anomodon viticulosus (*c*)
Barbula convoluta
B. fallax
B unguiculata

Bryum capillare
Campylopus flexuosus (*a*)
Ceratodon purpureus
Ctenidium molluscum (*c*)
Dicranum scoparium

(*a*) Species characteristic of acid pockets.
(*c*) Species characteristic of calcareous pockets.

Encalypta streptocarpa (c)
E. vulgaris (c)
Eurhynchium murale (c)
E. striatum (c)
Hypnum cupressiforme
Polytrichum piliferum (a)
Pottia lanceolata (c)
Rhacomitrium canescens (a)
R. lanuginosum (a)

Tortula intermedia (c)
T. muralis
T. ruralis (c)

Frullania tamarisci
Porella platyphylla (c)
Scapania aequiloba (c)
S. aspera (c)
S. gracilis (a)

(a) Species characteristic of acid pockets.
(c) Species characteristic of calcareous pockets.

7. *On calcareous walls and boulders* (*i.e. growing directly on the limestone rock itself*)

Barbula recurvirostra
B. revoluta
B. rigidula
B. trifaria
B. vinealis
Camptothecium lutescens
C. sericeum
Ctenidium molluscum
Encalypta streptocarpa
Grimmia apocarpa
G. pulvinata
Hypnum cupressiforme
Neckera complanata
N. crispa
Orthotrichum anomalum var.
 saxatile

O. cupulatum
Rhynchostegiella tenella
Tortella tortuosa
Tortula intermedia
T. muralis
T. ruralis
Trichostomum brachydontium
T. crispulum

Plagiochila asplenioides
Porella laevigata
P. platyphylla
Scapania aspera

8. *In chalk pits and quarries* (*many of these occur in comparable situations on limestone generally*)

Acrocladium cuspidatum
Barbula convoluta
B. cylindrica
B. fallax
B. unguiculata
Brachythecium rutabulum
Bryum capillare
Camptothecium lutescens
C. sericeum

Campylium chrysophyllum
Ctenidium molluscum
Dicranella varia
Eurhynchium striatum
E. swartzii
Fissidens adianthoides
F. cristatus
F. taxifolius
Funaria hygrometrica

Hypnum cupressiforme
Pseudoscleropodium purum
Rhytidiadelphus squarrosus
R. triquetrus
Seligeria calcarea

S. pusilla

Leiocolea turbinata
Plagiochila asplenioides
Porella platyphylla

9. *In chalk and limestone grassland*

Camptothecium lutescens
Campylium chrysophyllum
Ctenidium molluscum
Dicranum scoparium
Ditrichum flexicaule
Eurhynchium swartzii
Fissidens cristatus
Hylocomium splendens
Hypnum cupressiforme (including
  vars. lacunosum and tectorum)

Mnium undulatum
Pseudoscleropodium purum
Rhytidiadelphus squarrosus
R. triquetrus
Rhytidium rugosum (rather rare)
Thuidium abietinum
T. philiberti
T. tamariscinum

10. *On the ground or on banks in beech woods on the chalk*

Anomodon viticulosus (*c*)
Atrichum undulatum
Barbula cylindrica (*c*)
B. recurvirostra (*c*)
Brachythecium rutabulum
Cirriphyllum piliferum (*g*)
Ctenidium molluscum (*c*)
Dicranella heteromalla (*a*)
Dicranum scoparium
Cirriphyllum crassinervium
Eurhynchium praelongum
E. striatum (*c*)
Fissidens bryoides
F. taxifolius
Homalia trichomanoides
Hylocomium splendens (*g*)
Leucobryum glaucum (*a*)
Mnium hornum
M. longirostrum

M. undulatum (*g*)
Polytrichum formosum (*a*)
P. gracile (*a*)
Pseudoscleropodium purum (*g*)
Rhytidiadelphus squarrosus (*g*)
R. triquetrus
Thamnium alopecurum (*c*)
Thuidium tamariscinum
Tortula subulata (*c*)

Cephalozia bicuspidata (*a*)
Diplophyllum albicans (*a*)
Lejeunea cavifolia (*c*)
Lepidozia reptans
Lophocolea bidentata (*g*)
Nardia scalaris (*a*)
Plagiochila asplenioides
Porella platyphylla (*c*)
Scapania nemorosa

(*a*) Acid substratum, e.g. surface humus or leached layers.
(*c*) Calcareous ground.
(*g*) Among grass.

11. *On the ground in oak woods (e.g.* Quercus robur *woods of southern England*)

Atrichum undulatum
Brachythecium rutabulum
Dicranella heteromalla
Dicranum scoparium
Eurhynchium praelongum
E. striatum
Fissidens bryoides
F. exilis
F. taxifolius
Hylocomium splendens
Isopterygium elegans
Leucobryum glaucum
Mnium affine
M. hornum
M. longirostrum
M. undulatum
Plagiothecium denticulatum

P. silvaticum
P. undulatum
Pleuridium acuminatum
Polytrichum formosum
P. gracile
Rhytidiadelphus triquetrus
Thuidium tamariscinum

Cephalozia bicuspidata
Diplophyllum albicans
Lepidozia reptans
Lophocolea bidentata
Nardia scalaris
Plagiochila asplenioides
Plectocolea crenulata
Scapania irrigua
S. nemorosa

12. *On the ground in alder carr*

Acrocladium cuspidatum
Atrichum undulatum
Brachythecium rutabulum
Cratoneuron filicinum
Dicranella heteromalla
Eurhynchium praelongum
Fissidens taxifolius
Leptodictyum riparium
Mnium affine
M. longirostrum

M. punctatum
M. seligeri
M. undulatum
Plagiothecium denticulatum

Calypogeia trichomanis
Lophocolea bidentata
Pellia fabbroniana
Plagiochila asplenioides
Trichocolea tomentella

13. *On trees, stumps and logs in woodland*

Amblystegium serpens
Aulacomnium androgynum (r)
Brachythecium rutabulum
B. velutinum
Bryum capillare
Campylopus pyriformis (r)

Ceratodon purpureus
Cryphaea heteromalla (especially on elders) (e)
Dicranella heteromalla
Dicranoweisia cirrata (e)
Dicranum scoparium

(e) Characteristically epiphytic on trunk or branches.
(r) Characteristic of rotting stumps and logs.

Eurhynchium confertum
E. praelongum
Homalia trichomanoides
Hypnum cupressiforme
H. cupressiforme var. filiforme (e)
Isopterygium elegans (r)
Isothecium myosuroides
I. myurum
Neckera complanata
Orthodontium lineare (r)
Orthotrichum affine (e)
O. lyellii (e)
Plagiothecium denticulatum (r)
P. silvaticum (r)

Tetraphis pellucida (r)
Tortula laevipila (e)
Ulota bruchii (e)
U. crispa (e)
U. phyllantha (e)

Frullania dilatata
F. tamarisci
Lophocolea cuspidata (r)
L. heterophylla (r)
Metzgeria furcata (e)
Microlejeunea ulicina (e)
Radula complanata (e)

(e) Characteristically epiphytic on trunk or branches.
(r) Characteristic of rotting stumps and logs.

14. *Mountain oak wood* (e.g. Quercus petraea *woodland in steep-sided rocky valley of mountain stream*)

This list includes a selection of the very rich bryophyte flora to be found on boulders, banks and stumps, but excludes species that are confined to the stream and stream-side boulders (cf. list 19).

Brachythecium rutabulum
Bryum capillare
Dicranodontium denudatum
Dicranum majus
D. scoparium
Eurhynchium praelongum
E. striatum
Hylocomium brevirostre
H. splendens
Hypnum cupressiforme
Isopterygium elegans
Isothecium myosuroides
I. myurum
Leucobryum glaucum
Mnium hornum
M. longirostrum
M. undulatum
Plagiothecium denticulatum

P. undulatum
Pleurozium schreberi
Polytrichum formosum
Pseudoscleropodium purum
Rhytidiadelphus loreus
R. triquetrus
Thamnium alopecurum
Thuidium tamariscinum

Bazzania trilobata
Diplophyllum albicans
Nardia scalaris
Plagiochila asplenioides
P. spinulosa
Plectocolea crenulata
Saccogyna viticulosa
Scapania gracilis
S. nemorosa

15. *On the ground in pinewoods*

Campylopus flexuosus
C. pyriformis
Dicranella heteromalla
Dicranum scoparium
Hylocomium brevirostre (*h*)
H. splendens
Hypnum cupressiforme
Leucobryum glaucum
Plagiothecium undulatum
Pleurozium schreberi
Polytrichum formosum
Pseudoscleropodium purum

Ptilium crista-castrensis (*h*)
Rhytidiadelphus loreus (*h*)
R. squarrosus
R. triquetrus
Sphagnum palustre
Thuidium tamariscinum

Cephalozia bicuspidata
Diplophyllum albicans
Lepidozia reptans
Lophocolea bidentata

(*h*) Characteristic of Highland conifer woods—e.g. the indigenous *Pinus sylvestris* forest of north Britain.

16. *On heaths*

Brachythecium albicans
Bryum erythrocarpum
B. inclinatum
Campylopus flexuosus
C. pyriformis
Ceratodon purpureus
Dicranum scoparium
Funaria hygrometrica
Hylocomium splendens
Hypnum cupressiforme var. erice-
torum
Leucobryum glaucum
Pleurozium schreberi

Pohlia nutans
Polytrichum juniperinum
P. piliferum
Pseudoscleropodium purum
Rhytidiadelphus squarrosus

Cephalozia bicuspidata
Cephaloziella spp.
Diplophyllum albicans
Gymnocolea inflata
Lepidozia reptans
Nardia scalaris
Ptilidium ciliare

17. *In bogs (acid peat)*

Aulacomnium palustre
Breutelia chrysocoma
Campylopus atrovirens
Drepanocladus exannuiatus
D. fluitans
D. revolvens
Leucobryum glaucum

Polytrichum commune
P. alpestre
Scorpidium scorpioides
Sphagnum spp.
Splachnum ampullaceum (on
dung)

Calypogeia fissa
C. trichomanis
Cephalozia bicuspidata
C. connivens

Cephaloziella spp. (in *Sphagnum*, etc.)
Mylia anomala
Odontoschisma sphagni

18. *In marshes and fens*

Acrocladium cordifolium
A. cuspidatum
A. giganteum
Aulacomnium palustre
Brachythecium rutabulum
Bryum pseudotriquetrum
Campylium stellatum
Climacium dendroides
Cratoneuron commutatum
C. filicinum
Dicranum bonjeani
Drepanocladus aduncus
D. fluitans
D. revolvens (especially var. inter-medius)
Fissidens adianthoides

Leptodictyum riparium
Mnium punctatum
M. pseudopunctatum
M. seligeri
M. undulatum
Philonotis calcarea
Sphagnum plumulosum*
S. subsecundum*

Calypogeia fissa
C. trichomanis
Chiloscyphus polyanthus
Pellia fabbroniana
Riccardia multifida
R. pinguis

19. *On wet rocks in and by mountain streams*

Blindia acuta
Brachythecium plumosum
B. rivulare
Cinclidotus fontinaloides (*b*)
Cratoneuron commutatum (*b*) (*w*)
C. filicinum (*b*)
Dichodontium pellucidum
Dicranella squarrosa
Eucladium verticillatum (*b*) (*w*)
Eurhynchium riparioides
Fissidens adianthoides
Fontinalis antipyretica var. gracilis
F. squamosa

Grimmia alpicola var. rivularis
Heterocladium heteropterum
Hygroamblystegium fluviatile
H. tenax
Hygrohypnum eugyrium
H. luridum (*b*)
H. ochraceum
Hyocomium flagellare (*a*) (*w*)
Orthotrichum rivulare
Rhacomitrium aciculare (*a*)
R. aquaticum (*a*)
Thamnium alopecurum

(*b*) Especially on basic (calcareous) rocks.†
(*a*) Especially on acid, siliceous rocks.
(*w*) Especially about waterfalls.

* Both tend to be absent from typical fens.
† See note at the foot of p. 408.

Anthelia julacea
Chiloscyphus polyanthus
Conocephalum conicum (*b*)
Jungermannia cordifolia
J. tristis

Marsupella emarginata
Nardia compressa
Riccardia sinuata
Scapania undulata

(*b*) Especially on basic (calcareous) rocks.*

20. *On siliceous boulders in mountain country*

Andreaea rothii
A. rupestris
Dicranoweisia crispula
Grimmia doniana
G. trichophylla
Hedwigia ciliata
Hypnum cupressiforme
Isothecium myosuroides
Orthotrichum anomalum (typical
   form)
O. rupestre

Ptychomitrium polyphyllum
Rhacomitrium fasciculare
R. heterostichum
R. lanuginosum
Ulota hutchinsiae (rather rare)

Frullania tamarisci
Gymnomitrium crenulatum
G. obtusum
Marsupella emarginata

21. *Moorland and mountain slopes*

This list includes the species of peat cuttings, detritus, wet flushes and those which grow in the turf on mountain slopes, but it excludes the species of boulders, and mountain cliffs (see lists 20 and 22).

Acrocladium cuspidatum (*f*)
A. sarmentosum (*f*)
Andreaea alpina (*d*)
Anomobryum filiforme (*d*)

Aulacomnium palustre
Brachythecium rivulare (*f*)
Breutelia chrysocoma
Bryum alpinum (*d*)

(*d*) Especially on moist detritus.
(*f*) In flushes.

* It must be admitted that the implied contrast here, and the glossary definition of 'siliceous' are not wholly satisfactory; for all igneous rocks and sandstones contain some silica and hence are, in a sense, siliceous; further, a sharp line between calcareous (basic) rocks and siliceous (acid) rocks cannot always be drawn. Thus, whereas limestone and chalk are typical calcareous (and basic) sedimentary rocks and many sandstones are both siliceous and acid, some sandstones are in varying degree calcareous. Moreover, in the complex igneous rock systems of many mountain regions it is possible to find siliceous rocks that are basic in varying degree or that are acid on the whole but contain pockets of basic material. Thus gabbro, dolerite and basalt are, geologically, basic rocks in that they contain only 45–55% silica; and the acid volcanic rock rhyolite may contain pockets filled with the basic calcite. Hence a bryophyte is often an indicator, not so much of the geological character of the rock as a whole, but rather of the nature of the precise ledge or pocket where it happens to be growing.

B. pallens (*d*)
B. pseudotriquetrum (*f*)
Campylopus atrovirens
C. flexuosus (*p*) (*t*)
C. fragilis
C. pyriformis (*p*)
Ceratodon purpureus
Climacium dendroides
Dichodontium pellucidum (*d*)
Dicranella heteromalla (*p*)
D. squarrosa (*f*)
Dicranum fuscescens (*t*)
D. scoparium
Diphyscium foliosum
Ditrichum heteromallum (*d*)
Drepanocladus revolvens (*f*)
Funaria obtusa
Hylocomium splendens (*t*)
Hypnum cupressiforme
Leucobryum glaucum (*t*)
Oligotrichum hercynicum (*d*)
Orthodontium lineare (*p*)
Plagiothecium undulatum
Pleurozium schreberi (*t*)
Pohlia elongata
P. nutans (*p*)

Polytrichum aloides (*d*)
P. alpestre (*t*)
P. alpinum (*t*)
P. commune (*t*)
P. juniperinum
P. piliferum
P. urnigerum (*d*)
Pseudoscleropodium purum (*t*)
Rhacomitrium lanuginosum
Rhytidiadelphus loreus (*t*)
R. squarrosus (*t*)
Scorpidium scorpioides (*f*)
Sphagnum spp.
Splachnum ampullaceum
Thuidium tamariscinum (*t*)

Cephalozia bicuspidata (*p*)
Cephaloziella spp.
Diplophyllum albicans (*d*)
Gymnocolea inflata (*d*)
Lepidozia spp.
Nardia scalaris (*d*)
Odontoschisma sphagni
Plectocolea crenulata (*d*)
Ptilidium ciliare
Scapania undulata (*f*)

(*d*) Especially on moist detritus.
(*f*) In flushes.
(*p*) Colonizing bare peaty banks and cuttings.
(*t*) Especially in the turf of mountain slopes (i.e. among grass, etc.).

22. *Wet rock clefts and rock ledges on mountains*

Amphidium mougeotii
Andreaea spp. (*s*)
Anoectangium compactum
Barbula cylindrica (*b*)
B. recurvirostra (*b*)
B. spadicea (*b*)
Bartramia ithyphylla
B. pomiformis

Blindia acuta
Breutelia chrysocoma
Bryum alpinum
B. pseudotriquetrum
Campylopus atrovirens (*s*)
C. flexuosus (*s*)
C. schwarzii (*s*)
Ctenidium molluscum (*b*)

(*b*) Exclusively or principally on base-rich rock.
(*s*) Exclusively or principally on acid, siliceous rock.

Distichium capillaceum (*b*)
Encalypta ciliata (*b*)
Fissidens adianthoides
F. cristatus
F. osmundoides
Grimmia funalis
G. torquata
Gymnostomum aeruginosum (*b*)
Hookeria lucens
Isopterygium pulchellum
Isothecium myosuroides
Mnium hornum
M. marginatum (*b*)
M. punctatum
M. undulatum
Neckera crispa (*b*)
Orthothecium intricatum (*b*)
O. rufescens (*b*)
Plagiobryum zierii (*b*)
Plagiopus oederi (*b*)
Pohlia albicans
P. cruda

Polytrichum urnigerum (*s*)
Rhabdoweisia denticulata
Rhacomitrium ellipticum (*s*)
R. lanuginosum (*s*)
Sphagnum papillosum (*s*)
S. subsecundum (*s*)
Tortella tortuosa (*b*)
Trichostomum tenuirostre (*s*)

Bazzania tricrenata
Conocephalum conicum (*b*)
Diplophyllum albicans (*s*)
Herberta adunca
H. hutchinsiae
Marsupella emarginata
Nardia scalaris (*s*)
Orthocaulis floerkii
Pellia epiphylla (*s*)
Plagiochila asplenioides
P. spinulosa
Preissia quadrata (*b*)
Scapania undulata

(*b*) Exclusively or principally on base-rich rock.
(*s*) Exclusively or principally on acid, siliceous rock.

23. *On sand dunes*

(Moist hollows and dune slacks are dealt with separately in list 24.)

Barbula convoluta
B. fallax
Brachythecium albicans
Bryum argenteum
B. capillare
B. intermedium
B. pendulum
Camptothecium lutescens
Ceratodon purpureus
Climacium dendroides
Cratoneuron filicinum
Dicranum scoparium
Ditrichum flexicaule
Entodon orthocarpus

Hypnum cupressiforme
Mnium longirostrum
M. undulatum
Pseudoscleropodium purum
Rhytidiadelphus squarrosus
R. triquetrus
Thuidium philiberti
Tortella flavovirens
Tortula ruraliformis

Lophocolea bidentata
L. cuspidata
Scapania aspera

24. *Dune slacks (including pools with marsh vegetation and moist hollows with short turf)*

Acrocladium cordifolium
A. cuspidatum
A. giganteum
Barbula convoluta
B. tophacea
Bryum intermedium
B. pallens
B. pendulum
B. pseudotriquetrum
Campylium polygamum
C. stellatum
Cratoneuron commutatum
C. filicinum

Drepanocladus aduncus
D. lycopodioides
D. revolvens
Fissidens adianthoides
Mnium cuspidatum
M. punctatum
M. seligeri

Anthoceros laevis
Pellia fabbroniana
Preissia quadrata
Riccardia pinguis
R. sinuata

25. *On maritime rocks, liable to inundation by salt spray*

Grimmia maritima
Tortella flavovirens
Trichostomum brachydontium

T. crispulum
Ulota phyllantha

INDEX

(Main species appear in bold type; additional species appear in ordinary type; synonyms in italic.)

Acaulon muticum, 170
Acrocladium cordifolium, 279
Acrocladium cuspidatum, 278
Acrocladium giganteum, 279
Acrocladium sarmentosum, 279
Acrocladium stramineum, 279
Alicularia compressa, 355
Alicularia Geoscyphus, 355
Alicularia scalaris, 354
Aloina aloides, 164
Aloina ambigua, 164
Aloina rigida, 164
Amblystegium filicinum, 262
Amblystegium fluviatile, 270
Amblystegium irriguum, 269
Amblystegium serpens, 268
Amblystegium varium, 269
Amphidium lapponicum, 236
Amphidium mougeotii, 235
Anastrepta orcadensis, 352
Andreaea alpina, 108
Andreaea petrophila, 107
Andreaea rothii, 108
Andreaea rupestris, 107
Andreaeales, 107
Aneura multifida, 385
Aneura pinguis, 383
Aneura sinuata, 386
Anoectangium compactum,
  178
Anomobryum filiforme, 210
Anomodon viticulosus, 259
Anthelia julacea, 321
Anthelia juratzkana, 322
Anthoceros husnoti, 399
Anthoceros laevis, 398
Anthoceros punctatus, 397
Anthocerotales, 397
Antitrichia curtipendula, 251
Aplozia cordifolia, 352
Aplozia crenulata, 355
Aplozia pumila, 353
Aplozia riparia, 353
Aplozia sphaerocarpa, 353
Archidium alternifolium, 128
Atrichum crispum, 110
Atrichum undulatum, 109

Aulacomnium androgynum, 229
Aulacomnium palustre, 227

Barbilophozia barbata, 352
Barbula convoluta, 171
Barbula cylindrica, 177
Barbula fallax, 174
Barbula lurida, 176
Barbula recurvirostra, 175
Barbula revoluta, 176
Barbula rigidula, 176
Barbula rubella, 175
Barbula sinuosa, 182
Barbula tophacea, 176
Barbula trifaria, 176
Barbula unguiculata, 172
Barbula vinealis, 177
Bartramia halleriana, 231
Bartramia ithyphylla, 231
Bartramia Oederi, 231
Bartramia pomiformis, 230
Bazzania tricrenata, 327
Bazzania trilobata, 326
Blasia pusilla, 381
Blepharostoma trichophyllum, 323
Blindia acuta, 135
Brachythecium albicans, 284
Brachythecium caespitosum, 291
Brachythecium glareosum, 291
Brachythecium illecebrum, 291
Brachythecium plumosum, 290
Brachythecium populeum, 291
Brachythecium purum, 299
Brachythecium rivulare, 287
Brachythecium rutabulum, 286
Brachythecium velutinum, 288
Breutelia arcuata, 233
Breutelia chrysocoma, 233
Bryales, 109
Bryum, 211
Bryum alpinum, 218
Bryum argenteum, 216
Bryum atropurpureum, 217
Bryum bicolor, 217
Bryum bimum, 215
Bryum caespiticium, 215
Bryum capillare, 219

Bryum erythrocarpum, 221
Bryum filiforme, 210
Bryum inclinatum, 221
Bryum intermedium, 221
Bryum pallens, 212
Bryum pendulum, 221
Bryum pseudotriquetrum, 214
Bryum pseudotriquetrum var. **bimum**, 215
Bryum roseum, 221
Buxbaumia aphylla, 122
Buxbaumiales, 121

Calypogeia arguta, 332
Calypogeia fissa, 331
Calypogeia trichomanis, 330
Camptothecium lutescens, 283
Camptothecium sericeum, 282
Campylium chrysophyllum, 266
Campylium polygamum, 267
Campylium stellatum, 265
Campylopus atrovirens, 151
Campylopus brevipilus, 152
Campylopus flexuosus, 150
Campylopus fragilis, 153
Campylopus pyriformis, 149
Campylopus schwarzii, 153
Catharinea crispa, 110
Catharinea undulata, 109
Cephalozia bicuspidata, 332
Cephalozia connivens, 333
Cephalozia fluitans, 334
Cephalozia Francisci, 334
Cephalozia media, 334
Cephaloziella, 335
Cephaloziella byssacea, 336
Cephaloziella hampeana, 337
Cephaloziella rubella, 338
Cephaloziella Starkii, 336
Ceratodon purpureus, 133
Chiloscyphus pallescens, 343
Chiloscyphus pallescens var. fragilis, 343
Chiloscyphus polyanthus, 341
Cinclidium stygium, 227
Cinclidotus Brebissonii, 171
Cinclidotus fontinaloides, 170
Cinclidotus mucronatus, 171
Cirriphyllum crassinervium, 292
Cirriphyllum piliferum, 291
Cladopodiella fluitans, 334
Cladopodiella francisci, 334
Climacium dendroides, 249
Cololejeunea calcarea, 378
Colura calyptrifolia, 378
Conocephalum conicum, 389, Plate XVII
Cratoneuron commutatum, 263

Cratoneuron commutatum var. **falcatum**, 264
Cratoneuron filicinum, 262
Cryphaea heteromalla, 251
Ctenidium molluscum, 310, plate XIII
Cylindrothecium concinnum, 302
Cynodontium bruntonii, 140

Desmatodon convolutus, 164
Dichodontium flavescens, 141
Dichodontium pellucidum, 141
Dichodontium pellucidum var. **flavescens**, 141
Dicranales, 127
Dicranella cerviculata, 140
Dicranella heteromalla, 139
Dicranella rufescens, 140
Dicranella squarrosa, 136
Dicranella varia, 138
Dicranodontium denudatum, 153
Dicranodontium longirostre, 153
Dicranoweisia cirrata, 142
Dicranoweisia crispula, 143
Dicranum bonjeani, 146
Dicranum falcatum, 148
Dicranum fuscescens, 143
Dicranum majus, 144
Dicranum scoparium, 147
Dicranum scottianum, 148
Diphyscium foliosum, 121
Diplophyllum albicans, 362
Distichium capillaceum, 131
Ditrichum cylindricum, 131
Ditrichum flexicaule, 130
Ditrichum heteromallum, 129
Ditrichum homomallum, 129
Ditrichum tenuifolium, 131
Drepanocladus aduncus, 270
Drepanocladus exannulatus, 275
Drepanocladus fluitans, 271
Drepanocladus lycopodioides, 275
Drepanocladus revolvens, 273
Drepanocladus uncinatus, 274

Encalypta ciliata, 156
Encalypta streptocarpa, 155
Encalypta vulgaris, 154
Entodon orthocarpus, 302
Ephemerum serratum, 201
Eubryales, 206
Eucalyx hyalinus, 356
Eucalyx obovatus, 356
Eucladium verticillatum, 178
Eurhynchium confertum, 298
Eurhynchium crassinervium, 292

Eurhynchium murale, 299
Eurhynchium myosuroides, 281
Eurhynchium myurum, 279
Eurhynchium piliferum, 291
Eurhynchium praelongum, 294
Eurhynchium pumilum, 299
Eurhynchium riparioides, 296
Eurhynchium rusciforme, 296
Eurhynchium striatum, 293
Eurhynchium swartzii, 295
Eurhynchium tenellum, 299

Fissidens, 122
Fissidens adianthoides, 124, Plate III
Fissidens bryoides, 123
Fissidens crassipes, 126
Fissidens cristatus, 127
Fissidens decipiens, 127
Fissidens exilis, 126
Fissidens incurvus, 126
Fissidens osmundoides, 127
Fissidens pusillus, 126
Fissidens taxifolius, 124
Fissidentales, 122
Fontinalis antipyretica, 247
Fontinalis antipyretica var. **gracilis**, 247
Fontinalis squamosa, 249
Fossombronia dumortieri, 379
Fossombronia pusilla, 378
Fossombronia wondraczeki, 379
Frullania dilatata, 373
Frullania fragilifolia, 375
Frullania germana, 375
Frullania tamarisci, 372
Funaria attenuata, 200
Funaria ericetorum, 200
Funaria hygrometrica, 198
Funaria obtusa, 200
Funaria Templetoni, 200
Funariales, 198

Grimmiales, 184
Grimmia alpicola var. rivularis, 190
Grimmia apocarpa, 185
Grimmia apocarpa var. *rivularis*, 190
Grimmia decipiens, 191
Grimmia doniana, 186
Grimmia funalis, 190
Grimmia hartmanii, 191
Grimmia maritima, 184
Grimmia patens, 191
Grimmia pulvinata, 187, Plate V
Grimmia torquata, 190
Grimmia trichophylla, 189
Gymnocolea inflata, 348

Gymnomitrium adustum, 359
Gymnomitrium concinnatum, 359
Gymnomitrium crenulatum, 358
Gymnomitrium obtusum, 359
Gymnomitrium varians, 359
Gymnostomum aeruginosum, 177
Gymnostomum recurvirostrum, 178

Hedwigia ciliata, 250
Herberta adunca, 322
Herberta hutchinsiae, 322
Heterocladium heteropterum, 259
Homalia trichomanoides, 254
Hookeriales, 257
Hookeria lucens, 257
Hygroamblystegium fluviatile, 270
Hygroamblystegium tenax, 269
Hygrohypnum eugyrium, 276
Hygrohypnum luridum, 275
Hygrohypnum ochraceum, 276
Hylocomium brevirostre, 318
Hylocomium loreum, 315
Hylocomium rugosum, 318
Hylocomium splendens, 317
Hylocomium squarrosum, 314
Hylocomium triquetrum, 313
Hyocomium flagellare, 311
Hypnobryales, 258
Hypnum aduncum, 270
Hypnum chrysophyllum, 266
Hypnum commutatum, 263
Hypnum cordifolium, 279
Hypnum crista-castrensis, 311
Hypnum cupressiforme, 307, Plate XII
Hypnum cupressiforme var. *elatum*, 309
Hypnum cupressiforme var. **ericetorum**, 309
Hypnum cupressiforme var. **filiforme**, 308
Hypnum cupressiforme var. **lacunosum**, 309
Hypnum cupressiforme var. **resupinatum**, 308
Hypnum cupressiforme var. **tectorum**, 309
Hypnum cuspidatum, 278
Hypnum eugyrium, 276
Hypnum exannulatum, 275
Hypnum falcatum, 263
Hypnum fluitans, 271
Hypnum giganteum, 279
Hypnum lycopodioides, 275
Hypnum molluscum, 310
Hypnum ochraceum, 276
Hypnum palustre, 275
Hypnum patientiae, 309
Hypnum polygamum, 267

Hypnum revolvens, 273
Hypnum riparium, 267
Hypnum sarmentosum, 279
Hypnum Schreberi, 300
Hypnum scorpioides, 276
Hypnum stellatum, 265
Hypnum stramineum, 279
Hypnum uncinatum, 274

Isobryales, 234
Isopaches bicrenatus, 351
Isopterygium depressum, 303
Isopterygium elegans, 302
Isopterygium pulchellum, 303
Isothecium myosuroides, 281
Isothecium myurum, 279

Jubula hutchinsiae, 375
Jungermannia cordifolia, 352
Jungermannia pumila, 353
Jungermannia sphaerocarpa, 353
Jungermannia tristis, 353
Jungermanniales Acrogynae, 321
Jungermanniales Anacrogynae, 378
Jungermannineae, 321

Leiocolea muelleri, 348
Leiocolea turbinata, 347
Lejeunea cavifolia, 375
Lejeunea lamacerina, 376
Lejeunea patens, 376
Lejeunea planiuscula, 376
Lepidozia pearsoni, 329
Lepidozia pinnata, 329
Lepidozia reptans, 327
Lepidozia setacea, 328
Lepidozia trichoclados, 329
Leptobryum pyriforme, 206
Leptodictyum riparium, 267
Leptodontium flexifolium, 184
Leptoscyphus anomalus, 344
Leptoscyphus Taylori, 343
Leskea polycarpa, 258
Leucobryum glaucum, 153, Plate IV
Leucodon sciuroides, 251
Lophocolea alata, 341
Lophocolea bidentata, 338
Lophocolea cuspidata, 339
Lophocolea heterophylla, 340
Lophozia alpestris, 347
Lophozia attenuata, 350
Lophozia barbata, 352
Lophozia bicrenata, 351
Lophozia excisa, 347
Lophozia Floerkii, 349

Lophozia incisa, 347
Lophozia Muelleri, 348
Lophozia porphyroleuca, 347
Lophozia quinquedentata, 352
Lophozia turbinata, 347
Lophozia ventricosa, 346
Lunularia cruciata, 391, Plate XVIII

Madotheca Cordeana, 369
Madotheca laevigata, 369
Madotheca platyphylla, 368
Marchantiales, 387
Marchantia polymorpha, 387, Plates XV and XVI
Marchesinia mackayi, 378
Marsupella aquatica, 356
Marsupella emarginata, 356
Marsupella funckii, 358
Metzgeria conjugata, 383
Metzgeria furcata, 382
Metzgeria hamata, 383
Metzgeria pubescens, 383
Metzgerineae, 378
Microlejeunea ulicina, 376
Mnium affine, 227
Mnium affine var. *elatum*, 227
Mnium cuspidatum, 227
Mnium hornum, 223
Mnium longirostrum, 227
Mnium marginatum, 227
Mnium pseudopunctatum, 227
Mnium punctatum, 226
Mnium rostratum, 227
Mnium seligeri, 227
Mnium serratum, 227
Mnium stellare, 227
Mnium subglobosum, 227
Mnium undulatum, 224, Plate VII
Moerckia flotowiana, 382
Mylia anomala, 344
Mylia taylori, 343
Myurium hebridarum, 252

Nardia compressa, 355
Nardia geoscyphus, 355
Nardia scalaris, 354
Neckera complanata, 253
Neckera crispa, 252
Nowellia curvifolia, 334

Odontoschisma denudatum, 335
Odontoschisma sphagni, 334
Oligotrichum hercynicum, 110
Orthocaulis attenuatus, 350
Orthocaulis floerkii, 349

Orthodontium gracile, 207
Orthodontium gracile var. *heterocarpum*, 207
Orthodontium lineare, 207
Orthothecium intricatum, 302
Orthothecium rufescens, 302
Orthotrichum, 238
Orthotrichum affine, 240
Orthotrichum anomalum, 239
Orthotrichum anomalum var. **saxatile**, 239
Orthotrichum cupulatum, 244
Orthotrichum diaphanum, 243
Orthotrichum leiocarpum, 244
Orthotrichum lyellii, 241
Orthotrichum pulchellum, 244
Orthotrichum rivulare, 244
Orthotrichum rupestre, 244
Orthotrichum striatum, 244
Orthotrichum tenellum, 244

Pellia calycina, 381
Pellia epiphylla, 380
Pellia fabbroniana, 381
Pellia neesiana, 381
Phascum curvicollum, 170
Phascum cuspidatum, 168
Phascum floerkeanum, 170
Philonotis calcarea, 232
Philonotis fontana, 231
Physcomitrella patens, 201
Physcomitrium pyriforme, 200
Plagiobryum zierii, 210
Plagiochila asplenioides, 359, Plate XIV
Plagiochila asplenioides var. **major**, 361
Plagiochila punctata, 362
Plagiochila spinulosa, 361
Plagiopus oederi, 231
Plagiothecium denticulatum, 304
Plagiothecium depressum, 303
Plagiothecium elegans, 302
Plagiothecium pulchellum, 303
Plagiothecium silvaticum, 306
Plagiothecium undulatum, 305, Plate XI
Plectocolea crenulata, 355
Plectocolea hyalina, 356
Plectocolea obovata, 356
Pleuridium acuminatum, 127
Pleuridium alternifolium, 128
Pleuridium axillare, 128
Pleuridium subulatum, 128
Pleuridium subulatum (of Dixon, *Handbook*), 127
Pleurozia purpurea, 371
Pleurozium schreberi, 300

Pohlia albicans, 210
Pohlia annotina, 210
Pohlia cruda, 210
Pohlia delicatula, 209
Pohlia elongata, 210
Pohlia nutans, 207
Pohlia proligera, 210
Polytrichales, 109
Polytrichum, 112
Polytrichum aloides, 113
Polytrichum alpestre, 120
Polytrichum alpinum, 120
Polytrichum commune, 119, Plate II
Polytrichum formosum, 117
Polytrichum gracile, 120
Polytrichum juniperinum, 115, Plate I
Polytrichum nanum, 120
Polytrichum piliferum, 116
Polytrichum strictum, 120
Polytrichum urnigerum, 114
Porella cordeana, 369
Porella laevigata, 369
Porella platyphylla, 368
Porotrichum alopecurum, 255
Pottiales, 154
Pottia davalliana, 167
Pottia heimii, 168
Pottia intermedia, 168
Pottia lanceolata, 164
Pottia minutula, 167
Pottia recta, 168
Pottia truncata, 166
Pottia truncatula, 166
Preissia quadrata, 388
Pseudephemerum nitidum, 128
Pseudoscleropodium purum, 299, Plate X
Pterogonium gracile, 251
Pterygoneurum ovatum, 164
Pterygophyllum lucens, 257
Ptilidium ciliare, 322
Ptilidium pulcherrimum, 323
Ptilium crista-castrensis, 311
Ptychomitrium polyphyllum, 234
Pylaisia polyantha, 309

Radula aquilegia, 370
Radula complanata, 369
Radula lindbergiana, 370
Reboulia hemisphaerica, 392
Rhabdoweisia denticulata, 140
Rhacomitrium aciculare, 191
Rhacomitrium aquaticum, 192
Rhacomitrium canescens, 196
Rhacomitrium ellipticum, 198
Rhacomitrium fasciculare, 193

Rhacomitrium heterostichum, 194
Rhacomitrium heterostichum var. gracilescens, 194
Rhacomitrium lanuginosum, 197, Plate VI
Rhacomitrium protensum, 192
Rhacomitrium sudeticum, 194
Rhodobyrum roseum, 221
Rhynchostegiella pallidirostra, 299
Rhynchostegiella tenella, 299
Rhytidiadelphus loreus, 315
Rhytidiadelphus squarrosus, 314
Rhytidiadelphus triquetrus, 313
Rhytidium rugosum, 318
Riccardia multifida, 385
Riccardia pinguis, 383
Riccardia sinuata, 386
Riccia crystallina, 396
Riccia fluitans, 393
Riccia glauca, 395
Riccia sorocarpa, 396
Ricciocarpus natans, 392

Saccogyna viticulosa, 345
Scapania aequiloba, 367
Scapania aspera, 367
Scapania compacta, 367
Scapania curta, 367
Scapania dentata, 365
Scapania gracilis, 366
Scapania irrigua, 367
Scapania nemorosa, 363
Scapania subalpina, 367
Scapania undulata, 365
Schistostegales, 203
Schistostega osmundacea, 203
Schistostega pennata, 203
Scleropodium caespitosum, 291
Scleropodium illecebrum, 291
Scorpidium scorpioides, 276
Seligeria calcarea, 134
Seligeria pusilla, 135
Seligeria recurvata, 135
Sphaerocarpales, 396
Sphaerocarpus michelii, 396
Sphaerocarpus texanus, 396
Sphagnales, 100
Sphagnum, 100
Sphagnum *acutifolium* var. *rubellum*, 106
Sphagnum acutifolium var. *subnitens*, 104
Sphagnum auriculatum, 106
Sphagnum compactum, 106
Sphagnum crassicladum, 106
Sphagnum cuspidatum, 103
Sphagnum cymbifolium, 101
Sphagnum intermedium, 106

Sphagnum inundatum, 106
Sphagnum magellanicum, 105
Sphagnum medium, 105
Sphagnum palustre, 101
Sphagnum papillosum, 106
Sphagnum plumulosum, 104
Sphagnum recurvum, 106
Sphagnum rigidum, 106
Sphagnum rubellum, 106
Sphagnum subsecundum, 106
Sphenolobus exsectiformis, 352
Sphenolobus minutus, 352
Splachnum ampullaceum, 202
Splachnum ovatum, 203
Splachnum sphaericum, 203
Swartzia montana, 131

Targionia hypophylla, 392
Tetraphidales, 205
Tetraphis pellucida, 205
Tetraplodon mnioides, 203
Thamnium alopecurum, 255, Plate VIII
Thuidium abietinum, 261
Thuidium delicatulum, 262
Thuidium philiberti, 262
Thuidium tamariscinum, 260, Plate IX
Tortella flavovirens, 181
Tortella tortuosa, 179
Tortula aloides, 164
Tortula ambigua, 164
Tortula atrovirens, 164
Tortula intermedia, 159
Tortula laevipila, 160
Tortula latifolia, 164
Tortula marginata, 164
Tortula muralis, 162
Tortula mutica, 164
Tortula papillosa, 164
Tortula pusilla, 164
Tortula rigida, 164
Tortula ruraliformis, 158
Tortula ruralis, 157
Tortula subulata, 161
Trichocolea tomentella, 325
Trichostomum brachydontium, 181
Trichostomum crispulum, 182
Trichostomum flavovirens, 181
Trichostomum mutabile, 181
Trichostomum sinuosum, 182
Trichostomum tenuirostre, 182
Trichostomum tortuosum, 179
Tritomaria exsectiformis, 352
Tritomaria quinquedentata, 352

Ulota bruchii, 247

Ulota crispa, 245
Ulota hutchinsiae, 247
Ulota phyllantha, 244

Webera albicans, 210
Webera annotina, 210
Webera carnea, 209
Webera cruda, 210
Webera elongata, 210
Webera nutans, 207
Webera proligera, 210
Weissia controversa, 182

Weissia crispa, 183
Weissia curvirostris, 178
Weissia microstoma, 183
Weissia rupestris, 177
Weissia tortilis, 183
Weissia verticillata, 178
Weissia viridula, 182

Zygodon conoideus, 238
Zygodon lapponicus, 236
Zygodon Mougeotii, 235
Zygodon viridissimus, 236